continued on back

Clustering Algorithms

A WILEY PUBLICATION IN APPLIED STATISTICS

Clustering Algorithms

JOHN A. HARTIGAN,
Department of Statistics
Yale University

JOHN WILEY & SONS New York · London · Sydney · Toronto

Hartigan, John A   1937–
    Clustering algorithms.

    (Wiley series in probability and mathematical
statistics)
    Includes bibliographical references.
    1. Cluster analysis. 2. Electronic data processing
—Cluster analysis. I. Title.

QA278.H36   519.5′3   74-14573
ISBN 0-471-35645-X

Printed in the United States of America

10 9 8 7 6 5 4 3 2

To Pamela

Preface

In the last ten years, there has been an explosive development of clustering techniques and an increasing range of applications of these techniques. Yet there are at present only three books in English on clustering: the pioneering *Numerical Taxonomy* by Sokal and Sneath, oriented toward biology; *Cluster Analysis* by Tryon and Bailey, oriented toward psychology; and *Mathematical Taxonomy* by Jardine and Sibson. The principal contribution of statisticians has been in the area of discriminant analysis, the problem of assigning new observations to known groups. The more difficult, more important, and more interesting problem is discovery of the groups in the first place. Although modern clustering techniques began development in biological taxonomy, they are generally applicable to all types of data. They should be used routinely in early descriptions of data, playing the same role for multivaiate data that histograms play for univariate data.

The origin of this book was the general methodology lecture on clustering, given at the invitation of Dr. S. Greenhouse to the December 1970 meeting of the American Statistical Association. The notes for the lecture were used in a seminar series at Yale in early 1971 and later for a number of seminars given on behalf of the Institute of Advanced Technology, Control Data Corporation, at various places in the United States and overseas. These intensive two-day seminars required the preparation of more detailed notes describing algorithms, discussing their computational properties, and listing small data sets for hand application of the algorithms. After considerable evolution, the notes became the basis for the first draft of the book, which was used in a course at Yale University in 1973.

One difficulty of the two-day seminars was describing the various algorithms explicitly enough for students to apply them to actual data. A comprehensible but unambiguous description is needed before the algorithm can sensibly be discussed or applied. The technique eventually developed was the step-by-step description used in this book, an amalgamation of verbal description and Fortran notation. These descriptions form the skeleton of the book, fleshed out by applications, evaluations, and alternative techniques.

The book could be used as a textbook in a data analysis course that included some work on clustering or as a resource book for persons actually planning to do some clustering. The chapters are pretty well independent of each other, and therefore the one or two chapters containing algorithms of special interest may be read alone. On the other hand, the algorithms become increasingly complex as the book proceeds, and it is easier to work up to the later chapters via the early chapters.

Fortran programs implementing the algorithms described in the chapter are listed at the end of each chapter. An attempt has been made to keep these programs machine independent, and each program has been run on several different data sets, but my deficiencies as a programmer and comment card writer could make the programs tricky to use. The ideal user is an experienced Fortran programmer who is willing to adapt the programs to his own needs.

I am indebted to G. E. Dallal, W. Maurer, S. Schwager, and especially D. A. Meeter, who discovered many errors in facts and style in the first draft of the book. I am also indebted to Mrs. Barbara Amato, who cheerfully typed numerous revisions.

April 1974

JOHN A. HARTIGAN

Contents

Introduction

I.1 CLUSTERING

Table I.1 consists of a number of observations on minor planets. There are a very large number of such planets in orbits between Mars and Jupiter. The first minor planet is Ceres, discovered in 1801 by Piazzi and Gauss. In a photograph against the fixed stars, a minor planet sighting will show as a curved streak from which its orbital elements may be computed. Many astronomers see the minor planets as noise obscuring the observation of other interesting movements.

There are about 2000 named minor planets, and many thousands of sightings of these and perhaps other planets. An important problem in keeping track of the minor planets is deciding which sightings are of the same planet. In particular, if a new planet is claimed, it must be checked that the sighting does not agree with any planet already named.

Clustering is the *grouping* of *similar objects*. The naming of minor planets and the classification of sightings are typical clustering problems. The objects are the sightings. Two objects are similar if, considering measurement error, the sightings could plausibly be of the same planet. A group is a set of sightings of the same planet.

It is clear that such classifications occur constantly in thought and speech. Objects that differ in insignificant details are given the same name, can be treated the same, and can be expected to act the same. For example, a wife notices that the man coming in the door differs only in insignificant details from her husband that left in the morning, and so she expects him to answer to the same name.

The word clustering is almost synonymous with classification. In this book it is a general term for formal, planned, purposeful, or scientific classification. Other words that have been used for this purpose are numerical taxonomy, taximetrics, taxonorics, morphometrics, botryology, nosology, nosography, and systematics. General references are Blackwelder (1966) for animal taxonomy, Sokal and Sneath (1963) for the first and most important book in numerical taxonomy, Good (1965) for a classification of classification problems, Jardine and Sibson (1971) for a mathematical development, and Cormack (1971) for a recent survey of the literature.

I.2 EXAMPLES OF CLUSTERING

Naming is classifying. It is not necessary (or possible) that a naming scheme be best, but for effective communication it is necessary that different people give the same name to the same objects. The most familiar naming scheme is the taxonomy of animals and plants.

I.2.1 Animals and Plants

Formal classifications of animals and plants date back to Aristotle, but the modern system is essentially that of Linnaeus (1753). Each species belongs to a series of clusters of increasing size with a decreasing number of common characteristics. For example, man belongs to the primates, the mammals, the amniotes, the vertebrates, the animals.

1

Table I.1   Minor Planets[a]

NAME [b]	NODE[c] (DEGREES)	INCLINATION[d] (DEGREES)	AXIS[e] (A.U.)
1935RF	130.916	4.659	2.2562
1941FD	132.2	4.7	2.13
1955QT	130.07	4.79	2.1893
1940YL	338.333	16.773	2.7465
1953NH	339.625	16.067	2.7335
1930SY	80.804	4.622	2.1890
1949HM	80.804	4.622	2.1906
1929EC	115.072	2.666	3.1676
1948RO	89.9	2.1	3.35
1951AM	115.072	2.666	3.1676
1938DL	135.6	1.0	2.6
1951AX	153.1	6.5	2.45
1924TZ	59.9	5.7	2.79
1931DQ	69.6	4.7	2.81
1936AB	78.1	6.6	2.90
1952DA	55.144	4.542	3.0343
1948RB	194.6	1.8	3.0200
1948RH	164.1	10.0	1.93
1948TG	34.2	12.5	2.82

[a] The grouped sightings are some of those tentatively identified as being from the same planet in *Elements of Minor Planets*, University of Cincinnati Observatory (1961).
[b] The year of sighting and the initials of the astronomer.
[c] The angle, in the plane of the earth's orbit, at which the minor planet crosses the earth's orbit.
[d] The angle between the plane of the earth's orbit and the plane of the planet's orbit.
[e] The maximum distance of the minor planet from the sun, divided by the same quantity for the earth.

This tree, which was originally developed to name objects consistently, was given physical significance in the evolutionary theories of Darwin, which state that man, for example, has ancestors at the various levels of the tree. Man has an ancestor in common with a monkey, a rabbit, a frog, a fish, and a mosquito. If your naming scheme works exceptionally well, look for the reason why!

The tree is used in storing and disseminating knowledge. For example, vertebrates have a backbone, bilateral symmetry, four limbs, a head with two eyes and a mouth, a heart with blood circulation, a liver, and other common properties. Once you've seen one vertebrate, you've seen them all. It is not necessary to record these properties separately for each species. See Borradaile and Potts (1958) for accumulation of knowledge about an ant, moving down the tree through *Metazoa*, *Arthropoda*, *Insecta*, *Pterygota*, *Endopterygota*, *Hymenoptera*, and *Formicoidea*.

The techniques of taxonomy deserve imitation in other areas. The tree structure is now used as a standard clustering structure. The functions of naming objects and of storing information cheaply are also generalized to other areas. It is in the construction of the tree that it becomes difficult to generalize the methods of animal and plant taxonomy. New classifications of groups of species are frequently suggested and much disputed. The *principles* of classification, which might sometimes settle the disputes, are themselves very underdeveloped.

Modern clustering owes its development to high-speed computers and to a small group of *numerical taxonomists*. The stronghold of numerical taxonomy is the journal *Systematic Zoology*. Sokal and Sneath's book (1963) has had a very important seminal influence. Since computers will do exactly what they are told, it is necessary to have precise definitions of the meaning of a cluster, of the data type, and of the meaning of similarity before computers can be useful. The intervention of the computer has thus caused extensive development of clustering principles. On the other hand, traditional taxonomists suspect that their rich and instinctive knowledge of species in their field cannot be reduced in any substantial part to machine-readable form and are wisely wary of accepting classifications from the computer. For such a "traditionalist" view, see Darlington (1971). For some recent examples of numerical taxonomy, see Dupont and Hendrick (1971), Stearn (1971), and Small, Bassett, and Crompton (1971).

I.2.2 Medicine

The principal classification problem in medicine is the classification of disease. Webster's dictionary defines a disease as an impairment of the normal state of the living animal that interrupts or modifies the performance of the vital functions, being a response to environmental factors (such as malnutrition, industrial hazards, or climate), to specific infective agents (such as worms, bacteria, or viruses), or to inherent defects of the organism. Webster's has already classified diseases into three types. The World Health Organization produces a *Manual of the International Statistical Classification of Diseases, Injuries, and Causes of Death* (1965). This provides a standard nomenclature necessary to compile statistics, especially death statistics, comparable across different countries and times. General theories of classification of disease or general principles such as the evolutionary principle of taxonomy in biology are not known to the author. See Temkin (1970) for some history of classifications. As Feinstein (1972) remarks, nosology differs from biological taxonomy in being oriented to treatment; it is important to separate diseases that require different treatment.

Numerical techniques have had only a slight impact on the classification of disease, perhaps because medical data and especially medical histories are not easy to assimilate into the standard data structures that were first developed for biological taxonomy. In particular, since a disease is an abnormality, it is always important to scale observations against normal observations for the patient or for a population, as in Goldwyn et al. (1971). The data structure used in the following studies is a matrix of measurements on a number of patients for various characteristics relevant to the disease. A measure of similarity is usually computed between each pair of patients, and groups of similar patients are then constructed. Knusman and Toeller (1972) discover three subgroups of diabetes mellitus, using factor analysis. Winkel and Tygstrup (1971) identify two distinct subgroups of 400 cirrhosis patients, but leave 70 % of the patients unclassified. Baron and Fraser (1968), in an experimental test of clustering methods on 50 cirrhosis patients measured on 330 characteristics, show that the single-linkage algorithm conforms less well to previous diagnoses than the average-linkage algorithm. Hayhoe et al. (1964) identify four groups in 140 cases of leukemia and propose diagnostic criteria to distinguish the four groups. Manning and Watson (1966) divide 99 heart patients into three groups agreeing substantially with physicians' diagnoses of univalvular lesions, multivalvular lesions, and arteriosclerotic, hypertensive, or pulmonary disease.

Bouckaert (1971) uses single linkage to select clusters of patients from 85 persons presenting a goiter and also to select three syndromes (clusters of symptoms) that correspond to the common description of simple goiter, hyperthyroidism, and cancer. The usual practice in classification still follows the traditional biological taxonomists' technique of selecting one or two important variables by expert judgement and classifying according to these variables, as in Schrek et al (1972), where it is proposed to classify lymphocytic leukemia by the cross-sectional area of blood lymphocytes and by the percentage of smooth nuclei.

A particular type of classification within a disease is the identification of stages of severity—for example, for renal disease (1971). Various symptoms are grouped by expert judgement to make up ordered classes of severity in three categories. Goldwyn et al. (1971) use clustering techniques to stage critically ill patients.

For diseases that are caused by viruses and bacteria, the techniques of numerical taxonomy carry over, and there are many papers using such techniques. For example, Goodfellow (1971) measures 241 characters on 281 bacteria, some biochemical, some physiological, and some nutritional. He identifies seven groups substantially conforming to groups already known. However, the classifications of viruses in Wilner (1964) and Wildy (1971) and the classification of bacteria in Prevot (1966) are still based on picking important variables by expert judgement.

Stark et al. (1962) use clustering techniques to identify abnormal electrocardiograms.

There are other uses of classification in medicine besides the direct classification of disease. Blood group serology is classification of blood, which began with the discovery of the A and B antigens by Landsteiner in 1900 (see, for example, Boorman and Dodd (1970)). Numerical techniques, as in Chakraverty (1971), frequently use a square data matrix in which the rows are antigens and the columns are the corresponding antibodies.

Finally, in epidemiology, diseases may be clustered by their pattern of distribution in space and time. Burbank (1972) identifies ten clusters of tumors by the pattern of death rates over time and over the United States and postulates a common causal agent for tumors within the same cluster.

I.2.3 Psychiatry

Diseases of the mind are more elusive than diseases of the body, and the classification of such diseases is in an uncertain state. There is agreement on the existence of paranoia, schizophrenia, and depression (such categories can be seen in Kant's classification published in 1790), but clear diagnostic criteria are not available, as Katz et al. (eds., 1970) remark. Shakow (1970) reports a study in which, of 134 patients diagnosed manic depressive at Boston Psychopathic Hospital, 28% were so diagnosed at Boston State Hospital and 10% were so diagnosed at Worcester State Hospital. A characteristic difficulty of classification of mental illness is the subjective, subtle, and variable character of the symptoms.

Numerical techniques have gained more acceptance in this area than in medical diagnosis. One of the earliest known contributions to clustering, by Zubin (1938a, 1938b), discusses a method of discovering subgroups of schizophrenic patients. The algorithm, which is of course oriented to hand calculation, is not clearly described. His schizophrenic group were on the average closer to his normal group than to themselves, which illustrates the variability problem.

The standard numerical technique collects mental state and historical data on each patient. The original data might be quite voluminous, but they are usually reduced by careful selection, by subjective averaging over groups of items, or by factor analysis. Some papers such as those of Everitt et al. (1971) and of Paykel (1970) seek clusters of patients. Others, such as those of Hautaluoma (1971) and of Lorr et al. (1963), seek clusters of symptoms (syndromes). Perhaps a great victory awaits clustering in psychiatry. Certainly, the psychiatrists have become hardened to numerical techniques and computerniks by their collaboration in factor analysis with psychologists. Certainly, the present classes and diagnostic rules satisfy no one. But perhaps the real problem is the vague data, which no amount of clever arithmetic will repair.

I.2.4 Archaeology and Anthropology

The field worker finds large numbers of objects such as stone tools, funeral objects, pieces of pottery, ceremonial statues, or skulls that he would like to divide into groups of similar objects, each group produced by the same civilization. Clustering techniques are surveyed in Weiner and Huizinger (eds., 1972) and in Hodson et al. (eds., 1971). Some recent papers are by Boyce (1969), who studies a number of average-linkage techniques on 20 skulls, and by Hodson (1969, 1970), who considers a wide range of techniques on three interesting data sets—broaches, stone tools, and copper tools.

I.2.5 Phytosociology

Phytosociology concerns the spatial distribution of plant and animal species. It bears the same relation to taxonomy that epidemiology bears to the classification of disease. Typical data consist of counts of the number of species in various quadrats. Clustering detects similar quadrats as being of the same type of habitat. An article by Whittaker (1962) contains a survey of traditional approaches. Lieth and Moore (1970) reorder the data matrix so that similar species are close in the new ordering and similar quadrats are close. Clark and Evans (1954) suggest a significance test for the random distribution of individuals (such as trees) within a region.

I.2.6 Miscellaneous

Clustering has been applied in diverse fields. In economics, Fisher (1969) considers input–output matrices in which the rows and columns have the same labels, so that the clustering of rows and columns must occur simultaneously. In market research, Goronzy (1970) clusters firms by various financial and operating characteristics, while King (1966) does so by stock price behavior. Frank and Green (1968) review a number of interesting applications. In linguistics, Dyen et al. (1967) use the proportion of matched words over a list of 196 meanings as a measure of the distance between two languages, with the aim of reconstructing an evolutionary tree of languages. Kaiser (1966) and Weaver and Hess (1963) consider numerical methods for establishing legislative districts. Abell (1960) finds clusters of galaxies by searching photographic plates of all high galactic latitudes. He lists 2712 such clusters and demonstrates that the clusters are not randomly distributed but exhibit further clustering themselves. Psychological applications are less common because of the dominance of factor analysis and multidimensional scaling, which are frequently interpreted as classifications. Miller (1969) has 50 Harvard students divide 48 nouns into categories according to similarity of meaning; the nouns are clustered into 5 groups, measuring similarity between two nouns by the proportion of students who place them in the same category. Wiley (1967) uses a factor-analysis-like technique on a similar data set.

I.3 FUNCTIONS OF CLUSTERING

The principal functions of clustering are to name, to display, to summarize, to predict, and to require explanation. Thus all objects in the same cluster will be given the same name. Objects are displayed, in order that subtle differences may become more apparent, by physically adjoining all objects in the same cluster. Data are summarized by referring to properties of clusters rather than to properties of individual objects. If some objects in a cluster have a certain property, other objects in the cluster will be

Table I.2 Mammals' Milk

	WATER	PROTEIN	FAT	LACTOSE
HORSE	90.1	2.6	1.0	6.9
DONKEY	90.3	1.7	1.4	6.2
MULE	90.0	2.0	1.8	5.5
CAMEL	87.7	3.5	3.4	4.8
LLAMA	86.5	3.9	3.2	5.6
ZEBRA	86.2	3.0	4.8	5.3
SHEEP	82.0	5.6	6.4	4.7
BUFFALO	82.1	5.9	7.9	4.7
GUINEA PIG	81.9	7.4	7.2	2.7
FOX	81.6	6.6	5.9	4.9
PIG	82.8	7.1	5.1	3.7
RABBIT	71.3	12.3	13.1	1.9
RAT	72.5	9.2	12.6	3.3
DEER	65.9	10.4	19.7	2.6
REINDEER	64.8	10.7	20.3	2.5
WHALE	64.8	11.1	21.2	1.6

Selected animals have been clustered by similarity of percentage constituents in milk. [From *Handbook of Biological Data*, Spector, ed. (1956), Saunders.]

Table I.3 Mammal's Milk Summarized

	NUMBER	WATER	PROTEIN	FAT	LACTOSE
HORSE-MULE	3	90.1 ± 0.2	2.1 ± 0.5	1.4 ± 0.4	6.2 ± 0.7
CAMEL-ZEBRA	3	87.0 ± 0.8	3.5 ± 0.5	4.0 ± 0.8	5.2 ± 0.4
SHEEP-PIG	5	82.2 ± 0.6	6.5 ± 0.9	6.5 ± 1.4	3.8 ± 1.1
RABBIT-RAT	2	71.9 ± 0.6	10.8 ± 1.6	12.9 ± 0.3	2.6 ± 0.7
DEER-WHALE	3	65.3 ± 0.6	10.8 ± 0.3	20.5 ± 0.8	2.1 ± 0.5

expected to have the same property. Finally, clear-cut and compelling clusters such as clusters of stars or animals require an explanation of their existence and so promote the development of theories such as the creation of matter in space or the evolutionary theory of Darwin.

Some data on mammal's milk will be used to illustrate these functions. These data are displayed in Table I.2, where animals with similar proportions of the constituents of milk are sorted into contiguous groups. Names for the groups are given by looking for characteristic properties of the groups. For example, the deer-reindeer-whale group is a high-fat group, the horse-donkey-mule group is a high-lactose group, and the rabbit-rat group is a high-protein group. A summary of the data for five groups on four variables appears in Table I.3. Such a summary makes the data easier to understand and to manipulate. For example, it now becomes apparent that water and lactose tend to increase together, that protein and fat tend to increase together, and that these two groups of properties are inversely related, with high water and lactose corresponding to low fat and protein. As for manipulation, the mean water content in the original data requires a sum of 16 numbers yielding the mean 80.03. The mean water content in the summarized data requires a sum of five numbers yielding the mean 80.13.

Prediction might occur in two ways. First, if a new object is classified into one of these groups by some other means, the same values would be predicted for the variables. Thus a mouse would be expected to be in the rabbit-rat group and to have a protein content of about 10.8%. Secondly, a new measurement of a similar type should exhibit a similar grouping. Thus, if a horse had 1.3% of a special protein, it would be predicted that a donkey also had 1.3% approximately of that protein.

The clusters require explanation, especially those clusters which differ quite a bit from the accepted mammalian taxonomy. For example, the zebra is classified with the camel and llama rather than with the horse by using this data. It may be that the classical taxonomy is incorrect, but of course it is supported by much more compelling data. It may be that the milk constituents are more dependent on the eating habits and local environment than on evolutionary history. Another odd grouping is the deer-reindeer-whale group. The high percentage of fat perhaps is necessary for resistance to cold.

I.4 STATISTICS AND DATA ANALYSIS

Like factor analysis, clustering techniques were first developed in an applied field (biological taxonomy) and are rarely accompanied by the expected statistical clothing of significance tests, probability models, loss functions, or optimal procedures. Although clustering techniques have been used (with arguable effect) in many different fields, they are not yet an accepted inhabitant of the statistical world. As Hills remarks after Cormack (1971), "The topic . . . calls to mind, irresistibly, the once fashionable custom of telling fortunes from tea leaves. There is the same rather arbitrary choice of raw material, the same passionately argued differences in technique from one teller to another, and, above all, the same injunction to judge the success of the teller solely by whether he proves to be right." Could statistics itself be described in this way? Do fortune tellers passionately argue technique? The principal difference between traditional biological taxonomy and numerical taxonomy is that techniques in numerical taxonomy are sufficiently precisely stated to allow passionate argument. As Gower says after Cormack (1971),

"No doubt much 'numerical taxonomic' work is logically unsound, but it has acquainted statisticians with some new problems whose separate identities are only now beginning to emerge. If statisticians do not like the formulations and solutions proposed, they should do better, rather than denigrate what others have done. Taxonomists must find it infuriating that statisticians, having done so little to help them, laugh at their efforts. I hope taxonomists who have real and, I think, interesting problems find it equally funny that so much statistical work, although logically sound, and often mathematically complicated (and surely done for fun), has little or no relevance to practical problems."

Well then, clustering is vulnerable on two fronts; the first, that the classifications delivered are not sufficiently compelling to convince the experts, who believe that detailed knowledge is more important than fancy manipulation; the second, that the techniques themselves are not based on sound probability models and the results are poorly evaluated and unstable when evaluated.

It may be that statistical methods are not appropriate for developing clusters, because some classification is often a prerequisite for statistical analyses. For example, crime statistics are based on the classification of crimes by police officers and on the geographical division of a country into various areas. There is a standard metropolitan area for each city to be used routinely in the collection of statistics. The important thing about these classifications is that there be clear rules for assigning individuals to them, rather than that they be optimal. Demographers worry about the classification process, since discovery of a trend is frequently only discovery of a change in classification practice. The New Haven police were accused in 1973 of classifying crimes benevolently to show an improvement in the serious crime rate, which would justify receiving a Federal grant. Likewise, the Yale grading system shows a self-congratulatory tendency to give students higher and higher grades so that of the four grades fail, pass, high pass, and honors, high pass is now a disgrace and pass and fail are regarded as extreme and unusual punishment. Thus, without a stable and appropriate classification scheme, statistical analyses are in vain. On the other hand, clustering techniques require raw data from some initial classification structure also, so it is doubtful whether formal techniques are sufficient for organizing the initial data gathering structure. Perhaps a mixture of informal and formal classification techniques is required.

Correspondingly, probability judgements depend basically on classifications and similarity judgements. The probability that it will rain an hour from now requires identification of "rain" (and rules for when rain occurs) and identification of "an hour from now." The numerical computation of this probability will require looking into the past for weather circumstances similar to the present one and counting how frequently it rained an hour later than those circumstances. Thus probability judgements arise from the principle of analogy that "like effects likely follow from like causes."

In some areas, such as in clustering of stars or in clustering of animals, the appropriateness of clustering is not in question. What about clustering used as a general purpose tool for data analysis? For one-dimensional data, the analogous procedure is to compute histograms. So one meaning of clustering is the computation of multi-variate histograms, which may be useful and revealing even if there are no "real" clusters. Another humble and primitive use of clustering is the estimation of missing values, which may be more accurately done from the similar values in a cluster than from the complete set of data. The summary function of clustering may save much computer time in later more sophisticated analyses.

I.5 TYPES OF DATA

I.5.1 Similarities

A cluster is a set of similar objects. The basic data in a clustering problem consist of a number of similarity judgements on a set of objects. The clustering operations attempt to represent these similarity judgements accurately in terms of standard similarity structures such as a partition or a tree. Suppose that there are M objects, $1 \leq I \leq M$. The similarity judgements might come in many forms. For example,

(i)   There is a real-valued distance $D(I, J)$ for each I, J.

(ii)   The distance $D(I, J)$ takes only two values 0 and 1, and $D(I, I) = 0$, $1 \leq I \leq M$. Thus each pair of objects is either similar or dissimilar.

(iii)   The distance $D(I, J)$ takes only values 0 and 1, and $D(I, J) \leq D(I, K) + D(K, J)$. In this case, the similarity relation expressed in (ii) is transitive, and the set of all objects is partitioned into a number of clusters such that $D(I, J) = 0$ if and only if I and J are in the same cluster.

(iv)   Triadic similarity judgements of the form "I and J are most similar of the three pairs (I, J), (I, K), and (K, J)."

I.5.2 Cases by Variables

The standard data structure in statistics assumes a number of cases (objects, individuals, items, operational taxonomic units) on each of which a number of variables (properties, characters) is measured. This data structure is almost always assumed in the book.

The variables may, in principle, take values in any space, but in practice there are five types of variables classified according to arbitrariness of the underlying scale:

(i)   *Counts* (e.g., the number of eyes on an ant) with no scale arbitrariness.

(ii)   *Ratio scale* (e.g., the volume of water in a cup), which is determined only in ratio to a standard volume.

(iii)   *Interval scale* (e.g., the height of a mountain) which is determined only from a standard position (sea level, say) and in terms of a standard unit (feet, say).

(iv) *Ordinal scale* (e.g., socioeconomic status), determined only by an ordered classification that may be changed by any monotonic transformation.

(v) *Category scale* (e.g., religion), determined by a classification that may be changed by any one-to-one transformation.

A given set of data may be *mixed* (containing variables of different types), it may be *heterogeneous* (variables of the same type but of different scales, such as temperature, rainfall, and corn production), or it may be *homogeneous* (variables measured on the same scale, such as percentage Republican vote for President in various years). There are a number of techniques transforming variables of one type to another or converting all variables to the same scale.

One special type of data has an identity between cases and variables. An example is imports, in dollars, from one country to another or the scores in a football conference where each team plays every other team. A distance matrix (or similarity matrix) is of this type. A standard approach in clustering (for example, Sokal and Sneath, 1963 and Jardine and Sibson, 1971) computes a distance matrix on the cases and then constructs the clusters from the distance matrix (see Chapter 2). In the distance approach, the two questions of the computation of distances from the cases by variables matrix and the computation of clusters from the distance matrix are separated. This causes a serious dilemma in that variables must be weighted before distances can be computed, yet it is desirable to have the weights depend on the within-cluster variability. An aesthetic objection to the distance approach is that the distance matrix is just a middle step between the actual data and the final clustering structure. It would be better to have algorithms and models that directly connect the data and the desired clustering structure.

Sometimes cases are clustered and sometimes variables are clustered. The traditional classification schemes in biology explicitly connect clusters of animals and clusters of characters; thus vertebrates are animals with backbones, two eyes, bilateral symmetry, four limbs, etc. A number of "two-way" algorithms in this book simultaneously produce clusters of cases and clusters of variables without a once-and-for-all distance calculation.

I.5.3 Other Data Structures

The cases-by-variables structure may be extended by the use of "not applicable" or "missing" values to include very general types of data. There are types of data which do not fit easily into this structure. One difficulty is the homology problem, in which it is not clear how a variable is to be measured on different objects. For example, the wings of a bird correspond to the arms of a man. The wings of a bee do not correspond to the wings of a bird, but there are more basic variables, such as the amino acid sequences in cytochrome-*c* molecules, which do correspond between a bird and a bee. The homology question makes it clear that the selection of variables and their measurement from object to object requires intimate knowledge of the objects and perhaps a preliminary informal classification.

A second difficulty arises when numerous measurements are applicable if some other condition is met. For example, if an insect has wings, the pattern of venation, the angle of repose, the relative size of the first and second pairs, whether or not the wings hook together in flight, and so on, are all useful measurements. It would be desirable to indicate explicitly in the data structure that these measurements are applicable only to insects with wings.

I.6 CLUSTERING STRUCTURES AND MODELS

There are only two clustering structures considered in this book, *partitions* and *trees*. A *cluster* is a subset of a set of objects. A partition is a family of clusters which have the property that each object lies in just one member of the partition. Algorithms for constructing partitions are considered in Chapters 3–7.

A model for cases-by-variable data, given a partition of the cases, is that within each cluster each variable is constant over cases. This model is made probabilistic by allowing cases to deviate randomly from the constant value in each cluster. A partition model for distance data is that all distances between pairs of objects in the same cluster are less than distances between pairs of objects in different clusters.

A tree is a family of clusters, which includes the set of all objects and for which any two clusters are disjoint or one includes the other. A partition, with the set of all objects added, is a tree.

A model on a distance matrix, given a tree, requires that the distance be an *ultrametric*. Thus if I, J, K are three objects, $D(I, J) \leq \max [D(I, K), D(J, K)]$. An ultrametric uniquely determines a tree for which $D(I, J) \leq D(K, L)$ whenever the objects I, J lie in the smallest cluster containing K and L. Thus a tree may be constructed from a distance matrix by finding the ultrametric closest to the distance matrix in some sense (see Chapter 11).

A weaker model requires that $D(I, J) < D(K, L)$ whenever I and J lie in a cluster strictly included in the smallest cluster containing K and L. This is similar to a model for triads which requires that I and J are more similar than K if I and J lie in a cluster that excludes K. An algorithm using triads is given in Chapter 10.

A tree model that applies directly on the data matrix requires that each variable be constant within clusters of a partition conforming to the tree; the partition is possibly different for different variables (see Chapters 14 and 18).

For clusters of both cases and variables, the basic unit is the *block*, which is a submatrix of the data matrix. The row margin and column margin of this submatrix form a case cluster and a column cluster. There are now three trees to consider: the tree formed by the case clusters, the tree formed by the row clusters, and the tree formed by the blocks themselves. Within a block, a variety of models are available. For example, for homogeneous data all values within a block might be assumed equal, or nearly so, as in Chapters 14 and 15. Or, for heterogeneous data, it might be assumed that the homogeneous model holds for some scaling of the variables to be discovered during the course of the algorithm, as in Chapter 16.

I.7 ALGORITHMS

All clustering algorithms are procedures for searching through the set of all possible clusterings to find one that fits the data reasonably well. Frequently, there is a numerical measure of fit which the algorithm attempts to optimize, but many useful algorithms do not explicitly optimize a criterion. The algorithms will be classified here by mode of search.

I.7.1 Sorting

The data are in cases-by-variables form. Each variable takes a small number of distinct values. An important variable is chosen somehow, and the objects are partitioned according to the values taken by this variable. Within each of the clusters of the partition, further partitioning takes place according to further important variables. Examples of this type of algorithm are given in Chapters 3, 17, and 18. These algorithms are quick in execution but are unsatisfactory in handling many variables, since only a few variables affect the classification.

I.7.2 Switching

A number of objects are to be partitioned. An initial partition is given, and new partitions are obtained by switching an object from one cluster to another, with the algorithm terminating when no further switches improve some criterion. Algorithms of this type occur in Chapters 4 and 6. These algorithms are relatively quick in execution but suffer from the uncertainty of the initial partition. There is always the possibility that a different initial partition might lead to a better final partition.

I.7.3 Joining

Initially begin with a number of clusters each consisting of a single object. Find the closest pair of clusters and join them together to form a new cluster, continuing this step until a single cluster containing all the original objects is obtained. This type of algorithm appears in Chapters 11–13, 15, and 16. These algorithms have gained wide acceptance in numerical taxonomy. The search for closest pairs is rather expensive, so that the algorithms are only practicable for moderately large ($<$1000) numbers of objects. There is often an embarrassingly rich overflow of clusters in the output from the algorithm, which must be reduced in number, often by *ad hoc* methods.

I.7.4 Splitting

Inversely to joining algorithms, splitting algorithms begin by partitioning the objects into several clusters and then partitioning each of these clusters into further clusters, and so on. Sorting algorithms are a special subset of splitting algorithms. Other splitting algorithms occur in Chapters 6 and 14. These algorithms have found less acceptance than joining algorithms because it is difficult to decide on a compelling splitting rule. The splitting may also require an expensive search.

I.7.5 Adding

A clustering structure (a partition or a tree) already exists, and each object is added to it in turn. For example, a typical object is initially selected for each cluster, and each object is added to the cluster to whose typical object it is closest. Algorithms of this type appear in Chapters 3, 9, and 10.

I.7.6 Searching

It sometimes happens that mathematical considerations rule out many possible clusterings for a particular criterion, and it becomes feasible to search over the remaining clusterings for the optimal one. Such an algorithm appears in Chapter 6.

I.8 INTERPRETATION AND EVALUATION OF CLUSTERS

An algorithm has been applied to the data producing these many clusters. Suppose that the clusters are subsets of cases for cases-by-variables data. At this point, all that is known is that the cases in clusters are similar in some way, and the clustering is not

useful until the ways of similarity are made clear. It is therefore important to summarize the behavior of variables within clusters. (These summaries are an intrinsic
part of the two-way algorithms in Chapters 14–16. Summaries for category variables
appear in Chapter 13 and for continuous variables in Chapter 18.)

For continuous variables the summaries might be mean, standard deviation,
minimum, and maximum for each variable for each cluster. If the clustering structure
is a partition, an analysis of variance will reveal which variables distinguish well
between clusters and which distinguish poorly. This information is useful for weighting
variables in later analyses. More generally a discriminant analysis will reveal which
combinations of variables distinguish best between clusters. The usual significance
tests of analysis of variance are not valid because the clusters will have been chosen
usually to maximize approximately a between-cluster sum of squares. Approximate
significance tests are known only in very few cases (see Chapters 4 and 11).

For category variables, agreement of a single variable with a partition is expressed
by a contingency table in which one margin is the variable and the other margin is the
partition. The partition is nothing but a category variable itself which is an amalgamation of all the category variables in the data, a category version of the first principal
component for continuous variables. The measure of agreement should be the proportion of correct predictions from the partition to the variable, under the predicting
rule that each member of the partition predicts that value of the variable that occurs
most frequently within it.

Between two partitions, the above measure of agreement is undesirable because it
is asymmetrical. There are many measures of association between category variables
in Goodman and Kruskal (1959). Suppose that the first variable takes values $1, 2, \ldots,$
M and the second variable takes values $1, 2, \ldots, N$. Let $L(I, J)$ denote the number
of times the first variable takes the value I and the second variable takes the value J.
Let $L(I, 0)$ be the number of times the first variable takes the value I, let $L(0, J)$ be
the number of times the second variable takes the value J, and let $L(0, 0)$ be the total
number of observations. One measure of association is the information measure

$$\sum \{1 \leq I \leq M, 1 \leq J \leq N\} L(I, J) \log L(I, J)$$
$$-\sum \{1 \leq I \leq M\} L(I, 0) \log L(I, 0)$$
$$-\sum \{1 \leq J \leq N\} L(0, J) \log L(0, J)$$
$$+ L(0, 0) \log L(0, 0).$$

(Here, and throughout the book, $\sum \{1 \leq I \leq M\}$ denotes summation over the
index I between the values 1 and M.) This is just the log likelihood ratio of the general
multinomial hypothesis against the hypothesis of independence between variables.
A measure that is more directly interpretable is the proportion of disagreed similarity
judgements,

$$\sum \{1 \leq I \leq M\} L(I, 0)^2 + \sum \{1 \leq J \leq N\} L(0, J)^2$$
$$- 2 \sum \{1 \leq I \leq M, 1 \leq J \leq N\} L(I, J)^2$$

divided by $L(0, 0) [L(0, 0) - 1]$.

Between two trees, a similar measure is the proportion of disagreed triads, the proportion of triples I, J, K in which I is more similar to J than to K according to the
first tree, but I is more similar to K than to J by the second tree.

One method of assessing the stability of clusters is to estimate the probability that
an object belongs to a cluster. In Wiley (1967) and in the mixture model in Chapter 5,

these probabilities are components of the model. For category variables amalgamated into a partition, each value of a variable has a frequency distribution over clusters. For each variable, a case has a frequency distribution over clusters according to the value of the variable. For a particular case, the frequency distribution is averaged over variables to form an estimated probability distribution that the case belongs to various clusters. A similar technique works for trees.

Clusters will be stable if all variables conform to the same clustering and unstable if different sets of variables suggest quite different clusters. This idea of stability may be applied directly by repeating the clustering on different sets of variables. The choice of sets might be made on prior knowledge; for example, classify insects by mouth parts, or by wing venation, or by genitalia, and assess the similarity of the classifications. In the absence of such knowledge, it is plausible to select variables from the original set so that each variable has, independently, probability 0.5 of appearing in the trial set. The selection and clustering are repeated several times, and agreement between the various clusterings is assessed. Examination of stability is useful in elimination of the many, only slightly different, clusters produced by joining methods.

I.9 USING THIS BOOK

Every chapter in the book with the exception of Chapter 2 has the same organization around a main algorithm, as follows:

(**i**)   concept and purpose of algorithm;

(**ii**)   step-by-step description of algorithm;

(**iii**)   execution of algorithm on a small data set;

(**iv**)   discussion and evaluation of algorithm and sometimes description of alternative strategies and algorithms;

(**v**)   suggestions for applying the algorithm, and various exercises, further techniques, and unsolved problems;

(**vi**)   a discussion of relevant literature;

(**vii**)   listing of Fortran programs.

Each chapter is more or less self-contained; one type of user of this book will be looking for a particular algorithm for a particular problem and should need to look at only two or three chapters. A second type of user will be learning about clustering in general, perhaps in a course. The book gets steadily more difficult, but there are some ridges in Chapter 4, 5, and 11 that perhaps should best be avoided in the first reading. These are starred. This second type of user is strongly urged to apply the techniques to his own small data sets and to try some of the problems in the THINGS TO DO sections. The difficult and impossible problems are starred. A sequence of chapters covering nearly all areas is 1–4, 8, 11, 13, 15, 18.

REFERENCES

GENERAL REFERENCES

BLACKWELDER, R. A. (1966). *Taxonomy: A Text and Reference Book*, Wiley, New York. A survey of classical biological taxonomy.

CORMACK, R. M. (1971). "A review of classification." *J. Roy. Stat. Soc.*, A **134**, 321–367. This is a careful and complete survey of the recent literature of classification, with reference to over 200 papers. The author is somewhat skeptical about the uses of classification, beginning "The availability of computer packages of classification techniques has led to the waste of more valuable scientific time than any other 'statistical' innovation" and ending with a quotation from Johnson: "Anyone who is prepared to learn quite a deal of matrix algebra, some classical mathematical statistics, some advanced geometry, a little set theory, perhaps a little information theory and graph theory, and some computer technique, and who has access to a good computer, and who enjoys mathematics . . . will probably find the development of new taximetric method much more rewarding, more up-to-date, more 'general,' and hence more prestigious than merely classifying animals or plants or working out their phylogenies." Beneath this bitter skin lies sweet flesh. Cormack distinguishes between *identification* or *assignment*, which allocate an individual to known classes, and *classification*, which constructs the classes. He considers three types of classification— *hierarchical*, *partitioning*, and *clumping*—in which the classes or clumps may overlap. There is a brief (and rather unsatisfactory) discussion of the purposes of classification. Similarity measures are listed, and the difficult problems of scaling are discussed; in particular, if within-cluster measures are used to scale, "there appears to be a circularity in trying to transform data by properties of the cluster that it is hoped to determine." The homology problem is discussed, wherein "it is not clear what variable in one entity corresponds to a particular variable in another."

"There are many intuitive ideas, often conflicting, of what constitutes a cluster, but few formal definitions." Sometimes, "the resulting clusters are defined only by the algorithm by which they were obtained." He divides algorithms into three types: agglomerative (joining close pairs), divisive (successive splitting), and clustering (reallocation of individuals between members of a partition. Another technique is the formation of single clusters, each of which is completed before another is initiated. The few known comparative and evaluative studies are reported. Measures of distance between a tree and similarity structure are given.

Searching for modes is closely related to single linkage. Overlapping clusters are discussed, following the Jardine and Sibson generalization of single linkage. A number of papers treat entities and variables simultaneously.

GOOD, I. J. (1965). "Categorization of classification." *Mathematics and Computer Science in Biology and Medicine*, H. M. Stationery Office, London. Many interesting remarks on problems in classification. In particular, a method of two-way splitting is proposed for a data matrix $\{A(I, J), 1 \leq I \leq M, 1 \leq J \leq N\}$, according to row vectors $\{R(I), 1 \leq I \leq M\}$ and column vectors $\{C(J), 1 \leq J \leq N\}$ that take the values ± 1 and maximize $\sum R(I)C(J)A(I, J)$.

GOODMAN, L. A., and KRUSKAL, W. H. (1959). "Measures of association for cross classification II: Further discussion and references." *J. Am. Stat. Ass.* **54**, 123–163. An earlier paper introduced some measures of association, and some further measures are discussed in the first section of this paper. The major part of the paper is an extensive and careful survey of the ancient and diverse literature of measures of association.

JARDINE, N., and SIBSON, R. (1971). *Mathematical Taxonomy*, Wiley, New York. The first part of the book develops various measures of dissimilarity. The central measure is based on information theory; this leads to Mahalanobis' D with normal subpopulations. The second part of the book assumes a dissimilarity matrix given and regards a cluster method as a function from dissimilarity matrices to trees (or dendrograms, or equivalently to dissimilarity matrices satisfying the ultrametric inequality). The single-linkage method corresponds to a continuous function on the space of dissimilarity matrices, whereas the other joining techniques are discontinuous. Overlapping clusters are suggested as a generalization of the single-linkage technique, which avoids chaining. The book concludes with some remarks on applications, with particular attention to problems in biological taxonomy.

SOKAL, R. R., and SNEATH, P. H. A. (1963). *Principles of Numerical Taxonomy*, Freeman, London. The book is divided into three parts. The first part is a general introduction to numerical taxonomy, including criticism of then current taxonomic practices. The second part discusses data preparation including selection of organisms, characters, and coding of characters. In this part also, methods of measuring similarity are proposed, with particular attention given to category data. There is some discussion of weighting and scaling.

A technique for shading similarity matrices yields clusters visually. Average- and single-linkage algorithms are discussed.

In the third part of the book, implications of these techniques for developing evolutionary connections are considered, and also applications to classification in other areas.

AREA REFERENCES

Biology

BORRADAILE, L. A., and POTTS, F. A. (1958). *The Invertebrata: A Manual for the Use of Students*. Cambridge U. P., London.

Metazoa: At least two body layers. The inner layer, the endoderm, contains cells for digestion of food. The outer layer, the ectoderm, contains nerve cells, sensors, and cells forming a protective sheet.

Arthropoda: Bilaterally symmetrical, segmented. Paired limbs on the somites, of which at least one pair functions as jaws. A chitinous cuticle. A nervous system like the Annelida. The coelom in the adult much reduced and replaced as a previsceral space by enlargement of the haemocoele. Without true nephridia, but with one or more pairs of coelomoducts as gonoducts and often as excretory organs. Without cilia.

Insecta: The body is divided into three distinct regions, the head, thorax, and abdomen. The head consists of six segments and there is a single pair of antennae. The thorax consists of three segments with three pairs of legs and usually two pairs of wings. The abdomen has 11 segments typically and does not possess ambulatory appendages. Genital apertures are situated near the anus.

Pterygota: Young stage born with full complement of adult segments. No abdominal locomotory appendages. Simple metamorphosis. Malphigian tubes present. Mouth parts free.

Endopterygota: Wings folded in repose. Holometabolous. Endopterygote. Usually with few malphigian tubes.

Hymenoptera: Head orthognathous. Mouth parts for biting or fluid feeding. Mandibles. First abdominal segment fused to thorax. Four membranous wings, anterior wings linked to posterior wings by a groove and hooks. Tarsus pentamerous. Female with ovipositor. Numerous malphigian tubes. Legless larva with well-formed head capsule.

Formicoidea: Social polymorphic insects in which two segments form the abdominal petiole. Females have a sting.

DARLINGTON, P. J., JR. (1971). "Modern taxonomy, reality and usefulness." *Systematic Zoology* **20,** 341–364. The ultimate purpose of taxonomy is a classification useful in storing and retrieving information. Some numerical taxonomists seem to be "slipping off into a kind of scientism which brings an appearance of precision by the abandonment of reality." As an example of modern practical taxonomy, Lindroth's taxonomy of ground beetles is discussed in detail.

DUPONT, P. F., and HENDRICK, L. R. (1971). "Deoxyribonucleic acid base composition, and numerical taxonomy of yeasts in the genus trichosporon." *J. Gen. Microbiol.* **66,** 349–359. The DNA base was determined for each of 10 species of trichosporon, and the species were divided into 4 groups according to the percentage of quanine and cytosine in each species. This classification was compared to a "hands-off" numerical taxonomy of 25 strains of trichosporon, based on 81 characteristics. There was only slight agreement between the two classifications.

SMALL, E., BASSETT, I. J., and CROMPTON, C. W. (1971). "Pollen phylogeny in *Clarkia.*" *Taxon* **20,** 739–749. Electromicroscopy is used to measure 16 characteristics of 42 species of pollen. There are twelve continuous variables measuring geometric properties of the trilobed grains and four discrete variables measuring presence or absence of patterns, etc. This poses an interesting difficulty in the combination of different types of data.

STEARN, W. T. (1971). "A survey of the tropical genera *Oplonia* and *Psilanthele* (*Acanthaceae*)." *Bull. Brit. Museum* (*Nat. Hist.*) *Bot.* **4,** 261–323. Uses numerical taxonomy conservatively in a revision of the genera.

Medicine

BARON, D. N., and FRASER, P. M. (1968). "Medical applications of taxonomic methods." *Brit. Med. Bull.* **24,** 236–240. A number of clustering algorithms were tested on 50 patients with liver disease, with 330 characters recorded for each patient. In conforming to previous diagnoses of the patients, the single-linkage algorithm was less accurate than the average-linkage technique. Weighting characters to give more weight to rarely appearing characters also improved conformity to previous diagnoses.

BOORMAN, K. E., and DODD, B. E. (1970). *An Introduction to Blood Group Serology,* Churchill, London. In 1900, Landsteiner described the agglutination that occurred when red cells of one individual were exposed to the action of serum from another and so discovered the A, B, O blood group system. The four groups are determined by the presence or absence of the antigens A and B in the red blood cells. Corresponding to the antigens A, B are antibodies −A, −B which occur in the sera of individuals lacking the corresponding antigens. Thus a person of type AB has serum containing no antibodies, which is acceptable to all. A person of type O has serum containing antibodies −A and −B, acceptable only to another person of type O.

BOUCKAERT, A. (1971). "Computer diagnosis of goiters: I. Classification and differential diagnosis; II. Syndrome recognition and diagnosis; III. Optimal sub-symptomatologies." *J. Chronic Dis.* **24,** 299–310, 311–320, 321–327. The single-linkage and complete-linkage algorithms are applied to a grouping of 85 patients with goiters, using 30 symptoms. The complete-linkage algorithm gives results in accordance with conventional nosology. Distance between patients is measured in a number of ways; for example, one measure is the proportion of symptoms they have in common.

Symptoms are grouped into symptom classes called syndromes, measuring distance between symptoms by correlation. Using single linkage, three syndromes are obtained, which are more or less coincident with the common descriptions of simple goiter, hyperthyroidism, and cancer.

BURBANK, F. (1972). "A sequential space-time cluster analysis of cancer mortality in the United States: etiological implications." *Amer. J. Epidemiol.* **95,** 393–417. Tumor types are defined in the International Classification of Disease either by cell type or by anatomical site. The correlations between death rates in the 50 United States are used as a measure of similarity between types. The most highly correlated types are amalgamated, until no pair of types have correlation exceeding 0.45. This yields 10 clusters of tumors which have different patterns of geographical distribution. The time trends for different tumor types are also considered. A common causal agent is postulated for tumor types with similar geographical distribution and time trend.

CHAKRAVERTY, P. (1971). "Antigenic relationship between influenza B viruses." *Bull. World Health Org.* **45,** 755–766. Single-linkage clustering is used to determine whether antigenic groupings could be demonstrated among a wide variety of strains of influenza B virus. The data are characteristic immunological data, with a correspondence between the rows (viruses) and columns (antisera). No clearly distinct groups were discovered.

"Criteria for the evaluation of the severity of established renal disease" (1971). *Ann. Intern. Med.* **75,** 251–252. The severity of the disease is classified in each of three categories: signs and symptoms, renal function impairment, and level of performance. The class assigned to each category is the highest class within which the patient falls.

Signs and Symptoms

    I.   No symptoms directly referrable to renal disease plus one or more of (a)–(e)
- (a) fixed proteinuria
- (b) repeatedly abnormal urine
- (c) radiographic abnormality in upper genito-urinary tract
- (d) hypertension
- (e) biopsy proved parenchymal renal disease

   II.   Two or more of
- (a) symptoms referrable to renal disease
- (b) radiographic evidence of osteodystrophy
- (c) stable anemia
- (d) metabolic acidosis
- (e) severe hypertension

  III.   Two or more of
- (a) osteodystrophy
- (b) peripheral neuropathy

(c) nausea and vomiting

(d) limited ability to conserve or excrete sodium and water, tending to sodium depletion, dehydration, congestive heart failure

(e) impaired mentation

IV. Any two or more of

    (a) uremic pericarditis

    (b) uremic bleeding diathesis

    (c) asterixis and impaired mentation

    (d) hypocalcemic tetany

V. Coma

Renal Function Impairment

Glomerular filtration rate (secondarily, on serum creatinine with a different set of intervals)

(a) normal

(b) 50–100%

(c) 20–50%

(d) 10–20%

(e) 5–10%

(f) 0–5%

Performance

What the patient thinks he can do:

(a) All usual activity

(b) No strenuous activity

(c) No usual daily activity

(d) Severe limitation of usual activity

(e) Semicoma

FEINSTEIN, A. R. (1972). "Clinical biostatistics. 13. Homogeneity, taxonomy and nosography." *Clin. Pharmacol. Ther.* **13**, 114–129. One of a series of papers on the philosophy of classification. Homogeneity of a group must be established before statistical techniques are applicable. Measures of similarity must depend on final purposes. Nosology differs from traditional biological taxonomy because the taxonomy of disease must be oriented to diagnosis and treatment.

GOLDWYN, R. M., FRIEDMAN, H. P., and SIEGEL, J. H. (1971). "Iteration and interaction in computer data bank analysis—case study in physiological classification and assessment of the critically ill." *Comp. Biomed. Res.* **4**, 607–622. In assessing the physiological status of critically ill patients, three stages of the septic process are identified, each characterized by its own pattern of homeostatic compensation. The principal variables used are the following (measured in logarithms):

  (i)   cardiac output/body area,

 (ii)   mean blood pressure,

(iii)   arterial venous difference in oxygen,

(iv)   heart rate,

 (v)   ejection time, and

(vi)   central venous pressure.

A base group of patients not critically ill is used as a background. The Friedman-Rubin algorithm is used to maximize trace B/trace W, where B denotes the covariance matrix between groups, and W the covariance matrix within groups. The three groups discovered represent stages with markedly different prognosis. Transitions of patients between the stages support the ordering of the stages.

GOODFELLOW, N. (1971). "Numerical taxonomy of some nocardioform bacteria." *J. Gen. Microbiol.* **69**, 33–80. There are 283 bacteria and 241 characters, some biochemical, some physiological, and some nutritional. Using two measures of similarity and two joining algorithms (the single- and average-linkage methods), seven major clusters are detected. These correspond to groupings already known. Many new characters are shown to be suitable for identifying bacteria.

HAYHOE, F. G. J., QUAGLINO, D., and DOLL, R. (1964). "The cytology and cytochemistry of acute leukemias." *Medical Research Council Special Report Series No. 304*, H.M. Stationery Office, London. Single linkage identified four groups in 140 cases of acute leukemia. Diagnostic criteria are proposed for distinguishing the groups.

KNUSMAN, R., and TOELLER, M. (1972). "Attempt to establish an unbiased classification of diabetes mellitus." *Diabetologia* **8**, 53. Using factor analysis, three groups are discovered closely related to age.

MANNING, R. T., and WATSON, L. (1966). "Signs, symptoms, and systematics." *J. Amer. Med. Assoc.* **198**, 1180–1188. Clustering techniques were tested on 99 patients with heart disease characterized by 129 items. The three clusters obtained agree substantially with physicians' diagnoses of univalvular lesions, multivalvular lesions, and arteriosclerotic, hypertensive, or pulmonary disease.

Manual of the International Statistical Classification of Diseases, Injuries, and Causes of Death (1965). World Health Organization, Geneva, Vols. 1 and 2. "Classification is fundamental to the quantitative study of any phenomenon. It is recognized as the basis of all scientific generalization, and is therefore an essential element in statistical methodology . . . . There are many who doubt the usefulness of attempts to compile statistics of disease, or even causes of death, because of difficulties of classification. To these, one can quote Professor Major Greenwood: 'The scientific purist, who will wait for medical statistics until they are nosologically exact, is no wiser than Horace's rustic waiting for the river to flow away'."

There are seventeen principal categories of disease, beginning with "Infective and Parasitic Diseases" and ending with the inevitable leftovers, such as "Certain Causes of Perinatal Morbidity and Mortality," "Symptoms and Ill-defined Conditions," and "Accidents, Poisonings and Violence." Each disease is represented by a three-digit code. All diseases of the same type have the same first two digits. There may be several types of diseases in a category. For example, the first category includes the following types: Intestinal infectious diseases (00), Tuberculosis (01), Zoonotic bacterial diseases (02), Other bacterial diseases (03), through Other infective and parasitic diseases (13). The type Tuberculosis includes Silicotuberculosis (010), Pulmonary Tuberculosis (011), through Late effects of tuberculosis (019).

The accident category is treated somewhat differently, with a separate code: E908 is "accident due to cataclysm."

Finally, there is a fourth digit which further breaks down each of the disease categories. For example, Late effects of tuberculosis (019) is divided into Respiratory tuberculosis (019.0), Central nervous system (019.1), Genito-Urinary system (019.2), Bones and joints (019.3), and Other specified organs (019.9).

PREVOT, A. H. (1966). *Manual for the Classification and Determination of the Anaerobic Bacteria*, Lea and Febiger, Philadelphia. Establishes the following classification scheme:

Schizomycetes
  Eubacteria (no true branching)
    Non-spore-forming
      spherical
      cylindrical
      spiral
    Spore-forming
      spherical
      cylindrical
      spiral
  Actinomycetaceae (tendency to form branches)

SCHREK, R., KNOSPE, W. H., and DONNELLY, W. (1972). "Classification of chronic lymphocytic leukemia by electron microscopy." *Fed. Proc.* **31,** A629. Blood lymphocytes from 43 patients with chronic lymphocytic lymphosarcoma cell leukemia were examined. The cross-sectional areas were classified A and B. This yields four groups: I A, I B, II A, II B. Groups I A and I B were mostly lymphocytic, groups II A and II B frequently had lymphosarcoma, I A were benign, and II B were aggressive.

STARK, L., OKAJIMA, M., and WHIPPLE, G. M. (1962). "Computer pattern recognition techniques: Electrocardiographic diagnosis." *Commun. Assoc. Comp. Mach.* **6,** 527–532. An electrocardiogram has three phases, PQ, QRS, ST-T, corresponding to different stages in the electrical excitation of the heart. Abnormal patterns are associated with pathological conditions. Techniques are described for recognizing abnormal patterns by computation rather than by expert judgement.

TEMKIN, O. (1970). "The history of classification in the medical sciences," in Katz, M. M., Cole, J. O., and Barton, W. E., eds. *Classification in Psychiatry and Psychotherapy*, National Institutes of Mental Health, Chevy Chase, Md. The Egyptians divided diseases according to anatomical location, and medical practitioners specialized accordingly. A number of ancient classification chemes are sketched, the four humors of the Greeks, the function-oriented classification of Galen, and Boissier de Sauvage's classes-orders-genera scheme in 1790, which parallels Linnaeus' systematics for animals and plants. Boissier de Sauvages believed that "species and genera of disease ought to be distinguished on a purely factual basis in contrast to causal theories. . . . [A] disease was to be defined by the enumeration of symptoms which suffice to recognize its genus or species, and to distinguish it from others."

WILDY, P. (1971). *Classification and Nomenclature of Viruses: First Report of the International Committee on Nomenclature of Viruses*, Karger, New York. Provides a classification scheme for recording important data about viruses. The classification lies more in identifying types of properties of viruses than in identifying groups of viruses. Each virus type is represented by a cryptogram recording

  (i)   Type of nucleic acid

  (ii) Strandedness of nucleic acid

  (iii) Molecular weight of nucleic acid

(iv) Percentage of nucleic acid in infective particles

(v) Outline of particle

(vi) Outline of nucleocapsid

(vii) Kinds of host infected

(viii) Kinds of vector

WILNER, B. I. (1964). *A Classification of the Major Groups of Human and Other Animal Viruses*, Burgess, Minneapolis. Viruses are sorted by five variables, with the most important variables used first (see Table I.4).

Table I.4   Classification of Viruses (Wilner, 1964)

SORTING VARIABLES

Nucleic Acid Core	Capsidal Symmetry	Envelope	Viral Particle Size	Acid Lability	Family
RNA	cubic	naked	18-38 mμ	stable	picornaviruses enteroviruses rhinoviruses
				labile	
			70-77 mμ	stable	reoviruses
		enveloped	30-75 mμ	labile	arboviruses
	helical	enveloped	90-300 mμ	labile	myxoviruses
			37-218 mμ	labile	myxoviruses like
DNA	complex	enveloped	120-250 mμ	labile	herpesviruses
			70-80 mμ	stable	adenoviruses
		naked	40-57 mμ	stable	papovaviruses
			18-26 mμ	stable	parvoviruses

WINKEL, P., and TYGSTRUP, N. (1971). "Classification of cirrhosis: the resolution of data modes and their recovery in an independent material." *Comp. Biomed. Res.* **4**, 417–426. Sneath's single-linkage method and Wishart's mode-seeking method are used to classify 400 patients with cirrhosis. Both methods identified two groups with distinct characteristics. Both methods left 70% of the patients unclassified.

Psychiatry

EVERITT, B. S., GOURLAY, A. J., and KENDELL, R. E. (1971). Attempt at validation of traditional psychiatric syndromes by cluster analysis. *Brit. J. Psychiat.* **119**, 399–412. "In spite of all the elegant and complex manipulations to which innumerable sets of data have been subjected, the classifications used by psychiatrists remain much as they were before the computer was invented." The *K*-means algorithm and a mixture algorithm are applied to two different sets of patients. There are 250 American patients and 250 British patients on which are measured 45 mental state items and 25 history items. These measurements are reduced to 10 factors by principal components and factor analysis. The clusters constructed corresponded to standard syndromes such as depression, schizophrenia, mania, drug abuse, and depressive delusions.

HAUTALUOMA, J. (1971). Syndromes, antecedents, and outcomes of psychoses: a cluster analytic study. *J. Consult. Clin. Psychol.* **37**, 332–344. "Current psychiatric mental illness description and classifications have a questionable status . . . . The entities called illnesses have low interjudge reliabilities. Frequently symptoms are unclassifiable." Clusters of symptoms are discovered empirically on 333 patients and validated on two more sets each of 333 patients by checking that symptoms within the clusters remained highly correlated. The clusters agree with the Lorr syndromes of excitement, hostile belligerence, paranoid projection, grandiose expansiveness, perceptual distortions, anxious intropunitiveness, retardation and apathy, disorientation, motor disturbance, and conceptual disorganization. It is noted that antecedent symptoms and outcomes of treatment are not highly related to any one of the clusters.

KANT, I. (1964). *The classification of mental disorders*, Charles T. Sullivan, ed., Doylestown Foundation, Doylestown, Pa. It is difficult to impose an orderly classification on that which is essentially and irreparably disordered. Kant recognizes four categories:

(i) *senselessness*, the inability to bring one's thoughts into that coherence which is necessary even for the mere possibility of experience;

(ii) *madness*, subjective impressions of a falsely inventive imagination are taken for actual perceptions;

(iii) *absurdity*, a disturbed capacity for judgement, whereby the mind is given to excessive analogies;

(iv) *frenzy*, the disorder of a disturbed reason.

KATZ, M. M., COLE, J. O., and BARTON, W. E., eds. (1970). *The Role and Methodology of Classification in Psychiatry and Psychopathology*, National Institutes of Mental Health, Chevy Chase, Md. All aspects of classification are considered in this conference report. As part of the editors' summary (p. 4), "psychoanalysts found grave difficulties in assigning patients to specific diagnostic categories and preferred to describe the presenting symptoms, while hospital psychiatrists felt that the assigning of diagnosis, if carefully done, had real clinical meaning and was a necessary prerequisite to good clinical practice. Even at the end, the position was by no means that diagnosis in its present form dictated treatment, but that the study of the individual case necessary to determine and arrive at a diagnosis brought with it an increase in the clarity with which the case was understood, and this study, plus the diagnosis, suggested certain directions for treatment and management."

LORR, M., KLETT, C. J., and McNAIR, D. M. (1963). *Syndromes of Psychosis*, Pergamon, New York. Includes a comprehensive review of earlier studies. By clustering symptoms, suggests 10 syndromes or clusters of symptoms.

PAYKEL, E. S. (1970). "Classification of depressed patients—cluster analysis derived grouping." *Brit. J. Psychiat.* **118**, 275. There are 165 depressed patients on which are measured 29 symptom ratings from a clinical interview, and six historical variables— namely, age, previous depressions, alcoholism, length of illness, neuroticism score, and stress score. These variables are reduced to six principal components. The patients are divided into four groups to maximize det T/det W in the Friedman-Rubin technique. The four groups are characterized as psychotic, anxious, hostile, and young depressive.

SHAKOW, D. (1970). *Classification in Psychiatry and Psychotherapy*, Katz, M. M., Cole, J. O., and Barton, W. E. eds. National Institutes of Mental Health, Chevy Chase, Md. Notes that of 57 patients diagnosed to have dementia praecox at Boston Psychopathic Hospital, 12% were so diagnosed at Boston State Hospital, and 37% were so diagnosed at Worcester State Hospital. Of 134 patients diagnosed manic depressive at Boston Psychopathic Hospital, 28% were so diagnosed at Boston State Hospital, and 10% were so diagnosed at Worcester State Hospital.

ZUBIN, J. (1938a). "A technique for measuring like-mindedness." *J. Abn. Soc. Psychol.* **33,** 508–516. A group of 68 schizophrenic patients and 68 matched normal individuals are measured on 70 items. Individuals which agree among themselves on more than 45 items are assembled into one group. The remaining individuals are sorted to find a group agreeing on at least 40 items. This process is continued until all abnormal individuals are classified, revealing eight groups.

ZUBIN, J. (1938b). "Socio-biological types and methods for their isolation." *Psychiatry* **1,** 237–247. A group of 68 schizophrenic patients was matched with a group of 68 normal individuals, and each group was given a personality inquiry form consisting of 70 items. The normals were more similar to each other on average than the abnormals, where similarity between two individuals is the proportion of similar responses over the 70 items. Groups of individuals similar among themselves were analyzed to detect common response patterns. In this way, six subgroups of abnormals and three subgroups of normals were established.

Archaeology and Anthropology

BOYCE, A. J. (1969). "Mapping diversity," in *Numerical Taxonomy*, Cole, A. J. ed., Academic, New York. A number of average-linkage clustering techniques are evaluated by application to 20 skulls of apes. It was found necessary to standardize the measurements for each ape to avoid size effects. The various techniques give only slightly different clusters.

HODSON, F. R. (1969). "Searching for structure within multivariate archaeological data." *World Archaeol.* **1,** 90–105. Average-linkage cluster analysis is applied to two sets of data. The first consists of 50 assemblages of stone tools from southwestern France and central Europe. The number of tools in the assemblages ranges from 127 to over 4000. The tools are divided, by expert judgement, into 46 types, and the data used in clustering are then the proportions of each type in each assemblage. Similarity between assemblages is measured by euclidean distance computed on the square root of the percentages. The average-linkage analysis suggests eight clusters of assemblages, which are closely related to groups already hypothesized. Three distinct groups, also conforming to the existing taxonomy, are visually apparent when the first two principal components are plotted.

The second set of data consists of spectrographic measurements of metal components in 100 copper tools. The percentages of the elements Cu, Sn, Pb, As, Sb, Ag, Ni, Bi, Au, Zn, Co, and Fe are estimated in each tool. Since many of the percentages are zero and variability is quite different for the different metals, a standardized measure of distance is necessary. The 16 groups obtained agreed reasonably well with geographical clustering of the tools.

HODSON, F. R. (1970). "Cluster analyses and archaeology. Some new developments and applications." *World Archaeol.* **1,** 299–320. Four clustering techniques are applied to a set of 30 iron age fibulae (broaches) mainly from a cemetery near Berne. Three angles and 10 dimensions are measured on each broach. The dimensions are expressed as proportions of total length so that overall size has no effect. The proportions are logged and standardized to unit variance, and then euclidean distance is used to measure dissimilarity between pairs.

The single-linkage algorithm gives two distinct clusters of six and seven objects, but other objects are added one at a time to these central groups in a characteristic chaining pattern, which make it difficult to define a working taxonomy. The average linkage algorithm "prevents the formation of chain clusters and provides the kind of structure that is useful in taxonomy. However, in doing so, it may force the units into an overly arbitrary framework." In particular, two similar units may be widely separated by the clustering. A double-linkage algorithm is a generalization of a single-linkage algorithm in which, approximately, each object is linked to the two objects closest to it. Overlapping clusters are the outcome of this algorithm. These differ only slightly from the single-linkage clusters. Finally, a version of the K-means algorithm produces partitions of various sizes. The correct partition size was not indicated by a sharp drop of within-cluster sum of squares; six clusters were nevertheless suggested. An attraction of the K-means technique is that it works directly on the data matrix rather than with distances, and so it saves computation time and produces interesting summary statistics.

HODSON, F. R., KENDALL, D. G., and TAUTU, P., eds. (1971). *Mathematics in the Archaeological and Historical Sciences*, University Press, Edinburgh. This "Proceedings of the Anglo-Romanian Conference" contains many relevant articles divided into the areas taxonomy, multidimensional scaling, and seriation and evolutionary tree structures in historical and other contexts. Rao clusters Indian caste groups, using Mahalanobis distance based on physical measurements. Hodson uses K-means on 500 ancient handaxes. Rowlett and Pollnac reduce 104 measurements, at 77 archaeological sites of Marnian cultural groups, to three factors, which they use to generate four clusters of sites. Kidd uses gene frequencies to reconstruct evolutionary histories of cattle. Kruskal, Dyen, and Black use cognation between words for 200 meanings in 371 Malayo-Polynesian languages to reconstruct the evolutionary tree for the languages. Buneman, Haigh, and Nita (in separate articles) consider evolutionary sequences for ancient hand-copied manuscripts.

WEINER, J. S., and HUIZINGER, J., eds. (1972). *The assessment of population affinities in man*, Oxford U. P., New York. A collection of papers including various ways of measuring distance between human populations and various studies of human populations.

Phytosociology

CLARK, P. J., and EVANS, F. C. (1954). "Distance to nearest neighbour as a measure of spatial relationships in populations." *Ecology* **35**, 445–453. For N individuals distributed at random over a large area, the distance of each individual to its nearest neighbor, averaged over all individuals is approximately normal with mean $\frac{1}{2}\sqrt{p}$ and standard deviation $0.26136\sqrt{Np}$, where p is the density of individuals. Larger-than-random values of the average distance correspond to uniform packing, whereas smaller-than-random indicate clustering.

LIETH, H., and MOORE, G. W. (1970). "Computerized clustering of species in phytosociological tables and its utilization in field work," in *Spatial Patterns and Statistical Distributions*, Patil, G. P., Pielou, E. C., and Waters, W. E. eds., Penn State Statistics Series. Measures proportion of each species in various quadrats. The species are reordered to reduce distances between neighboring species. Quadrats are similarly reordered.

WHITTAKER, R. H. (1962). "Classification of natural communities." *Bot. Rev.* **28**, 1–239.

Miscellaneous

ABELL, G. O. (1960). "The distribution of rich clusters of galaxies." *Astrophys. J. Suppl.* **32**, Vol. 3, pp. 211–288. Clusters of galaxies are to be inferred by examination of their projections on photographic plates. The galaxies are relatively close in projection on the plate and in brightness. That is, they satisfy the following:

(i) *Richness criterion.* A cluster must contain at least 50 members that are not more than two magnitudes fainter than the third-brightest member.

(ii) *Compactness criterion.* A cluster must be sufficiently compact that its 50 or more members are within a given radial distance r of its center.

2712 rich clusters are listed. These clusters are not randomly distributed over the high galactic latitudes studied, which indicates further clustering.

DYEN, I., JAMES, A. T., and COLE, J. W. L. (1967). "Language divergence and estimated word retention rate." *Language* **43**, 150–171. The data studied consist of the words in 89 Austronesian languages for 196 meanings (nearly the same as the Swadesh list). A sophisticated probability model is employed in which it is assumed that each pair of languages is separated in time by an unknown constant (twice the time since divergence from a common ancestor), and each meaning has a different mutation rate. The waiting time to mutation is assumed exponential. The time separations and mutation rates are then estimated by maximum likelihood.

FISHER, W. D. (1969). *Clustering and Aggregation in Economics*, Johns Hopkins, Baltimore. A number of economic applications are discussed. An interesting special data type is the input–output table, which is handled by a "lockstep progressive merger." This is a joining algorithm which joins the two closest rows and the corresponding two columns simultaneously.

FRANK, R. E., and GREEN, P. E. (1968). "Numerical taxonomy in marketing analysis, a review article." *J. Market. Res.* **5**, 83–98. A number of different clustering techniques and measures of similarity are illustrated on several marketing research studies. For example, 88 cities are divided into five groups by using a K-means technique, with data consisting of 14 business and demographic variables. In another study, the proportion of viewers in various age groups is used to cluster TV programs. In a brand-loyalty study, the percentages of various brands (within the three product categories beverages, coffee, and cereals) by weight were taken for 480 households. Both households and brands are clustered. Again, 100 different computer models were characterized by 12 variables such as word length, execution time, and transfer rate. In another study, the basic data consisted of a zero–one matrix in which each physician is classified as a light or heavy reader of each of 19 medical journals.

GORONZY, F. (1970). "A numerical taxonomy on business enterprises," in *Numerical Taxonomy*, Cole A. J., ed., Academic, New York. A questionnaire is completed by 50 firms, some questions dealing with fixed assets, others with productivity, others with labor statistics. To reduce size effects, average linkage is applied to the array of correlations between firms, yielding four clusters.

KAISER, H. F. (1966). "An objective method for establishing legislative districts." *Mid. J. Polit. Sci.* **10**, 200–213.

KING, B. F. (1966). "Market and industry factors in stock price behavior." *J. Bus.* **39**, 139–190. Clusters of stocks are obtained by using a joining algorithm on stock price data.

MILLER, G. A. (1969). "A psychological method to investigate verbal concepts." *J. Math. Psychol.* **6**, 169–191. The data are obtained directly from 50 Harvard students, who were asked to divide 48 common nouns into piles according to similarity of meaning. The number of piles ranged from 6 to 26. Similarity between two nouns is measured by the number of times they are sorted into the same pile. If each student partitions the nouns consistently with some common tree, this measure of similarity corresponds to an ultrametric. The single- and complete-linkage algorithms recovered very similar clusters. The complete-linkage method suggests five clusters of nouns, living things, nonliving things, quantitative terms, social interaction, and emotions.

The correlations between the original similarity measure and the ultrametrics corresponding to single and complete linkage were about 0.95. For random data, the corresponding correlations were about 0.25.

WEAVER, J. B., and HESS, S. W. (1963). "A procedure for nonpartisan districting—development of computer techniques." *Yale Law J.* **73**, 288–308.

WILEY, D. E. (1967). Latent partition analysis. *Psychometrika* **32**, 183–193. Each of 127 elementary school teachers divided 50 verbs into categories similar in the way in which they facilitated learning. A model is presented in which an underlying partition is postulated, and each member of a class in the partition has a set of probabilities of being assigned to the various clusters constructed by a teacher, with the set of probabilities being the same for all members of the class. The observed data matrix is thus to be approximated by the product of two matrices, which are estimated by factor analysis. The factors here are the classes in the latent partition.

CHAPTER 1

Profiles

1.1 INTRODUCTION

In the crime data in Table 1.1, the categories of crime (murder, rape, robbery, etc.) will be called "variables," and the cities will be "cases." (The values in the table are directly comparable within categories.) First analyses consist of single-variable summaries such as the minimum, maximum, mean, median, mode, standard deviation, and skewness; then come single-variable plots such as histograms and normal plots, and finally, plots of the variables taken two at a time.

The profile technique simultaneously plots several variables. It is useful in giving a feeling for the numbers without commitment to any mode of analysis. It is especially useful, in clustering, in informally suggesting possible clusters of similar cases and

Table 1.1 City Crime

	MURDER MANSLAUGHTER	RAPE	ROBBERY	ASSAULT	BURGLARY	LARCENY	AUTO THEFT
ATLANTA	16.5	24.8	106	147	1112	905	494
BOSTON	4.2	13.3	122	90	982	669	954
CHICAGO	11.6	24.7	340	242	808	609	645
DALLAS	18.1	34.2	184	293	1668	901	602
DENVER	6.9	41.5	173	191	1534	1368	780
DETROIT	13.0	35.7	477	220	1566	1183	788
HARTFORD	2.5	8.8	68	103	1017	724	468
HONOLULU	3.6	12.7	42	28	1457	1102	637
HOUSTON	16.8	26.6	289	186	1509	787	697
KANSAS CITY	10.8	43.2	255	226	1494	955	765
LOS ANGELES	9.7	51.8	286	355	1902	1386	862
NEW ORLEANS	10.3	39.7	266	283	1056	1036	776
NEW YORK	9.4	19.4	522	267	1674	1392	848
PORTLAND	5.0	23.0	157	144	1530	1281	488
TUCSON	5.1	22.9	85	148	1206	756	483
WASHINGTON	12.5	27.6	524	217	1496	1003	739

From the United States Statistical Abstract (1970) per 100,000 population.

also clusters of similar variables. It is sometimes necessary, before clustering, to decide the weights to be given to the different variables, and profiles may suggest reasonable weights.

Profiles are best described as histograms on each variable, connected between variables by identifying cases (usually the case name is ignored in plotting a single histogram). As with other plotting techniques, the imprecise objectives of the profiles technique make it difficult to give completely explicit instructions for constructing profiles. There seems to be an unavoidable subjective element in the scaling and positioning of the variables. Three techniques are described: The first is an iterative one which demands judgement in rescaling and repositioning the variables; the second uses ranks to solve the scaling problem and a measure of complexity of profiles to be minimized in ordering the variables; the third uses the first two eigenvectors of the data matrix.

1.2 PROFILES ALGORITHM

STEP 1.   Choose a symbol for each case, preferably only one or two characters and preferably mnemonic so that the case can readily be identified from its symbol.

STEP 2A.   For each variable, plot the cases along a horizontal line, identifying each case by its symbol. If a number of cases have identical values, their symbols should be placed vertically over this value as in a histogram.

STEP 2B.   The horizontal scale for each variable is initially set so that the minima for different variables coincide and the maxima coincide, approximately.

STEP 2C.   The vertical positions of the horizontal scales for each variable are assigned so that "similar" variables are in adjacent rows. This may be a subjective decision, or more formal measurements of similarity between variables such as correlation might be used.

STEP 3.   A *profile* for each case is drawn by connecting the symbols for the case in the various horizontal scales, one for each variable. For many cases, drawing all the profiles gives you spaghetti. So just draw a few at a time, with different sets of profiles on different copies of the horizontal scales, or draw different sets in different colors.

STEP 4.   Rescale and reposition the variables to make the case profiles smoother. For example, minimize the number of crossings of profiles. An important option in this step is to reverse a variable, using a negative scale factor.

STEP 5.   Clusters of cases will correspond to profiles of similar appearance, and clusters of variables will be positioned closely together. The final scaling of the variables might be an appropriate weighting in later distance calculations, and several variables positioned close together might each be down-weighted. Finally, transformations of the variables might be suggested.

1.3 PROFILE OF CITY CRIME

STEP 1.   Two symbols will be used for each city, although a single symbol code is possible for 16 cases. There are six ambiguities in a first-letter code, but the two-letter code consisting of the first and last letter in each name, except for two-name cities where it consists of the first letters in the two names, is unambiguous.

ATLANTA	AA	BOSTON	BN
DENVER	DR	DETROIT	DT
HOUSTON	HN	KANSAS CITY	KC
NEW YORK	NY	PORTLAND	PD
CHICAGO	CO	DALLAS	DS
HARTFORD	HD	HONOLULU	HU
LOS ANGELES	LA	NEW ORLEANS	NO
TUCSON	TN	WASHINGTON	WN

STEP 2. The initial scales appear in Table 1.2. A 20-cell histogram is drawn for each variable. It is not necessary to agonize at this stage over the scaling or positioning of the variables. Grossly rounded minima and maxima are adequate. The periods marking the boundaries of the cells are useful in visually separating the symbols.

Table 1.2   Initial Profiles of City Crime

```
MURDER     0|  .  .HD.BN.PD|  .DR.  .NY.IA|CO.WN.DT.  .  |  .AA.DS.  .  |  20
                    HU TN              NO KC              HN

RAPE       0|  .  .  .HD.BN|  .  .NY.PD.AA|HN.  .DS.DT.  |DR.DC.  .  .IA|  50
                       HU           TN CO WN           NO

ROBBERY    0|  .HU.HD.AA.BN| PD.DS.  .  .  |KC.HN.  .CO.  |  .  .  .  .DT| 500
                   TN            DR          NO IA                      NY
                                                                       WN

ASSAULT    0|HU.  .  .  .BN|  .AA.  .HN.DR|DT.CO.NY.NO.DS|  .  .IA.  .  |  400
                    HD        PD            KC
                    TN                      WN

BURGLARY   0|  .  .  .  .  |  .  .  .CO.BN|AA.TN.  .  .DR|DT.DS.  .IA.  |2000
                                    HD             HU     NY
                                    NO             HN
                                                   KC
                                                   PD
                                                   WN

LARCENY  500|  .  .BN.HD.TN|HN.  .AA.  .KC|NO.HU.  .DT.  |PD.DR.IA.  .  |1500
                    CO           DS    WN                        NY

AUTO THEFT 0|  .  .  .  .  |  .  .  .HD.AA|  .DS.CO.HN.KC|DR.IA.  .BN.  |1000
                                    PD         HU     WN DT NY
                                    TN                 NO
```

STEP 3. Profiles are drawn for selected cities (Figure 1.1). It will be seen that there is a distinct low-personal-crime group (HD, BN, PD, HU, and TN) and that this group is also low, though less distinctly, in property crime. Except for robbery and murder, LA is high on most crimes. The cities WN, DT, and NY have a similar pattern, moderate except in robbery.

STEP 4. Burglary needs rescaling to range from 500 to 2000. Auto theft needs rescaling to range from 500 to 1000. This will bring the low-crime groups into co-incidence on all variables. A slight increase in smoothness of profiles is obtained by

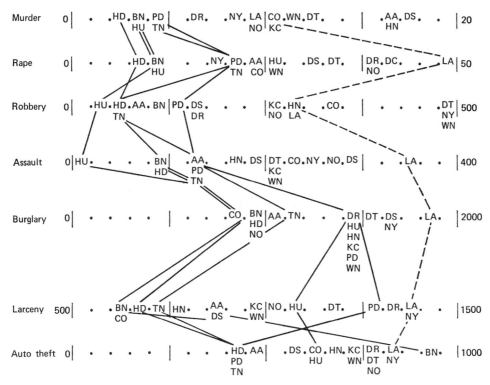

Figure 1.1 Initial profiles of city crime.

Table 1.3 Rescaled Crime Data

	MURDER	RAPE	ASSAULT	ROBBERY	BURGLARY	LARCENY	AUTO THEFT
HARTFORD	−	−	−	−	−	−	−
TUCSON	−	·	−	−	−	−	−
BOSTON	−	−	−	−	−	−	+
HONOLULU	−	−	−	−	·	·	·
PORTLAND	−	·	−	−	·	+	−
ATLANTA	+	·	−	−	·	·	−
CHICAGO	·	·	·	·	−	−	·
DENVER	−	+	·	−	·	+	·
NEW ORLEANS	·	+	·	·	−	·	·
KANSAS CITY	·	·	·	·	·	·	·
DALLAS	+	·	·	−	·	·	·
HOUSTON	+	·	·	·	·	−	·
DETROIT	·	·	·	+	·	·	·
WASHINGTON	·	·	·	+	·	·	·
NEW YORK	·	·	·	+	·	+	·
LOS ANGELES	·	+	+	·	+	+	·

In each crime category, cities are divided into three clusters, low, average, and high denoted by −, ·, and +, respectively. (See Table 1.2 for the basis of the clusters.)

interchanging robbery and assault, and this is good on outside grounds because robbery is both a personal and property crime and should be the border between the personal and property variables.

STEP 5.   The histograms for each variable seem to be multimodal (an easy thing to be deceived about for a small number of cases). There is a very pronounced low-personal-crime cluster, which fragments for property crimes. The final scales suggest the following weights for variables: murder (50), rape (20), robbery (2), assault (2), burglary (1), larceny (1), and auto theft (1).

It would be worthwhile to cluster cities separately on personal and property crimes to check for discordances between the clusterings.

A complete rescaling of each variable could be effected by measuring high, average, or low in each crime category. This procedure is suggested by the three modes appearing in nearly every category. There is no suggestion that a log transformation (almost routine for rate data like this) would be useful. The rescaled data are given in Table 1.3.

1.4 RANK PROFILES ALGORITHM

Preliminaries.   A data matrix $\{A(I, J), 1 \leq I \leq M, 1 \leq J \leq N\}$ is given with M cases and N variables. The rank profiles algorithm transforms the values of each variable to ranks and in this way solves the scaling problem. (The solution is a draconian one, since the multimodality characteristic of well-clustered variables is not visible when the variables are transformed to ranks.) The second important decision is the positioning of the variables. For a given order of variables, define a *crossing* to take place if $A(I, J) > A(K, J)$ but $A(I, J + 1) < A(K, J + 1)$. The total number of crossings for all I, K $(1 \leq I < K \leq M)$ and all J $(1 \leq J < N)$ is a measure of complexity of the profiles. [This measure is equivalent to $-\sum \{1 \leq J < N\}$ $\tau(J, J + 1)$, where $\tau(J, J + 1)$ denotes Kendall's rank correlation τ between variables J and $J + 1$.] The algorithm orders the variables to minimize the total crossings. Maurer suggested some parts of this algorithm.

STEP 1.   Define a name for each case I $(1 \leq I \leq M)$.

STEP 2.   To avoid ties, take a very small random number U and add $U \times I$ to each data value $A(I, J)$. Then define $R(K, J)$, for $1 \leq K \leq M, 1 \leq J \leq N$, to be the name of the case which is greater than exactly $K - 1$ of $A(I, J)$ $(1 \leq I \leq M)$.

STEP 3.   Define a distance $D(J, K)$ between each pair of variables J, K for $1 \leq J$, $K \leq N$, to be the number of pairs I, L with $1 \leq I, L \leq M$ and $A(I, J) > A(L, J)$ and $A(I, K) < A(L, K)$.

STEP 4.   The array NR states for each variable which variable is to its right in the final order. Initially, $NR(J) = 0$ $(1 \leq J \leq N)$. It is convenient to invent bounding variables 0 and $N + 1$ with $D(0, J) = D(N + 1, J) = 0$.

STEP 5.   Find variables $J(1)$ and $J(2)$ whose distance is minimum. Set $NR(0) = J(1)$, $NR[J(1)] = J(2), NR[J(2)] = N + 1, NR(N + 1) = 0$.

STEP 6. For each variable I $(1 \leq I \leq N)$ with NR$(I) = 0$ and for each variable $J(0 \leq J \leq N)$ with $NR(J) \neq 0$, compute $E(I, J) = D(I, J) + D[I, NR(J)] - D[J, NR(J)]$. This is the increase in sum of distances due to adding I between J and $NR(J)$. The quantities $E(I, J)$ need not all be recomputed after each addition.

STEP 7. Find I, J minimizing $E(I, J)$. Set NR$(I) =$ NR(J), then set NR$(J) = I$. If any variables I $(1 \leq I \leq N)$ with $NR(I) = 0$ remain, return to Step 6.

STEP 8. Define $K(1) =$ NR(0), and for each J $(2 \leq J \leq N)$ define $K(J) = NR[K(J - 1)]$. Then $K(1), K(2), \ldots, K(N)$ is the final order of the variables.

STEP 9. The transformed array is printed as follows:

$$\{R[1, K(L)], 1 \leq L \leq N\} \qquad \text{(first row)}$$
$$\{R[2, K(L)], 1 \leq L \leq N\} \qquad \text{(second row)}$$

$$.$$
$$.$$
$$.$$

$$\{R[I, K(L)], 1 \leq L \leq N\} \qquad \text{(Ith row)}.$$

The rank profiles algorithm, applied to the crime data, is shown in Table 1.4. Note that the procedure for ordering variables does not guarantee that the number of crossings is minimal. For small numbers of variables (<7) it is feasible to minimize exactly by examining all permutations.

Table 1.4 Rank Profiles of Crime Data

	MURDER	RAPE	ROBBERY		ASSAULT		BURGLARY		LARCENY		AUTO THEFT	
HARTFORD	HD	HD HD	HD	HD	HU		HU		CO		CO		HD	HD
HONOLULU	HU	HU	HD	HD	BN	BN	BN	BN	BN	BN	TN	
BOSTON	BN	BN BN	BN	TN		HD	HD	HD	HD	HD	HD	PD	
PORTLAND	PD	NY	AA		PD		NO		TN		AA	
TUCSON	TN	TN	BN	BN	AA		AA		HN		DS	
DENVER	DR	PD	PD		TN		TN		DS		HU	
NEW YORK	NY	CO	DR		HN		HU		AA		CO	
LOS ANGELES	LA	AA	DS		DR		KC		KC		HN	
NEW ORLEANS	NO	HN	KC		WN		WN		WN		WN	
KANSAS CITY	KC	WN	NO		DT		HN		NO		KC	
CHICAGO	CO	DS	LA		KC		PD		HU		NO	
WASHINGTON	WN	DT	HN		CO		DR		DT		DR	
DETROIT	DT	NO	CO		NY		DT		PD		DT	
ATLANTA	AA	DR	DT		NO		DS		DR		LA	
HOUSTON	HN	KC	NY		DS		NY		LA		NY	
DALLAS	DS	LA	WN		LA		LA		NY		BN	BN

Boston and Hartford emphasized to show that Bostonians are good, except they can't resist an unattended automobile.

The computer program given for reordering the data values is a slow one for large numbers of objects, requiring $M(M - 1)$ comparisons when an efficient sorting program takes $O(M \log M)$. The program given should not be used for $M > 50$.

1.5 LINEARLY OPTIMAL PROFILES ALGORITHM

Preliminaries. In drawing profiles, the scaling and positioning of the variables are determined to make the profiles for all cases as "smooth" as possible. The definition of "smooth" used in the rank profile technique is the number of crossings of the profiles. However, this technique becomes prohibitively expensive for many variables (say, more than six) and, since it only orders the variables, does not reveal clusters of very similar variables.

Another definition of smoothness requires that the profiles be as nearly linear as possible. Ideally, if $A(I, J)$ denotes the value of the Ith case for the Jth variable $(1 \leq I \leq M, 1 \leq J \leq N)$, then

$$B(J) + C(J)A(I, J) = D(I) + E(I)F(J).$$

Here $B(J)$ and $C(J)$ are the scale parameters for the Jth variable, $F(J)$ is the position of the Jth variable, and $D(I)$ and $E(I)$ determine the straight-line profile of the Ith case (see Figure 1.2). In fitting real data to such a model, it is necessary to define thresholds for each variable, proportional to the desired error variance of the observed values from the fitted values. An initial estimate of such thresholds is the raw variance for each variable. The weighting decision is somewhat subjective; a variable that you

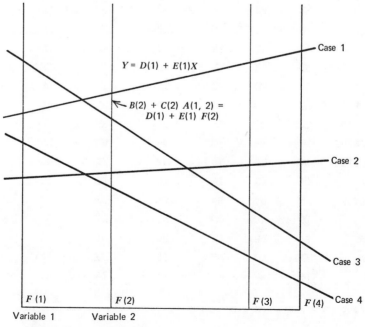

Figure 1.2 Linear profiles.

want to have little influence is given a large threshold. In the following, all variables will be presumed prestandardized to have equal thresholds.

The problem is to choose $B(J)$, $C(J)$, $F(J)$, $E(I)$, and $D(I)$ to minimize the sum of squared errors,

$$\sum \{1 \leq I \leq M, 1 \leq J \leq N\} \left(A(I, J) - \frac{[E(I)F(J) + D(I) - B(J)]}{C(J)} \right)^2.$$

This eigenvector problem is discussed in Section 1.7. The coefficients $B(J)$, $C(J)$, $F(J)$, $E(I)$, and $D(I)$ are not uniquely determined by the requirement that they minimize the sum of squares. Ignore the profile parameters $E(I)$ and $D(I)$ and concentrate on the scale and position parameters $B(J)$, $C(J)$, and $F(J)$. Let $\bar{A}(J)$ denote the mean of the Jth variable, and let $\{E1(J)\}$ and $\{E2(J)\}$ denote the first two eigenvectors of the covariance matrix $(1/M) \sum \{1 \leq I \leq M\} [A(I, J) - \bar{A}(J)][A(I, K) - \bar{A}(K)]$. Then

$$\frac{B(J)}{C(J)} = -\bar{A}(J),$$

$$\frac{F(J)}{C(J)} = E2(J),$$

$$\frac{1}{C(J)} = E1(J)$$

is a solution of the "linearly optimal" fitting problem.

Thus the variables are shifted to have mean zero, and the values for the Jth variable are divided by $E1(J)$, the coefficient of the Jth variable in the first eigenvector. The Jth variable is positioned at $E2(J)/E1(J)$. This technique is useful in providing precise scaling and positioning instructions, but the linearity of the model implies that the data can be represented in two dimensions, which is a rare occurrence. Of course, using the first two eigenvectors, both cases and variables may be represented on the plane, the variables by plotting $E1(J)$ and $E2(J)$ and the cases by plotting the corresponding eigenvectors $\sum A(I, J)E1(J)$ and $\sum A(I, J)E2(J)$. The "linearly optimal" profile technique is a new way of using the first two eigenvectors, in which the variables and cases are represented by lines rather than points. It is possible to represent the data values exactly in this new technique.

STEP 1.   Prestandardize variables so that an error of size 1 is equivalent across all variables.

STEP 2.   Compute the covariance matrix S,

$$S(J, K) = \frac{1}{M} \sum \{1 \leq I \leq M\}[A(I, J) - \bar{A}(J)][A(I, K) - \bar{A}(K)],$$

where

$$\bar{A}(J) = \sum \{1 \leq I \leq M\} \frac{A(I, J)}{M}.$$

Let $\{E1(J), 1 \leq J \leq N\}$ be the first eigenvector of S, that solution to the equations

$$\sum \{1 \leq K \leq N\} S(J, K)E1(K) = CE1(J) \qquad (1 \leq J \leq N),$$

which maximizes C. Let $\{E2(J), 1 \leq J \leq N\}$ be the second eigenvector of S.

STEP 3. Replace $A(I,J)$ by $[A(I,J) - \bar{A}(J)]/E1(J)$. Position the variable J at $E2(J)/E1(J)$.

STEP 4. The fitted profile, for the Ith case, is the line

$$y = F1(I) + F2(I)\,x,$$

where

$$F1(I) = (\sum \{1 \leq J \leq N\}\, A(I,J)E1(J))(\sum \{1 \leq J \leq N\}\, E1(J)^2)^{-1},$$

and

$$F2(I) = (\sum \{1 \leq J \leq N\}\, A(I,J)E2(J))(\sum \{1 \leq J \leq N\}\, E2(J)^2)^{-1}.$$

Thus the fitted value of the Ith case and the Jth variable is $F1(I) + F2(I)E2(J)/E1(J)$ since the Jth variable is in position $E2(J)/E1(J)$.

1.6 LINEARLY OPTIMAL PROFILES OF CRIME DATA

In Table 1.5, scale and position parameters are computed with thresholds equal to variances. Standardization then requires that the eigenvectors be computed on the correlation matrix of the variables. The scale for the Jth variable transforms $A(I,J)$ to $[A(I,J) - \bar{A}(J)]/SD(J)E1(J)$ and the position is $E2(J)/E1(J)$, where $\bar{A}(J)$ and $SD(J)$ are the mean and standard deviation of the Jth variable. The corresponding case eigenvectors $F1(I)$ and $F2(I)$ determine the line $y = F1(I) + F2(I)\,x$, which corresponds to the Ith case; see Table 1.6.

Profiles based on these eigenvectors are given in Table 1.7. The corresponding point representation based on the first two eigenvectors is given in Figure 1.3. The scales are not much different from those guessed at crudely from the minima and maxima.

Table 1.5 Scale and Position Based on Eigenvectors

VARIABLE	MEAN	SD	1ST EIGENV	2ND EIGENV	SCALE	POSITION
MURDER	10	5	.28	.61	(X-10)/1.4	2.2
RAPE	28	12	.44	.06	(X-28)/5.2	0.1
ROBBERY	244	150	.38	.19	(X-244)/57	0.5
ASSAULT	196	80	.46	.27	(X-196)/37	0.6
BURGLARY	1378	300	.39	-.40	(X-1378)/117	-1.0
LARCENY	1004	250	.35	-.59	(X-1004)/87	-1.7
AUTO THEFT	689	150	.31	-.12	(X-689)/46	- .4

Note: The scale divisor is the product of the standard deviation and the

first eigenvector. The position is the 2nd eigenvector divided by the

first.

Thresholds are raw variances; the eigenvectors therefore are computed from the correlation matrix.

Table 1.6 Case Eigenvectors Determining Profiles

CASE	F1	F2	CASE	F1	F2
ATLANTA	-1.2	1.2	HOUSTON	0.3	1.2
BOSTON	-2.2	-0.2	KANSAS CITY	1.1	0.2
CHICAGO	-0.9	2.2	LOS ANGELES	3.5	-1.0
DALLAS	1.2	1.3	NEW ORLEANS	0.8	0.7
DENVER	1.0	-1.5	NEW YORK	2.0	-0.9
DETROIT	1.9	0.1	PORTLAND	-0.8	-1.6
HARTFORD	-3.5	- .2	TUCSON	-2.1	-0.0
HONOLULU	-2.3	-2.0	WASHINGTON	1.2	0.6

The intercept is the first eigenvector, the slope is the second; the profile for the Ith case is $y = F1(I) + F2(I)\,x$.

Table 1.7 Profiles Using Eigenvectors

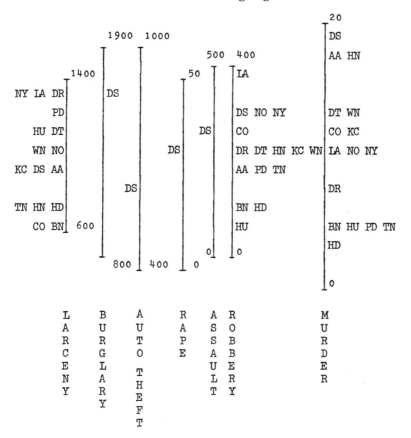

Positions and scales of variables determined by the first two eigenvectors of the correlation matrix. To avoid overwhelming you with a welter of symbols, only Dallas is recorded throughout.

37

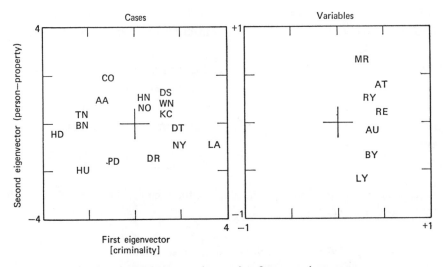

Figure 1.3 Cases and variables as points, using first two eigenvectors.

The positions of the variables are interesting; murder is separated from all other variables, assault and robbery are quite close. Possibly the crimes are ordered by increasing severity? All personal crimes are together, and all property crimes are together.

1.7 THINGS TO DO

1.7.1 Eigenvalues

The data matrix $A(I, J)$ is to be approximated by a sum $\sum \{1 \leq K \leq L\}\, R(I, K)C(J, K)$. If the sum of squared residuals

$$\sum \{1 \leq I \leq M, 1 \leq J \leq N\}\, (A(I, J) - \sum R(I, K)C(J, K))^2$$

is minimized, then $R(I, K)$ and $C(J, K)$ satisfy the equations

$$\sum \{1 \leq I \leq M\}\, R(I, K)A(I, J) = C(J, K)\sum \{1 \leq I \leq M\}\, R^2(I, K)$$

and

$$\sum \{1 \leq J \leq N\}\, C(J, K)A(I, J) = R(I, K)\sum \{1 \leq J \leq N\}\, C^2(J, K).$$

The *eigenvalues* $E(K) = [\sum R^2(I, K) \sum C^2(J, K)]^{1/2}$ $(1 \leq K \leq L)$ are unique up to a permutation. If $E(K)$ differs from other eigenvalues, eigenvectors $\{R(I, K), 1 \leq I \leq M\}$ and $\{C(J, K), 1 \leq J \leq N\}$ are unique up to a scale factor. Two row eigenvectors $\{R(I, K)\}$ and $\{R(I, KK)\}$ with $E(K) \neq E(KK)$ are orthogonal with $\sum \{1 \leq I \leq M\}$ $R(I, K)R(I, KK) = 0$, and similarly for two column eigenvectors. Now let $L = \min(M, N)$. Even if some two eigenvalues are equal, it is possible to write

$$\underset{M \times N}{A} = \underset{M \times L}{R} \; \underset{L \times L}{D} \; \underset{L \times N}{C'},$$

where R and C have orthogonal columns and D is a diagonal matrix. The eigenvalues of A are the diagonal elements of D. The row eigenvectors R are also eigenvectors of AA', and the column eigenvectors C are also eigenvectors of $A'A$.

For $L = 2$, the eigenvalues $E(1)$ and $E(2)$ are the two largest eigenvalues of A. The fitted-data matrix $\{R(I, 1)C(J, 1) + R(I, 2)C(J, 2)\}$ may be regarded as M vectors of dimension N, the vectors $R(I, 1)\{C(J, 1), 1 \leq J \leq N\} + R(I, 2)\{C(J, 2),$

$1 \leq J \leq N$}. These vectors lie in the plane spanned by {$C(J, 1), 1 \leq J \leq N$} and {$C(J, 2), 1 \leq J \leq N$}. Regarding the data as M points in N-dimensional space, this plane is closest to the data.

The M fitted points on the plane are obtained by orthogonal projection of the observed points onto the plane. Setting $\sum${$1 \leq J \leq N$} $C(J, 1)^2 = 1$ and $\sum${$1 \leq J \leq N$} $C^2(J, 2) = 1$, these points are (after a rotation) the points {$[R(I, 1), R(I, 2)], 1 \leq I \leq M$}. The orientation of the plane in N-dimensional space is best seen by projecting the vectors $(1, 0, \ldots, 0), (0, 1, \ldots, 0), \ldots, (0, \ldots, 1)$, corresponding to the axes, onto the plane. The ends of these projected axes are the points {$C(J, 1), C(J, 2), 1 \leq J \leq N$}. Thus plotting the two eigenvectors for each row and each column gives a view of the data from an angle where they most appear to lie in a plane (equivalently, they appear to have the largest covariance matrix from this angle). The distribution of points is determined by the row eigenvectors; the angle of view, by the column eigenvectors.

It is usually desirable to remove column means, so that the best fitting plane is not required to pass through zero but through the mean point. If the columns are measured on different scales, they should be rescaled to have variance one or to reflect a desired weighting. It is sometimes sensible with many variables, all measuring the same thing, to remove a row mean also.

1.7.2 Lines

This technique is the transpose of the profiles method. A horizontal line is drawn for each case. An upper and lower value is given for each variable, corresponding to the ends of the line. For each case, variables are plotted on the line according to their values for the case. A different symbol is used for each variable. The positioning of the cases now becomes important. Similar optimality criteria to those used in "linearly optimal profiles" might determine the positioning of the cases.

1.7.3 Bilines

In the profiles and lines methods, first the variables are represented as parallel lines with the cases as general lines, the cases are then represented as parallel lines with the variables as general lines. Obviously, both cases and variables might be general lines laid down on a plane. Two marks are set on each variable line. The value of a case is then determined by the intersection with a case line, relative to these marks. Theoretical and computational aspects need work.

1.7.4 Faces

Each case is a face and the variables are features such as mouth, nose, eye position, and eye width. Extreme values of variables are assigned to extreme values of features. This technique was invented by Chernoff [1973].

1.7.5 Boxes

If three variables are measured on each case, the case may be represented by a box whose three sides correspond to the three variables. This is a blander version of "faces" that is easy to implement on a printer using the symbols |, −, and /. For more than three variables, divide the variables into three similar sets and assign each set of variables to a side of the box. The visual effect is of a box with strings wrapped around it parallel to the edges.

1.7.6   Block Histograms

A block histogram may be computed in two (or more) dimensions by finding the smallest rectangular block that contains (say) 20 points, then the smallest rectangular block that contains 20 of the remaining points (discounting area included in the first block), and so on.

REFERENCES

CHERNOFF, H. (1973). Using Faces to Represent Points in k-Dimensional Space Graphically. *J. Am. Stat. Ass.* **68,** 361–368. Six measurements of the dimensions of 88 fossil specimens of the Nummulitidae shellfish group were represented by the width and height of a face, the eccentricity of the upper face, the vertical position and curvature of the month, and the size of the eyes. One face is given for each specimen. A compelling clustering of the 88 specimens into 3 distinct groups is then possible by visual sorting of the drawn faces.

TRYON, R. C., and BAILEY, D. C. (1970). *Cluster Analysis*, McGraw-Hill, New York. In this book there is a technique of representing correlations between variables by profiles, with one profile giving the correlations between a single variable and all other variables. Variables are permuted to smooth the profiles, and similar profiles signal clusters of variables.

PROGRAMS

IN	reads a data matrix, with row and column labels. The data format is specified by the user.
OUT	prints data matrix with row and column labels.
PROF	draws profiles, given positions and scales for the variables.
RANK	prints row labels in the order of each variable, one column of labels for each variable.
LINE	represents each case by symbols distributed over an interval, one symbol for each variable.
PLOT	plots several pairs of variables on the same page.
EIGEN	computes first two row and column eigenvectors of a matrix using a crude but quick iterative method.
HIST	computes and prints univariate histograms.
KC	discretizes a value, used in plotting.
RANGE	finds minimum and maximum of vector.
PMANY	draws many pairwise plots on the same page.
TABL	computes contingency table between two variables.
MODAL	finds blocks of minimum area containing given number of points.
PLACE	used by MODAL for identifying blocks already removed.
DENSITY	used by MODAL to evaluate a given rectangular block.
BDRAW	draws bivariate block histograms.
TRY1	finds best set of multivariate blocks, changing one at a time.
PICK	changes a given side of a given block.
LOB	computes log likelihood for a given set of blocks.
MHIST	represents multivariate block histogram.
LBOX	draws vertical, horizontal, or sloped line in constructing boxes.
BOXES	represents each case as a box, with several variables assigned to each side of the box.

```
      SUBROUTINE IN(A,M,N)
C......................................................................20 MAY 1973
C.... READS IN BORDERED ARRAY IN STANDARD FORMAT
C.... M = NUMBER OF ROWS, INCLUDING ONE ROW FOR COLUMN LABELS
C.... N = NUMBER OF COLUMNS, INCLUDING ONE COLUMN FOR ROW LABELS
C.... A = BORDERED ARRAY, FIRST ROW CONTAINS TITLE AND COLUMN LABELS,
C         FIRST COLUMN CONTAINS TITLE AND ROW LABELS.
C.... DATA SHOULD BE ARRANGED AS FOLLOWS....
C.... FIRST CARD  TITLE,NUMBER OF ROWS, NUMBER OF COLUMNS, IN FORMAT(A5,2I5)
C     THE NUMBERS OF ROWS AND COLUMNS REQUESTED IN THE SUBROUTINE CALL MAY BE
C     LESS THAN THESE NUMBERS.
C.... SECOND CARD  FORMAT FOR ONE ROW OF THE DATA MATRIX,FIRST WORD BEING THE
C     ROW LABEL.  FOR EXAMPLE, (A5,2F10.6).
C.... THIRD CARD(S)  COLUMN LABELS IN FORMAT 16(1X,A4)
C.... NEXT CARDS  DATA ACCORDING TO FORMAT IN SECOND CARD
C......................................................................
      DIMENSION A(M,N)
      DIMENSION FMT(20)
    1 FORMAT(1X,A4,2I5)
      WRITE(6,6)A(1,1)
      READ(5,1) A(1,1),MM,NN
    6 FORMAT(20H READING DATA MATRIX   ,A5)
      WRITE(6,7) MM,M
    7 FORMAT(15H ACTUAL ROWS = ,I5,18H REQUESTED ROWS = ,I5)
      WRITE(6,8) NN,N
    8 FORMAT(18H ACTUAL COLUMNS = ,I5,21H REQUESTED COLUMNS = ,I5)
      IF(M.GT.MM.OR.N.GT.NN) RETURN
      READ(5,2)(FMT(I),I=1,20)
    2 FORMAT(20A4)
      READ(5,4)(A(1,J),J=2,N)
    4 FORMAT(16(1X,A4))
      WRITE(6,9)(A(1,J),J=2,N)
    9 FORMAT(1X,10(8X,A4))
      DO 20 I=2,M
      READ(5,FMT)(A(I,J),J=1,N)
      WRITE(6,5)(A(I,J),J=1,N)
    5 FORMAT(1X,A4,10G12.4/(5X,10G12.4))
   20 CONTINUE
      RETURN
      END




      SUBROUTINE OUT(A,M,N)
C......................................................................20 MAY 1973
C     PRINTS OUT A BORDERED ARRAY A.
C.... M = NUMBER OF ROWS
C.... N = NUMBER OF COLUMNS
C.... A = M BY N BORDERED ARRAY
C......................................................................
      DIMENSION A(M,N)
      WRITE(6,1) A(1,1)
    1 FORMAT(6H ARRAY,1X,A4)
      WRITE(6,2)(A(1,J),J=2,N)
    2 FORMAT(10X,5A10,2X,5A10)
      DO 20 I=2,M
      WRITE(6,3)(A(I,J),J=1,N)
    3 FORMAT(1X,A4,5X,5F10.4,2X,5F10.4/(10X,5F10.4,2X,5F10.4))
   20 CONTINUE
      RETURN
      END
```

```
      SUBROUTINE PROF(M,N,A,B)
C...............................................................20 MAY 1973
C.... PROFILE OF BORDERED ARRAY A,ACCORDING TO SCALE AND POSITION VECTOR B
C.... M = NUMBER OF ROWS
C.... N = NUMBER OF COLUMNS
C.... A = M BY N BORDERED ARRAY
C.... B = 3 BY N POSITION VECTOR
C          A(I,J) IS PLOTTED AT (B(3,J)   ,   B(1,J)+B(2,J)*A(I,J)  )
C..............................................................................
      DIMENSION A(M,N),B(3,N)
      DIMENSION AA(25)
      DATA BB,BL/1H',1H /
      DATA YMIN,YMAX,XMIN,XMAX/4*0./
C.... MINIMUM AND MAXIMUM
      DO 10 I=2,M
      DO 10 J=2,N
      KK=I*J-4
      YY=B(1,J)+B(2,J)*A(I,J)
      XX=B(3,J)
      IF(YY.LT.YMIN.OR.KK.EQ.0) YMIN=YY
      IF(YY.GT.YMAX.OR.KK.EQ.0) YMAX =YY
      IF(XX.LT.XMIN.OR.KK.EQ.0) XMIN=XX
      IF(XX.GT.XMAX.OR.KK.EQ.0) XMAX=XX
   10 CONTINUE
C.... SET SCALES
      NP=M/20+1
      FY=(NP*50)/(YMAX-YMIN+.000001)
      FX=25/(XMAX-XMIN+.000001)
C.... WRITE TITLES
      WRITE(6,1) A(1,1)
    1 FORMAT(13H1PROFILES OF ,A4)
      WRITE(6,2)
    2 FORMAT(32H0VARIABLE      MINIMUM      MAXIMUM    )
      DO 30 J=2,N
      Y1=(YMAX-B(1,J))/B(2,J)
      Y2=(YMIN-B(1,J))/B(2,J)
      WRITE(6,3) A(1,J),Y2,Y1
    3 FORMAT(3X,A4,4X,2F10.4)
   30 CONTINUE
C.... WRITE PROFILES
      IL=-1
   45 IL=IL+1
C.... INITIALIZE LINE
      DO 40 I=1,25
   40 AA(I)=BL
      DO 43 J=2,N
      JJ=(B(3,J)-XMIN)*FX+1
      AA(JJ)=BB
   43 IF(IL.EQ.0) AA(JJ)=A(1,J)
      IF(IL.EQ.0) GO TO 44
C.... FILL IN VALUES ON LINE
      DO 41 J=2,N
      DO 41 I=2,M
      K=(B(1,J)+B(2,J)*A(I,J)-YMIN)*FY+1
      IF(K.NE.IL) GO TO 41
      JJ=(B(3,J)-XMIN)*FX+1
      IF(JJ.LT.1.OR.JJ.GT.25) GO TO 41
      AA(JJ)=A(I,1)
      AA(JJ)=A(I,1)
   41 CONTINUE
   44 CONTINUE
      WRITE(6,5) IL,(AA(I),I=1,25)
    5 FORMAT(1X,I3,1X,25(1X,A4))
      IF(IL.LE.NP*50) GO TO 45
      RETURN
      END
```

```
      SUBROUTINE RANK(M,N,A,SC)
C..................................................................20 MAY 1973
C.... ORDERS EACH COLUMN OF A AND REPLACES VALUES BY ROW NAMES
C.... M = NUMBER OF ROWS
C.... N = NUMBER OF COLUMNS
C.... A = M BY N BORDERED ARRAY
C.... SC = M BY 1 SCRATCH VECTOR
C..................................................................
      DIMENSION A(M,N),SC(M)
      DO 20 J=2,N
C.... ORDER VALUES OF THE JTH COLUMN
      DO 10 I=2,M
      SC(I)=A(I,1)
      DO 10 II=2,I
      IF(A(I,J).GE.A(II,J)) GO TO 10
      C=A(I,J)
      A(I,J)=A(II,J)
      A(II,J)=C
      C=SC(I)
      SC(I)=SC(II)
      SC(II)=C
   10 CONTINUE
C.... REPLACE VALUES BY CORRESPONDING ROW NAMES
      DO 11 I=2,M
   11 A(I,J)=SC(I)
   20 CONTINUE
C...... WRITE ARRAY OF NAMES
      WRITE(6,1) A(1,1)
    1 FORMAT(18H1RANKING OF ARRAY ,A4)
      WRITE(6,2)(A(1,J),J=2,N)
    2 FORMAT(25(1X,A5))
      WRITE(6,3)
    3 FORMAT(1H0)
      DO 40 I=2,M
      WRITE(6,2)(A(I,J),J=2,N)
   40 CONTINUE
      RETURN
      END
```

```fortran
      SUBROUTINE LINE(M,N,A,R,S)
C.....................................................................20 MAY 1973
C.... PRINTS EACH CASE AS A SET OF SYMBOLS, ONE FOR EACH VARIABLE
C.... M = NUMBER OF ROWS
C.... N = NUMBER OF COLUMNS
C.... A = BORDERED DATA MATRIX
C.... R = N BY 2 ARRAY.  FOR ITH VARIABLE, R(I,1) IS MAPPED INTO 0,
C          R(I,2) IS MAPPED INTO 100.
C          IF R(1,1) = 0 ON INPUT, THE PROGRAM COMPUTES THE ARRAY R.
C.... S = N BY 1 SYMBOL ARRAY, COMPUTED IN PROGRAM IF S(1)=0 ON INPUT.
C.....................................................................
      DIMENSION A(M,N),R(N,2),S(N)
      DIMENSION SA(36),AA(105)
      DATA P1,P2/1H(,1H)/
      DATA SA/1H1,1H2,1H3,1H4,1H5,1H6,1H7,1H8,1H9,
     *        1HA,1HB,1HC,1HD,1HE,1HF,1HG,1HH,1HI,
     *        1HJ,1HK,1HL,1HM,1HN,1HO,1HP,1HQ,1HR,
     *        1HS,1HT,1HU,1HV,1HW,1HX,1HY,1HZ,1H./
C.... COMPUTE RANGES AND SYMBOLS IF REQUESTED
      IF(R(1,1).NE.0) GO TO 30
      DO 20 J=2,N
      R(J,1)=A(2,J)
      R(J,2)=A(2,J)
      DO 21 I=2,M
      IF(A(I,J).LT.R(J,1)) R(J,1)=A(I,J)
   21 IF(A(I,J).GT.R(J,2)) R(J,2)=A(I,J)
      IF(R(J,1).EQ.R(J,2)) R(J,1)=R(J,1)-1
   20 CONTINUE
   30 IF(S(1).NE.0) GO TO 40
      DO 31 J=2,N
      JJ=J-((J-1)/35)*35
      JJ=JJ-1
   31 S(J)=SA(JJ)
   40 CONTINUE
C.... TITLES
      WRITE(6,2) A(1,1)
    2 FORMAT(21H1LINE PRINT OF ARRAY ,A4)
      WRITE(6,3)
    3 FORMAT(40H0VARIABLE   SYMBOL        MIN        MAX
      WRITE(6,4)(A(1,J),S(J),R(J,1),R(J,2),J=2,N)
    4 FORMAT(3X,A4,5X,A1,5X,2F10.4)
      WRITE(6,5)
    5 FORMAT(1H0)
C.... PRINT CASES
      DO 49 I=2,M
      DO 41 J=1,105
   41 AA(J)=SA(36)
      DO 42 J=2,N
      JJ=100*(A(I,J)-R(J,1))/(R(J,2)-R(J,1))+3
   44 IF(JJ.LT.3) JJ=1
      IF(JJ.GT.103) JJ=105
      IF(AA(JJ).EQ.SA(36)) GO TO 42
      IF(JJ.LT.51) JJ=JJ+1
      IF(JJ.GT.51) JJ=JJ-1
      IF(JJ.NE.51) GO TO 44
   42 AA(JJ)=S(J)
      AA(2)=P1
      AA(104)=P2

      WRITE(6,1) A(I,1),(AA(J),J=1,105)
    1 FORMAT(1X,A4,5X,105A1)
   49 CONTINUE
      RETURN
      END
```

```
      SUBROUTINE PLOT(X,NN,N)
C.......................................................................20 MAY 1973
C.... PLOTS A BORDERED 3 BY N ARRAY,X.
C       THE NAMES OF THE POINTS ARE IN THE FIRST ROW, THE X VALUES ARE IN THE
C       SECOND ROW, THE Y VALUES ARE IN THE THIRD ROW.
C       SEVERAL ARRAYS MAY BE PLOTTED ON THE SAME PAGE BY SETTING N NEGATIVE.
C       SET RANGES BY CALLING PLOT WITH N = 0.
C.... NN = DIMENSION OF ARRAY.
C.... N = CONTROL PARAMETER.
C.... X = 3 BY N ARRAY.
C.......................................................................
      DIMENSION X(3,NN)
      DIMENSION T(20),A(51,26)
      DATA B,IC/1H ,0/
      DIMENSION E(7,10)
C.... INITIALIZE PLOT MATRIX A.
      IF(IC.NE.0) GO TO 15
      IU=51
      JU=26
      DO 11 I=1,IU
      DO 11 J=1,JU
   11 A(I,J)=B
   15 CONTINUE
C.... SET RANGES
      IF(N.NE.0.AND.IC.NE.0) GO TO 20
      X1=X(2,2)
      X2=X(2,3)
      Y1=X(3,2)
      Y2=X(3,3)
      IF(NN.EQ.3) GO TO 20
      DO 10 J=2,NN
      IF(X(2,J).LT.X1) X1=X(2,J)
      IF(X(2,J).GT.X2) X2=X(2,J)
      IF(X(3,J).LT.Y1) Y1=X(3,J)
      IF(X(3,J).GT.Y2) Y2=X(3,J)
   10 CONTINUE
      IF(N.EQ.0) RETURN
   20 CONTINUE
C.... PLOT POINTS IN X ARRAY ON A MATRIX
      DO 21 J=2,NN
      II=(X(2,J)-X1)*(JU-1)/(X2-X1)+1
      IF(II.LT.1) II=1
      IF(II.GT.26) II=26
      JJ=(X(3,J)-Y1)*(IU-1)/(Y2-Y1)+1
      IF(JJ.LT.1) JJ=1
      IF(JJ.GT.51) JJ=51
      JJ=52-JJ
   21 A(JJ,II)=X(1,J)
      IC=IC+1
      E(1,IC)=X(1,1)
      E(2,IC)=X(2,1)
      E(3,IC)=X1
      E(4,IC)=X2
      E(5,IC)=X(3,1)
      E(6,IC)=Y1
      E(7,IC)=Y2
      IF(N.LT.0) RETURN
C... PRINT A MATRIX
      WRITE(6,1)
    1 FORMAT(
     *62H1PLOTS OF ARRAY X-AXIS    MIN      MAX    Y-AXIS    MIN      MAX)
      WRITE(6,2)((E(I,J),I=1,7),J=1,IC)
    2 FORMAT(11X,A4,2X,A4,2F10.4,2X,A4,2F10.4)
      WRITE(6,4)
    4 FORMAT(1H0)
      WRITE(6,6)
    6 FORMAT(10X,106(1H-))
      WRITE(6,5)((A(I,J),J=1,26),I=1,51)
    5 FORMAT(10X,1H',26A4,1H')
      WRITE(6,6)
      IC=0
      RETURN
      END
```

```fortran
      SUBROUTINE EIGEN(M,N,A,X,Y)
C.............................................................20 MAY 1973
C.... FINDS FIRST TWO EIGENVECTORS OF A BY CRUDE BUT QUICK ITERATION
C.... M = NUMBER OF ROWS
C.... N = NUMBER OF COLUMNS
C.... A = BORDERED ARRAY
C.... X = BORDERED 3 BY M ARRAY, FIRST TWO ROW EIGENVECTORS
C.... Y = BORDERED 3 BY N ARRAY, FIRST TWO COLUMN EIGENVECTORS,MSSQ =1.
C.........................................................................
      DIMENSION A(M,N),X(3,M),Y(3,N)
      DIMENSION BB(6)
C.... INSERT X AND Y BORDERS
      DATA BB/4HEIGC,4HCOL1,4HCOL2,4HEIGR,4HROW1,4HROW2/
      DO 12 J=1,3
      Y(J,1)=BB(J)
   12 X(J,1)=BB(J+3)
      DO 10 J=2,N
   10 Y(1,J)=A(1,J)
      DO 11 I=2,M
   11 X(1,I)=A(I,1)
C.... INITIALIZE COLUMNS
      DO 20 J=2,N
      DO 20 K=2,3
   20 Y(K,J)=0
      DO 21 J=2,M
      DO 21 K=2,3
   21 X(K,J)=J**K
      DO 60 IT=1,5
C.... ITERATE ON Y
      DO 32 K=2,3
      KK=5-K
      DO 50 J=2,N
      Y(K,J)=0
      DO 50 I=2,M
   50 Y(K,J)=Y(K,J)+(A(I,J)-X(KK,I)*Y(KK,J))*X(K,I)
C.... STANDARDIZE Y
      SY=0
      DO 30 J=2,N
   30 SY=SY+Y(K,J)**2
      SY=(SY/(N-1))**(0.5)
      DO 31 J=2,N
   31 IF(SY.NE.0) Y(K,J)=Y(K,J)/SY
      SX=0
      DO 33 J=2,M
   33 SX=SX+X(K,J)**2
      DO 34 J=2,M
   34 IF(SX.NE.0) X(K,J)=X(K,J)*SY/SX
   32 CONTINUE
C.... ITERATE X
      DO 40 K=2,3
      KK=5-K
      DO 40 I=2,M
      X(K,I)=0
      DO 40 J=2,N
   40 X(K,I)=X(K,I)+(A(I,J)-X(KK,I)*Y(KK,J))*Y(K,J)/(N-1)
   60 CONTINUE
      RETURN
      END
```

46

```
      SUBROUTINE HIST(N,X,T,X1,X2,NI,P)
C.........................................................................20 NOV 1973
C.... SUBROUTINE FOR PRINTING HISTOGRAMS
C.... N = NUMBER OF DATA VALUES
C.... X = VECTOR TO BE COUNTED
C.... T = TITLE
C.... X1 = MINIMUM
C.... X2 = MAXIMUM
C        IF X1 .LE.X2 THESE ARE COMPUTED FROM X
C.... NI = NUMBER OF INTERVALS( O IS CHANGED TO 100)
C.... P = DIVISOR FOR VERTICAL SCALE( O CHANGED TO 1.)
C.........................................................................
      DIMENSION X(N)
      DIMENSION A(102),AA(102)
      DATA B,C,H,V/1H ,1HO,1H_,1HI/
C.... DEFAULT OPTIONS
      IF(P.EQ.0) P=1
      IF(NI.EQ.0) NI=100
      CALL RANGE(X,N,X1,X2)
      DO 20 I=1,102
   20 A(I)=0
C.... COMPUTE FREQUENCIES
      DO 30 J=1,N
      K=KC(X(J),X1,X2,101)
   30 A(K)=A(K)+1
      L=0
      L=100/NI
      IF(L.LT.1) L=1
C.... ADJUST FOR INTERVALS
      DO 40 J=1,N,L
      JU=J+L-1
      S=0
      IF(JU.GT.N) JU=N
      DO 41 K=J,JU
   41 S=S+A(K)
      DO 42 K=J,JU
   42 A(K)=S/(P*L)
   40 CONTINUE
C.... FIND MAXIMUM VALUE
      AM=0
      DO 50 K=1,102
   50 IF(A(K).GT.AM) AM=A(K)
C.... PRINT HISTOGRAM
      MM=AM
      WRITE(6,1) T
    1 FORMAT(1H ,A4,' HISTOGRAM')
      WRITE(6,2)
    2 FORMAT(1H+,19X,102(1H_))
      DO 60 K=1,MM
      AA(1)=V
      DO 61 I=1,102
      IF(I.GT.1) AA(I)=B
      AA(102)=V
   61 IF(A(I)+K.GT.MM) AA(I)=C
      KK=MM-K+1
   60 WRITE(6,3) KK,(AA(I),I=1,102)
    3 FORMAT(10X,I5,5X,102A1)
      WRITE(6,2)
      WRITE(6,5) X1,X2
    5 FORMAT(F25.4,5X,8(1HI,9X),1HI,F14.4)
      RETURN
      END
```

```
      FUNCTION KC(A,A1,A2,NC)
      KC=(A-A1)*NC/(A2-A1)+1.00001
      IF(KC.LT.1) KC=1
      IF(KC.GT.NC) KC=NC
      RETURN
      END
```

```
      SUBROUTINE RANGE(X,N,X1,X2)
      DIMENSION X(N)
      X1=X(1)
      X2=X(1)
      DO 20 I=1,N
      IF(X(I).LT.X1) X1=X(I)
   20 IF(X(I).GT.X2) X2=X(I)
      RETURN
      END
```

47

```
      SUBROUTINE PMANY(A,M,N,A1,A2,NC)
C..................................................................20 NOV 1973
C.... PLOTS MANY PAIRS OF VARIABLES SIMULTANEOUSLY
C.... M = NUMBER OF ROWS
C.... N = NUMBER OF COLUMNS
C.... A = M BY N BORDERED ARRAY
C.... A1 = 1 BY N ARRAY OF MINIMA FOR EACH VARIABLE
C.... A2 = 1 BY N ARRAY OF MAXIMA
C....IF A1.GE.A2 COMPUTE ACTUAL RANGE
C.... NC = 1 BY N ARRAY OF CELL NUMBERS ( (1 = 0 CHANGED TO NC =10)
C..................................................................
      DIMENSION A(M,N),A1(N),A2(N),NC(N),AA(122),AH(122)
      DATA B,H,V,D,T,P/1H ,1H_,1H|,1H.,1H:,1H+/
C.... ADJUST RANGES
      DO 10 J=2,N
      IF(NC(J).LE.0) NC(J)=10
      IF(A1(J).LT.A2(J)) GO TO 10
      A1(J)=A(2,J)
      A2(J)=A(2,J)
      DO 20 I=2,M
      IF(A(I,J).LT.A1(J)) A1(J)=A(I,J)
   20 IF(A(I,J).GT.A2(J)) A2(J)=A(I,J)
      IF(A1(J).EQ.A2(J)) RETURN
   10 CONTINUE
      WRITE(6,1) A(1,1)
    1 FORMAT(16H ALL PAIRS FROM ,A4)
C.... SET UP STARTING POINT OF BLOCKS
      NC(1)=2
      DO 60 J=2,N
   60 NC(J)=NC(J-1)+NC(J)+2
C.... HORIZONTAL LINE OF CHARACTERS
      DO 31 I=1,122
   31 AH(I)=B
      N1=N-1
      DO 33 J=1,N1
      JL=NC(J)
      JU=NC(J+1)-3
      DO 33 JJ=JL,JU
   33 AH(JJ)=H
C.... PLOT A LINE AT A TIME
      DO 30 JC=2,N
      WRITE(6,3)(A(1,J),J=2,N)
    3 FORMAT(10X,10(4X,A4,4X))
      WRITE(6,6)(AH(I),I=1,122)
C.... FIND VALUES CONDITIONAL ON JC
      JU=NC(JC)-NC(JC-1)-2
      DO 50 JP=1,JU
      DO 53 K=1,122
   53 AA(K)=0
      DO 51 I=2,M
      IF(KC(A(I,JC),A1(JC),A2(JC),JU).NE.JP) GO TO 51
      DO 52 J=2,N
         JJ=NC(J)-NC(J-1)-2
      K=KC(A(I,J),A1(J),A2(J),JJ)+NC(J-1)- 1
   52 AA(K)=AA(K)+1
   51 CONTINUE
C.... RECODE AA
      JL=NC(JC-1)
      L=AA(JL+JP-1)
      L=L+JL-1
      JJU=NC(JC)-3
      DO 63 K=JL,JJU
   63 AA(K)=0
      DO 61 KK=1,3
      DO 62 K=JL,JJU
   62 IF(K.LE.L) AA(K)=AA(K)+1
   61 L=L-JJU+JL-1
      DO 54 K=1,122
      IF(AA(K).GT.3) AA(K)=3
       IF(AA(K).EQ.0) AA(K)=B
      IF(AA(K).EQ.1) AA(K)=D
      IF(AA(K).EQ.2) AA(K)=T
   54 IF(AA(K).EQ.3) AA(K)=P
      AA(1)=V
      DO 55 J=2,N
      AA(NC(J)-1)=V
   55 AA(NC(J)-2)=V
      AA(NC(N)-1)=B
```

```
C.... TYPE AA
      IF(JP.EQ.1) WRITE(6,4) A1(JC),(AA(J),J=1,122)
      IF(JP.EQ.JU) WRITE(6,4) A2(JC),(AA(J),J=1,122)
      IF(JP.EQ.JU/2+1) WRITE(6,2) A(1,JC),(AA(J),J=1,122)
      IF(JP.NE.1.AND.JP.NE.JU.AND.JP.NE.JU/2+1)
     *WRITE(6,5) (AA(J),J=1,122)
    2 FORMAT(5X,A4,1X,122A1)
    4 FORMAT(F10.3,122A1)
    5 FORMAT(10X,122A1)
   50 CONTINUE
      WRITE(6,6)(AH(I),I=1,122)
    6 FORMAT(1H+,9X,122A1)
      IF(4*((JC-1)/4).EQ.JC-1) WRITE(6,7)
    7 FORMAT(1H1)
   30 CONTINUE
      RETURN
      END




      SUBROUTINE TABL(M,N,A,N1,N2,B,II,JJ,X1,X2,Y1,Y2)
C....................................................................20 NOV 1973
C.... COMPUTES CONTINGENCY TABLE BETWEEN II  AND JJ COLUMNS OF ARRAY A
C.... M = ROWS OF DATA ARRAY A
C.... N = COLS OF DATA ARRAY A

C.... A = DATA ARRAY A
C.... N1 = ROWS OF CONTINGENCY TABLE
C.... N2 = COLUMNS OF CONTINGENCY TABLE
C.... B = CONTINGENCY TABLE
C.... II = DATA COLUMN TO BE ROW OF TABLE
C.... JJ = DATA COLUMN TO BE COLUMN OF TABLE
C.... X1 = MINIMUM POINT OF II
C.... X2 = MAXIMUM POINT OF II(IF X1.GE.X2 COMPUTE)
C.... Y1 = MINIMUM POINT OF JJ
C.... Y2 = MAXIMUM POINT OF JJ
C....................................................................
      DIMENSION A(M,N),B(N1,N2)
      M1=M-1
      CALL RANGE(A(2,II),M1,X1,X2)
      CALL RANGE(A(2,JJ),M1,Y1,Y2)
      DO 30 I=1,N1
      DO 30 J=1,N2
   30 B(I,J)=0
      DO 40 I=2,M
      KI=KC(A(I,II),X1,X2,N1)
      KJ=KC(A(I,JJ),Y1,Y2,N2)
   40 B(KI,KJ)=B(KI,KJ)+1
      RETURN
      END
```

```
      SUBROUTINE MODAL(M,N,F,PT,B,NB,K)
C.......................................................... 31 DECEMBER 1973
C.... FINDS BLOCKS OF MINIMUM AREA CONTAINING FREQUENCY PT.
C.... M = NUMBER OF ROWS
C.... N = NUMBER OF COLUMNS
C.... F = M BY N ARRAY OF CELL FREQUENCIES
C.... PT = FREQUENCY THRESHOLD
C.... NB = BLOCK BOUNDARIES
C           NB(1,K)= FIRST ROW
C           NB(2,K) = FIRST COLUMN
C           NB(3,K) = LAST ROW
C           NB(4,K) = LAST COLUMNS
C           B(1,K)= FREQUENCY
C           B(2,K) = AREA
C           B(3,K)= DENSITY
C..........................................................................
      DIMENSION F(M,N),NB(4,50),B(3,50)
C.... FIND TOTAL FREQUENCY
      PP=0
      PTOT=0
      DO 9 I=1,M
      DO 9 J=1,N
    9 PTOT=PTOT+F(I,J)
      K=0
C.... INITIALISE BLOCK SIZE
      FM=0
      DO 70 I=1,M
      DO 70 J=1,N
   70 IF(F(I,J).GT.FM) FM=F(I,J)
      IF(FM.EQ.0) RETURN
      LI=PT/FM
      IF(LI.LT.1) LI=1
      LJ=1
   10 K=K+1
C.... INITIALISE BLOCK PARAMETERS
      NB(1,K)=1
      NB(2,K)=1
      NB(3,K)=M
      NB(4,K)=N
      B(1,K)=0
      B(2,K)=M*N
      B(3,K)=0
      IF(PP+PT.GT.PTOT) RETURN
   52 CONTINUE
      DO 21 I1=1,M
      I2=I1+LI-1
      IF(I2.GT.M) GO TO 21
      DO 20 J1=1,N
C.... FIND BLOCK EXCEEDING THRESHOLD OF MINIMUM AREA
      J2=J1+LJ-1
      IF(J2.GT.N) GO TO 20
      CALL DENSTY(I1,I2,J1,J2,M,N,F,P,A)
      IF(P.LT.PT) GO TO 20
      IF(P/A.LT.B(3,K).OR.A.GT.B(2,K)) GO TO 23
      NB(1,K)=I1
      NB(2,K)=J1
      NB(3,K)=I2
      NB(4,K)=J2
      B(1,K)=P
      B(2,K)=A
      B(3,K)=P/A
   23 CONTINUE
   20 CONTINUE
   22 CONTINUE
   21 CONTINUE
C.... REPLACE DENSITIES BY BLOCK DENSITY
      IF(B(1,K).NE.0) GO TO 51
      II=LI
      JJ=LJ
      MIN=M*N
      DO 50 I=1,M
      DO 50 J=1,N
      IF(I*J.LT.LI*LJ.OR.I*J.GT.MIN)     GO TO 50
      IF(I*J.EQ.LI*LJ.AND.J.LE.LJ) GO TO 50
      II=I
      JJ=J
      MIN=I*J
   50 CONTINUE
```

50

```
      LI=II
      LJ=JJ
      IF(MIN.LT.M*N) GO TO 52
   51 CONTINUE
      CALL PLACE(NB(1,K),NB(3,K),NB(2,K),NB(4,K),M,N,F,B(3,K))
      PP=PP+B(1,K)
      IF(K.LT.50) GO TO 10
      RETURN
      END




      SUBROUTINE PLACE(I1,I2,J1,J2,M,N,F,D)
C........................................................ 31 DECEMBER 1973
C.... REPLACES DENSITIES IN BLOCK BY -D
C.... SEE DENSTY FOR ARGUMENTS
C.........................................................................
      DIMENSION F(M,N)
      DO 10 I=I1,I2
      DO 10 J=J1,J2
      IF(F(I,J).LT.0) GO TO 10
      F(I,J)=-D
   10 CONTINUE
      RETURN
      END




      SUBROUTINE DENSTY(I1,I2,J1,J2,M,N,F,P,A)
C........................................................ 31 DECEMBER 1973
C.... FINDS PROBABILITY AND AREA FOR A BLOCK,WITH PREVIOUS BLOCKS INDICATED BY
C     A NEGATIVE FREQUENCY.   IF A .NE.0 ON INPUT IT WILL REPLACE DENSITIES BY
C     A NEGATIVE BLOCK DENSITY.
C....I1 = FIRST ROW OF BLOCK
C.... I2 = SECOND ROW OF BLOCK
C.... J1 = FIRST COLUMN OF BLOCK
C.... J2 = SECOND COLUMN OF BLOCK
C.... M = NUMBER OF ROW CELLS
C.... N = NUMBER OF COLUMN CELLS
C.... F = M BY N ARRAY OF CELL FREQUENCIES
C.... P = TOTAL BLOCK FREQUENCY
C.... A = AREA OF BLOCK
C........................................................ 31 DECEMBER 1973
      DIMENSION F(M,N)
      A=0
      P=0
      DO 10 I=I1,I2
      DO 10 J=J1,J2
      IF(F(I,J).LT.0) GO TO 10
      A=A+1
      P=P+F(I,J)
   10 CONTINUE
      RETURN
      END
```

```
      SUBROUTINE BDRAW(NB,K,A,M,N,NP)
C........................................................... 31 DECEMBER 1973
C.... DRAWS OVERLAPPING BLOCKS IN GIVEN ORDER
C...........................................................................
      DIMENSION A(M,N)
      DIMENSION NB(4,K)
      DATA B,H,V/1H ,1H_,1H|/
C.... CONSTRUCT BLOCKS IN REVERSE ORDER
      DO 5 I=1,M
      DO 5 J=1,N
    5 A(I,J)=B
      DO 10 KK=1,K
      L=K-KK+1
      I1=NB(1,L)*2-1
      I2=NB(3,L)*2
      J1=NB(2,L)*2-1
      J2=NB(4,L)*2+1
      DO 20 I=I1,I2
      DO 20 J=J1,J2
      IF(I.LT.I2.AND.I.GT.I1.AND.J.LT.J2.AND.J.GT.J1) A(I,J)=B
      IF(I.EQ.I1.AND.(J.EQ.J1.OR.J.EQ.J2)) GO TO 20
      IF(I.EQ.I1.OR.I.EQ.I2) A(I,J)=H
      IF(J.EQ.J1.OR.J.EQ.J2) A(I,J)=V
   20 CONTINUE
   10 CONTINUE
      M1=M-1
      DO 30 I=1,M1
   30 WRITE(6,3)(A(I,J),J=1,N)
    3 FORMAT(5X,120A1)
      RETURN
      END

      SUBROUTINE TRY1(A,M,N,B,K,L,XLL,V,D,NV)
C.............................................................23 DECEMBER 1973
C.... FINDS BEST SET OF K BLOCKS, CHANGING ONE AT A TIME
C.... A = M BY N BORDERED ARRAY
C.... M = NUMBER OF ROWS
C.... N = NUMBER OF COLUMNS
C.... B = 2 BY N BY K BLOCK ARRAY
C.... L = BLOCK OPTIMIZED BY PROGRAM
C.... XLL = LOG LIKELIHOOD AFTER OPTIMIZATION
C.... V = 2 BY K SCRATCH ARRAY
C.... D = 1 BY N SCRATCH ARRAY
C.... NV = 1 BY N ARRAY DEFINING ORDER OF VARIABLES
C...........................................................................
      DIMENSION NV(N)
      DIMENSION A(M,N),B(2,N,K)
      DIMENSION V(2,K),D(N)
      DIMENSION XX(5)
      CALL LOB(A,M,N,B,K,XM,V)
      DO 51 LD=1,3
      DO 20 JJ=2,N
      J=NV(JJ)
      B1=B(1,J,L)
      B2=B(2,J,L)
      VM=0
      KM=1
      DO 30 KK=1,5
      CALL PICK(KK,B1,B2,B(1,J,L),B(2,J,L))
      CALL LOB(A,M,N,B,K,XX(KK),V)
      IF(V(2,L).NE.0) V(2,L)=V(1,L)/V(2,L)
      IF(V(2,L).LE.VM) GO TO 30
      KM=KK
      VM=V(2,L)
   30 CONTINUE
      B(1,J,L)=B1
      B(2,J,L)=B2
      IF(XX(KM).LT.XM) GO TO 20
      XM=XX(KM)
      CALL PICK(KM,B1,B2,B(1,J,L),B(2,J,L))
   20 CONTINUE
   51 CONTINUE
   52 CONTINUE
      RETURN
      END
```

```
      SUBROUTINE PICK(K,B1,B2,BB1,BB2)
C............................................................... 31 DECEMBER 1973
C.... SUBROUTINE TO CHANGE RANGE OF INTERVAL WHILE SEARCHING FOR BEST BLOCK
C.... K = INDEX OF TYPE OF CHANGE
C.... B1 = ORIGINAL MINIMUM
C.... B2 = ORIGINAL MAXIMUM
C.... BB1 = NEW MINIMUM
C.... BB2 = NEW MAXIMUM
C.........................................................................
      BB=(B2-B1)/2.
      IF(K.EQ.1.OR.K.EQ.2) BB1=B1+BB
      IF(K.EQ.3.OR.K.EQ.4) BB2=B2-BB
      IF(K.EQ.4.OR.K.EQ.5) BB1=B1-BB
      IF(K.EQ.1.OR.K.EQ.5) BB1=B2+BB
      IF(K.EQ.3) BB1=B1
      IF(K.EQ.2) BB2=B2
      IF(K.EQ.5) BB1=B1+BB/2
      IF(K.EQ.5) BB2=B2-BB/2
      RETURN
      END


      SUBROUTINE LOB(A,M,N,B,K,XLL,V)
C............................................................... 23 DECEMBER 1973
C.... COMPUTES LOG LIKELIHOOD FIT FOR GIVEN SET OF BLOCKS
C.... M = NUMBER OF ROWS
C.... N = NUMBER OF COLUMNS
C.... A = M BY N BORDERED ARRAY
C.... B = 2 BY N BY K ARRAY DEFINING BLOCKS
C.... XLL = LOG LIKELIHOOD
C.... V = 2 BY K SCRATCH ARRAY
      DIMENSION A(M,N),B(2,N,K),V(2,K)
C.... COUNT NUMBER IN EACH BLOCK
C.........................................................................
      DO 20 KK=1,K
   20 V(1,KK)=0
      DO 30 I=2,M
      DO 32 KK=1,K
      DO 31 J=2,N
      IF(A(I,J).GT.B(2,J,KK).OR.A(I,J).LT.B(1,J,KK)) GO TO 32
   31 CONTINUE
      GO TO 30
   32 CONTINUE
   30 V(1,KK)=V(1,KK)+1
C.... COUNT VOLUME IN EACH BLOCK
      DO 33 KK=1,K
      V(2,KK)=1
      DO 34 J=2,N
   34 V(2,KK)=V(2,KK)*(B(2,J,KK)-B(1,J,KK))
C.... ADJUST FOR TWO BLOCKS OVERLAPPING
      DO 35 K1=1,KK
      IF(K1.EQ.KK) GO TO 33
      XV=1
      DO 36 J=2,N
      XL=AMAX1(B(1,J,KK),B(1,J,K1))
      XU=AMIN1(B(2,J,KK),B(2,J,K1))
      IF(XL.GE.XU) GO TO 35
   36 XV=XV*(XU-XL)
      V(2,KK)=V(2,KK)-XV
      DO 37 KK1=1,K1
      IF(KK1.EQ.K1) GO TO 35
      XV=1
      DO 38 J=2,N
      XL=AMAX1(B(1,J,KK),B(1,J,K1),B(1,J,KK1))
      XU=AMIN1(B(2,J,KK),B(2,J,K1),B(2,J,KK1))
      IF(XL.GE.XU) GO TO 37
   38 XV=XV*(XU-XL)
      V(2,KK)=V(2,KK)+XV
   37 CONTINUE
   35 CONTINUE
   33 CONTINUE
C.... COMPUTE LOG LIKELIHOOD
      XLL=0
      DO 50 KK=1,K
      IF(V(1,KK).EQ.0.OR.V(2,KK).EQ.0) GO TO 50
      XLL=XLL+V(1,KK)*ALOG(V(1,KK)/V(2,KK))
   50 CONTINUE
      RETURN
      END
```

53

```
            SUBROUTINE MHIST(A,B,NB,M,N,NC,X1,X2,KM,NM,NV)
C..........................................................12 DECEMBER 1973
C..... DRAWS MULTIVARIATE HISTOGRAM BASED ON BLOCKS IN B
C.... A = M BY N BORDERED ARRAY
C.... B = 2 BY N BY K ARRAY : B(1,J,K);B(2,J,K) IS RANGE FOR JTH VAR. , KTH BLK.
C.... M = NUMBER OF ROWS
C.... N = NUMBER OF COLUMNS
C.... NC = NUMBER OF CELLS
C.... X1 = 1 BY N ARRAY : X1(J) = LOWER BOUND OF JTH VARIABLE
C.... X2 = 1 BY N ARRAY : X2(J) = UPPER LIMIT OF JTH VARIABLE
C.... KM = NUMBER OF MODES
C.... NM = 1 BY M SCRATCH ARRAY
C.... NV = 1 BY N ARRAY DEFINING ORDER OF VARIABLES
C.............................................................................
      DIMENSION AA(112)
      DIMENSION NV(N)
      DIMENSION NB(2,N,KM)
      DIMENSION A(M,N),B(2,N,KM),X1(N),X2(N),NM(M)
      DATA BL,V,H/1H ,1H¦,1H_/
      DATA P/1H./
      WRITE(6,1) A(1,1),(A(1,J),X1(J),X2(J),J=2,N)
    1 FORMAT(21H1 MODAL HISTOGRAM OF ,A4/
     *60H INTERVALS ...¦    ¦.... FOR EACH VARIABLE DEFINE MODAL BLOCKS/
     *61H RECTANGLES TO RIGHT HAVE AREA PROPORTIONAL TO NUMBER IN MODE//
     *16H VARIABLE RANGES/(1X,A4,F9.3,2X,16(1H.),1X,F10.3))
      WRITE(6,2)
    2 FORMAT(1H+,15X,10(1H-))
      DO 9 I=2,M
    9 NM(I)=0
      DO 10 K=1,KM
C.... COUNT NUMBER SATISFYING CONDITIONS
      NK=0
      DO 20 I=2,M
      IF(NM(I).NE.0) GO TO 20
      DO 31 J=2,N
      IF(A(I,J).GT.B(2,J,K)) GO TO 30
      IF(A(I,J).LT.B(1,J,K)) GO TO 30
   31 CONTINUE
      NM(I)=1
      NK=NK+1
   30 CONTINUE
   20 CONTINUE
C.... GENERATE HEIGHT
      NK=NK+19
      DO 40 I=1,112
   40 AA(I)=BL
      DO 41 I=3,NK
   41 IF(I.NE.17.AND.I.NE.18.AND.I.NE.NK) AA(I)=H
      WRITE(6,5) (AA(I),I=1,112)
    5 FORMAT(15X,112A1)
      DO 42 I=19,NK
   42 AA(I)=BL
C.... WORK ON PROFILE
      DO 50 JJ=2,N
      J=NV(JJ)
      DO 53 I=1,18
   53 AA(I)=BL
      AA(19)=V
      AA(NK)=V
      K1=NB(1,J,K)+1
      K2=NB(2,J,K)+1
      AA(K1)=V
      AA(K2)=V
      J1=JJ+1
      IF(JJ.EQ.N) J1=JJ
      J1=NV(J1)
      K3=NB(1,J1,K)+1
      K4=NB(2,J1,K)+1
      DO 51 KK=2,17
      IF(KK.EQ.K1.OR.KK.EQ.K2) GO TO 51
      IF(KK.LT.K1) AA(KK)=P
      IF(KK.GT.K2) AA(KK)=P
      IF((KK-K3)*(KK-K1).LT.0) AA(KK)=H
      IF((KK-K4)*(KK-K2).LT.0) AA(KK)=H
   51 CONTINUE
C.... PRINT A LINE
      WRITE(6,6) A(1,J),(AA(I),I=1,112)
    6 FORMAT(10X,A4,1X,112A1)
```

54

```
   50 CONTINUE
      DO 61 I=3,NK
   61 IF(I.NE.17.AND.I.NE.18.AND.I.NE.NK) AA(I)=H
   10 WRITE(6,7)(AA(I),I=1,112)
    7 FORMAT(1H+,14X,112A1)
      WRITE(6,3)
    3 FORMAT(1H1)
      RETURN
      END




      SUBROUTINE LBOX(I,J,NL,A,IS)
C......................................................23 NOV 1973
C.... DRAWS LINE IN MATRIX A
C.... I = START ROW
C.... J = STARTING COLUMN
C.... NL = LENGTH OF LINE
C.... A = 21 BY 132 DATA MATRIX
C.... IS = 1:LEFT
C          2:DOWN
C          3:SLOPE
C......................................................................
      DIMENSION A(21,132)
      DATA D,V,H,S/1H.,1H|,1H_,1H//
      IF(I.LT.1.OR.I.GT.132) RETURN
      IF(J.LT.1.OR.J.GT.132) RETURN
      IF(A(I,J).NE.V) A(I,J)=S
      GO TO (10,20,30),IS
   10 DO 11 K=1,NL
      II=I
      JJ=J-K
   11 IF(JJ.GT.0) A(II,JJ)=H
      RETURN
   20 DO 21 K=1,NL
      JJ=J
      II=I+K
   21 IF(II.GT.0) A(II,JJ)=V
      RETURN
   30 DO 31 K=1,NL
      II=I-K
      JJ=J+K
   31 IF(K.NE.NL) A(II,JJ)=S
      RETURN
      END
```

```
      SUBROUTINE BOXES(A,M,N,A1,A2,NC,NN)
C.......................................................................23 NOV 1973
C.... PRODUCES ONE BOX PER CASE WITH VARIABLES DIVIDED INTO 3 SETS
C.... A = M BY N BORDERED DATA ARRAY
C.... M = NUMBER OF ROWS
C.... N = NUMBER OF COLUMNS
C.... A1 = 1 BY N LOWER LIMITS FOR EACH VARIABLE
C.... A2 = 1 BY N UPPER LIMITS FOR EACH VARIABLE
C.... NC = 1 BY N CELL SIZES
C.... N C = 1 BY N NUMBER OF INTERVALS FOR EACH VARIABLE
C.... NN = 3 BY N SIDE DIVISIONS: NN(K,I) IS THE ITH VARIABLE IN THE KTH DIMENSI
C.... NN = 3 BY N SIDE DIVISIONS: NN(K,I) IS THE ITH VARIABLE IN THE KTH SIDE
C.......................................................................
      DIMENSION A(M,N),A1(N),A2(N),NC(N),NN(3,N)
      DIMENSION NCS(3)
      DIMENSION P(21,132),NW(3)
C.... DIMENSIONS OF BOX
      DATA B/1H /
      N1=M-1
      DO 10 J=2,N
   10 CALL RANGE(A(2,J),N1,A1(J),A2(J))
      DO 11 K=1,3
      NW(K)=0
      DO 11 I=1,N
   11 IF(NN(K,I).NE.0) NW(K)=NW(K)+NC(K)
C.... WRITE RANGE AND BOX ASSIGNMENTS
      WRITE(6,5)
    5 FORMAT(22H1 VARIABLE ASSIGNMENTS)
      WRITE(6,6)
    6 FORMAT(5H     Z/4H X-//4H   I/4H   Y)
   17 FORMAT(12H VARIABLE : ,A4,13H   MINIMUM : ,F8.2,13H   MAXIMUM : ,
     * F8.3,25H   NUMBER OF INTERVALS : ,I4)
      DO 13 K=1,3
    8 FORMAT(//18H X VARIABLE RANGES/1H ,17(1H-))
    9 FORMAT(//18H Y VARIABLE RANGES/1H ,17(1H-))
    7 FORMAT(//18H Z VARIABLE RANGES/1H ,17(1H-))
      IF(K.EQ.1) WRITE(6,8)
      IF(K.EQ.2) WRITE(6,9)
      IF(K.EQ.3) WRITE(6,7)
      DO 14 J=1,N
      IF(NN(K,J).EQ.0) GO TO 13
      JL=NN(K,J)
      WRITE(6,17) A(1,JL), A1(JL),A2(JL),NC(JL)
   14 CONTINUE
   13 CONTINUE
      WRITE(6,3) A(1,1)
    3 FORMAT(1H1)
      NDD=NW(3)+NW(2)+1
      IF(NDD.GT.60) RETURN
      NR=132/(NW(1)+NW(3)+1)
      IF(NR.EQ.0) RETURN
C.... LOOP THROUGH CASES
      NLIN=0
      DO 20 II=2,M,NR
      IL=II
      IU=II+NR-1
      IF(IU.GT.M) IU=M
      DO 30 I=1,21
      DO 30 J=1,132
   30 P(I,J)=B
      NDD=0
      DO 31 I=IL,IU
      JS=(I-IL+1)*(NW(1)+NW(3)+1)-NW(3)
      IS=NW(3)+1
      DO 41 K=1,3
      NCS(K)=0
      DO 41 J=1,N
      IF(NN(K,J).EQ.0) GO TO 41
      JN=NN(K,J)
      NCH=KC(A(I,JN),A1(JN),A2(JN),NC(JN))
      NCS(K)=NCS(K)+NCH
   41 CONTINUE
      NNI = NCS(2)+NCS(3)+2
      IF(NNI.GE.NDD) NDD=NNI
      DO 32 K=1,3
   32 CALL LBOX(IS,JS,NCS(K),P,K)
C.... INSERT BOX FOR CASE I
      DO 40 KB=1,3
```

56

```
                  K=4-KB
                  ISS=IS
                  JSS=JS
                  DO 50 J=1,N
                  IF(NN(K,J).EQ.0) GO TO 40
                  JN=NN(K,J)
                  NCH=KC(A(I,JN),A1(JN),A2(JN),NC(JN))
                  IF(K.EQ.1) JSS=JSS-NCH
                  IF(K.EQ.2) ISS=ISS+NCH
                  IF(K.EQ.3) JSS=JSS+NCH
                  IF(K.EQ.3) ISS=ISS-NCH
                  DO 51 KK=1,3
                  IF(KK.EQ.K) GO TO 51
                  CALL LBOX(ISS,JSS,NCS(KK),P,KK)
               51 CONTINUE
               50 CONTINUE
               40 IF(K.EQ.3) P(ISS,JSS)=B
               31 CONTINUE
                  NLIN=NLIN+NDD
        C.... FINALLY, PRINT OUT ARRAY
                  DO 70 I=1,21
                  DO 71 J=1,132
                  IF(P(I,J).NE.B) GO TO 72
               71 CONTINUE
                  GO TO 70
               72 WRITE(6,1)(P(I,J),J=1,132)
               70 CONTINUE
                1 FORMAT(1X,132A1)
                  WRITE(6,2)(A(I,1),I=IL,IU)
                2 FORMAT(4(16X,A4,11X))
                  IF(NLIN+NW(2)+NW(3).LE.100) GO TO 20
                  NLIN=0
                  WRITE(6,3) A(1,1)
               20 CONTINUE
                  WRITE(6,3) A(1,1)
                  RETURN
                  END
```

CHAPTER 2

Distances

2.1 INTRODUCTION

From Table 2.1, it is desired to select clusters of similar Southern States. What does similar mean? No one thing. If the interest is in evaluating a government health program, the variables age, births, income, doctors, and infant mortality might be selected to base the clustering on and the others ignored. If the interest is in locating an industrial plant, the variables might be manufacturing, cars, telephones, and school years completed. Different measures of similarity and different clusters will be appropriate for different purposes.

A standard way of expressing similarity is through a set of distances between pairs of objects. Many clustering algorithms assume such distances given and set about constructing clusters of objects within which the distances are small. The choice of distance function is no less important than the choice of variables to be used in the study. A serious difficulty in choosing a distance lies in the fact that a clustering structure is more primitive than a distance function and that knowledge of clusters changes the choice of distance function. Thus a variable that distinguishes well between two established clusters should be given more weight in computing distances than a "junk" variable that distinguishes badly (see Friedman and Rubin, 1967).

2.2 EUCLIDEAN DISTANCES

There are M objects or cases, N variables, and the Jth variable has value $A(I, J)$ for the Ith case. Consider, for example, the demographic data for cases Alabama, Arkansas, Delaware, and Florida and for variables age and altitude. Here $M = 4$, $N = 2$, and a typical data value is $A(3, 2) = 6$, the value of the second variable for the third case (see Table 2.2).

Definition. The *euclidean distance* between case I and case K is

$$D(I, K) = (\sum \{1 \leq J \leq N\} [A(I, J) - A(K, J)]^2)^{1/2}.$$

In one, two, or three dimensions, this is just "straight-line" distance between the vectors corresponding to the Ith and Kth cases. For the data in Table 2.2, consider the distance between the first case, Alabama, and the third case, Delaware:

$$D(1, 3) = \{[A(1, 1) - A(3, 1)]^2 + [A(1, 2) - A(3, 2)]^2\}^{1/2}$$
$$= [(26 - 29)^2 + (50 - 6)^2]^{1/2}$$
$$= 44.1.$$

58

Table 2.1 Demographic Data for the South (1960–1965)

ALT : Mean altitude above sea level, in tens of feet
TEM : Mean annual temperature, in degrees fahrenheit
PRE : Mean annual precipitation in inches
POP : Number of persons per square mile
NEG : Number of Negroes per 100 of total population
AGE : Median age
URB : Urban population as percentage of total
BIR : Number of births per 1000 population
RUR : Rural farm population as percentage of total
MAN : Employment in manufacturing as percentage of total
CAR : Automobiles per 100 population
TEL : Telephones per 100 population
INC : Income in hundreds of dollars
FED : Federal revenue per 100 dollars of state and local revenue
LAW : Lawyers per 100,000 persons
SCH : School years completed, in tenths of a year
EDX : Expenditure on education in tens of dollars per pupil per year
DOC : Physicians per 100,000 persons
INF : Infant mortality of whites per 1000 births
HOU : Percentage of houses with sound plumbing
R60 : Percentage Republican votes in Presidential election 1960
R64 : Percentage Republican votes in Presidential election 1964
GOV : Percentage Republican votes in 1962–64 state governor elections

	ALT	TEM	PRE	POP	NEG	AGE	BIR	URB	RUR	MAN	CAR	TEL	INC	FED	LAW	DOC	INF	SCH	EDX	HOU	R60	R64	GOV
AL	50	68	68	64	30	26	22	55	12	31	39	33	19	22	82	79	25	89	28	54	42	70	4
AS	65	62	49	34	22	29	22	43	17	29	33	30	18	24	79	91	23	87	32	48	43	43	43
DE	6	54	45	226	14	29	23	66	4	37	40	53	33	15	115	135	20	108	54	80	49	39	49
FL	10	72	56	91	18	31	20	74	2	15	45	45	24	11	150	142	23	106	41	78	52	49	44
GA	60	62	47	68	29	26	24	55	9	32	37	36	22	19	125	102	23	88	33	58	37	54	0
KY	75	56	41	76	7	28	22	45	17	27	38	32	20	21	108	95	26	85	32	53	54	36	49
LA	10	70	63	72	32	25	25	63	7	17	32	36	21	21	128	114	21	86	42	61	29	57	39
MD	35	58	44	314	17	29	20	73	31	25	37	48	30	11	175	158	22	10	50	81	46	35	44
MI	30	65	49	46	42	24	24	38	23	31	29	25	16	21	101	76	23	86	27	45	25	87	38
MO	80	56	35	63	9	32	21	67	12	28	37	47	26	17	72	149	21	93	45	66	50	36	38
NC	70	60	44	93	25	26	22	40	16	42	35	31	20	15	77	100	22	85	32	57	48	44	43
SC	35	64	47	79	35	23	22	41	14	43	35	28	18	16	72	80	22	84	28	54	49	59	0
TE	90	61	47	85	17	28	22	52	15	35	35	35	20	22	116	113	24	86	30	57	53	45	49
TS	170	67	29	37	12	27	22	75	7	20	42	41	23	15	144	111	26	101	40	69	49	37	26
VA	95	59	44	100	21	27	21	56	9	27	34	38	24	16	114	108	24	92	38	66	52	46	36
WV	150	56	44	77	5	29	20	38	7	27	31	34	20	18	97	103	26	87	33	57	47	32	45

Notice that the variable *altitude* makes the major contribution to all the distances in Table 2.2 and that almost the same distance matrix would be obtained if age were ignored entirely. This distance calculation is a silly one because the squared distance is in units of squared years plus squared feet, and why should one year equal one foot? Age would dominate the calculations if it were measured in fine enough units, say months or weeks.

It is necessary then, when variables are measured in different units, to prescale the

Table 2.2 Euclidean Distances

| | DATA | | | DISTANCE | | |
	AGE (YRS)	ALTITUDE (FT)	AL	AS	DE	FL
ALABAMA	26	50	0	15.3	44.1	40.3
ARKANSAS	29	65	15.3	0	59.0	55.0
DELAWARE	29	6	44.1	59.0	0	4.5
FLORIDA	31	10	40.3	55.0	4.5	0

		WEIGHTED DISTANCE			
		AL	AS	DE	FL
VAR (AGE) = 4.25 YRS^2	AL	0	1.6	2.1	2.8
VAR (ALT) = 857 FT^2	AS	1.6	0	2.0	2.1
W(1) = 1/4	DE	2.1	2.0	0	1.0
W(2) = 1/900	FL	2.8	2.1	1.0	0

variables to make their values comparable or, equivalently, to compute a *weighted euclidean distance*

$$D(I, K) = (\sum \{1 \leq J \leq N\} \, W(J)[A(I, J) - A(K, J)]^2)^{1/2}.$$

This form of distance is not necessary if all variables are measured on the same scale—for example, the percentage Republican in the three elections. But even in this case, weights might be used to increase or decrease the importance of some variable, perhaps on subjective or *a priori* grounds. For example, if the next presidential election is the reason for organizing the data, the two presidential votes may be given more weight than the gubernatorial vote.

There are a number of unsatisfactory methods of choosing the weights $W(J)$.

(i) *Subjective.* Just as the investigator selected the variables to be used in the data, he is free to weight the variables according to considerations outside the data. He thinks to himself, 1 yr in age is equivalent to about 10 ft in altitude, and this implies a weight 1 yr^{-2} for age and a weight $\frac{1}{100}$ ft^{-2} for altitude. He expects variation *within* clusters to be about 1 yr in age to 10 ft in altitude.

(ii) *Measurement error.* Weights are chosen inversely proportional to measurement variance. (The units of weights are always the inverse square of measurement units.) For example, in the demographic data, the measurement variance for each variable is the average variance within states. Thus altitude might have a measurement variance of 45^2 on average, age about 15^2, and the relative weights of age to altitude would be 9 to 1. The measurement error is estimated from "replications" of the cases—that is, from a subjective preclustering of cases. This weighting scheme is used in the coefficient of racial likeness (Pearson, 1926). It may happen that there are no replications for the cases, so that this technique is unavailable. Even with the preclustering, there may be a tendency to overweight variables that are measured with great precision but are useless for clustering.

(iii) *Equal variance scale.* Weights are inversely proportional to variance and are thus chosen from the data. In Table 2.2,

VAR(AGE) = $\frac{1}{3}$[26^2 + 29^2 + 29^2 + 31^2 − $\frac{1}{4}$(26 + 29 + 29 + 31)^2] = 4.25 yrs^2

and VAR(ALT) = 857 ft²; therefore, by rounding, WT(1) = $\frac{1}{4}$ and WT(2) = $\frac{1}{900}$. Doing one calculation in detail,

$$D(1, 3) = [\tfrac{1}{4}(26 - 29)^2 + \tfrac{1}{900}(50 - 6)^2]^{1/2} = 2.1.$$

With this variance weighting, the distances are invariant under change of the units of measurement, and all variables make the same average contribution to the squared distances (see Sokal and Sneath, 1963).

Look at Table 2.3 and Figure 2.1 to see a serious defect in scaling to equalize variances. The percent Republican for president is taken for Georgia, Louisiana, South Carolina, West Virginia, Missouri, Kentucky, and Maryland in 1960 and 1964.

Table 2.3 Bad Weighting by Variances of Percent Republican for President

DATA

	1960	1964
GA	37	54
LA	29	57
SC	49	59
WV	47	32
MO	50	36
KY	54	36
MD	46	35

UNWEIGHTED DISTANCES

	GA	LA	SC	WV	MO	KY	MD
GA	0	9	13	24	22	25	21
LA	9	0	20	31	30	33	28
SC	13	20	0	27	23	24	24
WV	24	31	27	0	5	8	3
MO	22	30	23	5	0	4	4
KY	25	33	24	8	4	0	8
MD	21	28	24	3	4	8	0

WEIGHTED DISTANCES

VAR (1960) = 74
VAR (1964) = 141
W(1) = 2
W(2) = 1

	GA	LA	SC	WV	MO	KY	MD
GA	0	12	18	26	26	30	23
LA	12	0	28	36	36	42	33
SC	18	28	0	27	23	24	24
WV	26	36	27	0	6	11	3
MO	26	36	23	6	0	6	6
KY	30	42	24	11	6	0	11
MD	23	33	24	3	6	11	0

MAHALANOBIS DISTANCES

COV (1960, 1964) = -62
W(1, 1) = 2
W(1, 2) = W(2, 1) = 1
W(2, 2) = 1

	GA	LA	SC	WV	MO	KY	MD
GA	0	9	21	16	14	17	13
LA	9	0	30	19	21	25	18
SC	21	30	0	29	22	19	27
WV	16	19	29	0	8	13	2
MO	14	21	22	8	0	6	6
KY	17	25	19	13	6	0	12
MD	13	18	27	2	6	12	0

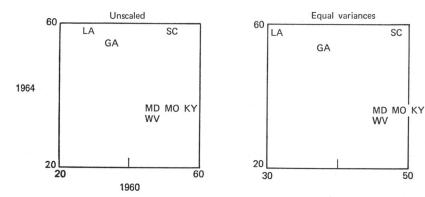

Figure 2.1 Decreased clarity of clusters after rescaling to equal variances—Republican voting data.

By using the unscaled data or the unweighted distances, two very distinct clusters are seen, GA, LA, SC and WV, MO, KY, MD. All the within-cluster distances (just) are smaller than the between-cluster distances. If the data are rescaled to have equal variances (or, equivalently, if weights are inversely proportional to variances in the distance calculation), the clusters become much less clear cut. The ratio of between- to within-cluster average distance is decreased and seven of the between-cluster distances become less than the within-cluster distance for SC and LA. The correct location of SC becomes uncertain.

To see the problem, suppose the variables were initially scaled so that within-cluster variances are equal. If all variables are now rescaled to have equal variances, those variables that have relatively large between-cluster variances will be reduced in importance. Therefore overall between-cluster variance will be reduced relative to within-cluster variance, which means that the clusters become less distinct.

Ideally, the scaling should be done so that the within-cluster variances are approximately equal. (The measurement error approach equalizes variances within a preclustering of the cases.) There is a basic circularity here:

(i) In order to cluster objects, it is necessary to propose a measure of distance between objects.

(ii) In order to define distance, it is necessary to weight the variables.

(iii) In order to weight the variables, it is necessary to know the clusters of objects so that within-cluster variances can be equalized.

2.3 RELATIONS BETWEEN VARIABLES

Suppose that the only data available were age, income, infant mortality, and the three voting variables. Even after standardization (say to equalize variances), the political effect has three times the weight of each of the demographic variables. It would be reasonable to give each of the voting variables weight $\frac{1}{3}$ and each of the others weight 1.

More generally, if it is noticed that several variables are highly correlated, it might be desirable to down-weight each. Similarity between variables thus affects the computation of similarity between cases (and vice versa). To accommodate relations between variables, a *generalized euclidean distance* is defined as

$$D(I, K)^2 = \sum \{1 \leq J, L \leq N\} \, W(J, L)[A(I, J) - A(K, J)][A(I, L) - A(K, L)].$$

The matrix $W(J, L)$ of weights, one for each pair of variables, must be positive semidefinite; that is, the above expression must be nonnegative for every possible A.

The choice of weights W is even more difficult than the previous case when only the diagonal elements were nonzero. There is a natural extension of the "equal variance" scale, in which W is the inverse of the matrix with (J, L) element:

$$\sum \{1 \leq I, K \leq M\} \, [A(I, J) - A(K, J)][A(I, L) - A(K, L)].$$

Equivalently, W is the inverse of the covariance matrix of the variables. (The covariance matrix may be computed over a reference class of cases other than the full data set—for example, within an initial clustering.) The distance is called the Mahalanobis distance, and arises naturally in multivariate normal theory (see Rao, 1948). The distance is invariant under any linear transformation of the original variables, including transformations of the type (ALT, AGE) → (2ALT + AGE, 3ALT − AGE). Invariance under this general class of linear transformations seems much less compelling than invariance under the change of measuring units of each of the variables.

The Mahalanobis distance based on the full data covariance matrix is even worse than the "equal variance" scale in decreasing the clarity of clusters. For the voting data in Table 2.3, the variances and covariances are VAR(1960) = 74, VAR(1964) = 141, and COV(1960, 1964) = −62. And so the weight matrix, proportional to the inverse of the covariance matrix is $W(1, 1) = 2$, $W(1, 2) = W(2, 1) = 1$, and $W(2, 2) = 1$. To do one distance computation in detail,

$$D(\text{LA, SC}) = [2(29 - 49)^2 + (29 - 49)(57 - 59)$$
$$+ (29 - 49)(57 - 59) + (57 - 59)^2]^{1/2} = 30.$$

Notice that the within-cluster distances are increased compared to the between-cluster distances, even from the already bad equalized variance case. The equalized variance technique is not too harmful to the clarity of clusters if all variables have about the same ratio of within- to between-cluster variance. The Mahalanobis distance reduces the clarity of clusters even in this case. Suppose there are two distinct clusters, equally well separating all the variables. The first eigenvector is approximately the vector between the mean of the cases in the first cluster and the mean of the cases in the second cluster. If the clusters are well separated, the first eigenvector will have a large variance compared to the other eigenvectors. Stomping out the brush fire, the Mahalanobis distance standardizes all eigenvectors to have equal variance, thus reducing the between-cluster component in the distances principally contributed by the first eigenvector (see Cronbach and Gleser, 1953).

The weights should be estimated from the within-cluster covariance of the variables (as with the equalized variance procedure). Of course, this involves the same circularity as before. The original treatments of Mahalanobis (1936) and Rao (1948) define the distance between two groups, given the means and common covariance matrix of the groups. These groups are a preclustering of the data.

Friedman and Rubin (1967) have considered a clustering technique that maximizes the average distance between clusters relative to the within-cluster covariance matrix.

This technique embraces the circularity by not assuming that distances are given once and for all.

2.4 DISGUISED EUCLIDEAN DISTANCES

(i) The distance between variables J and L is defined to be $1 - \text{CORR}(J, L)$, where

$$\text{CORR}(J, L) = \frac{(\sum \{1 \leq I \leq M\}[A(I, J) - \bar{A}(J)][A(I, L) - \bar{A}(L)])}{(\sum \{1 \leq I \leq M\}[A(I, J) - \bar{A}(J)]^2 \sum \{1 \leq I \leq M\}[A(I, L) - \bar{A}(L)]^2)^{1/2}},$$

where $\bar{A}(J) = \sum \{1 \leq I \leq M\} A(I, J)/M$. This rather horrible looking expression is just the usual definition of correlation.

Let A^* denote the matrix standardized to have mean zero and variance one for every variable. Let A^{*T} denote the transposed standardized matrix. Then the euclidean distance between the Jth and Lth row of this matrix is $(2M)^{1/2}[1 - \text{CORR}(J, L)]^{1/2}$. Thus many of the considerations relevant to computing distances between cases may be transferred to computing correlations. For example, the cases should be weighted, perhaps.

(ii) Suppose that the responses to a variable are not numerical but categorical— say, the religion or sex of a person. Define the distance between two cases I and K as the number of variables J for which $A(I, J) \neq A(K, J)$. This distance can also be cast as a euclidean distance by defining a new data matrix A^* with number of variables equal to the total number of categories (over all variables) in the original data matrix. Each of the new variables corresponds to a single category of one of the old variables. The new variable takes value 1 if case I occurs in this category and value 0 otherwise. The euclidean distance between cases is then equal to the square root of the distance just defined.

Weighting considerations apply as before, if desired, separately to each category.

2.5 OTHER DISTANCES

First consider distances for a single variable. A distance for the Jth variable, between cases I and K, will be denoted by $D(I, K \mid J)$.

(i)
$$D(I, K \mid J) = |A(I, J) - A(K, J)|.$$

This is the usual univariate euclidean distance, and is only appropriate for numerically valued variables.

(ii)
$$D(I, K \mid J) = |A(I, J) - A(K, J)| \qquad \text{if } |A(I, J) - A(K, J)| \leq T(J)$$
$$= T(J) \qquad \text{if } |A(I, J) - A(K, J)| > T(J).$$

This distance is a modification of euclidean distance that prevents any one variable from having too much weight in a particular comparison.

(iii)
$$D(I, K \mid J) = 0 \qquad \text{if } A(I, J) = A(K, J)$$
$$= 1 \qquad \text{if } A(I, J) \neq A(K, J).$$

This is a special case of (ii), appropriate for category variables.

There are a number of ways of combining distances over variables, each with its own way of weighting variables to make equal contribution to the distance.

(i)

$$D(I, K) = \sum \{1 \leq J \leq N\}\, D(I, K \,|\, J) W(J).$$

In this case, a choice of weights analogous to the equal variance choice selects $W(J)$, so that

$$\sum \{1 \leq I, K \leq M\}\, D(I, K \,|\, J) W(J) = 1.$$

Thus each variable makes the same average contribution to the distances $D(I, K)$.

(ii)

$$D(I, K) = [\sum \{1 \leq J \leq N\}\, D(I, K \,|\, J)^2 W(J)]^{1/2}.$$

This gives euclidean distance if $D(I, K \,|\, J) = |A(I, J) - A(K, J)|$.

(iii)

$$D(I, K) = \max \{1 \leq J \leq N\}\, D(I, K \,|\, J) W(J).$$

The weighting procedure here chooses $W(J)$ so that

$$\max \{1 \leq I, K \leq M\}\, D(I, K \,|\, J) W(J) = 1.$$

All three of these methods of combining distance are members of the Minkowski family,

$$D(I, K) = [\sum \{1 \leq J \leq N\}\, D(I, K \,|\, J)^p W(J)]^{1/p}$$

with $p = 1, 2, \infty$, respectively.

2.6 PLOTTING DISTANCES TO DETECT CLUSTERS

If the cases are distributed uniformly in M-dimensional space, the number of objects N within euclidean distance d of a given object will be Kd^M. Thus, if $d(N)$ is the distance of the object Nth closest to the given object, approximately

$$\log N = a + M \log d(N).$$

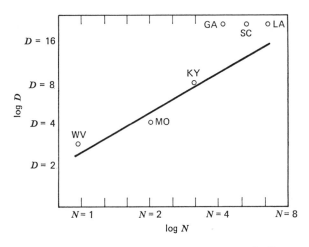

Figure 2.2 Clusters in plots of distances (based on Table 2.3, distances from Maryland).

If N_0 objects form a cluster near the given object, the distances $d(N)$ for $N > N_0$ will be greater than expected from the plot of objects within the cluster. Thus a break in the plot occurs at the boundary of the cluster. In Figure 2.2, distances are taken from Maryland using Table 2.3. The objects WV, MO, and KY are in a cluster about Maryland. The distances of GA, SC, and LA are greater than expected for a uniform distribution of objects.

2.7 THINGS TO DO

2.7.1 Distances between Variables

Complications arise because the variables lie on different scales or are of different types.

For two real variables measured on different scales, an appropriate rescaling gives them mean 0 and variance 1, and then euclidean distance is proportional to 1-correlation.

For two ordered variables, a plausible distance is

$$P(X < X^* \mid Y < Y^*, X \neq X^*),$$

where X, Y and X^*, Y^* are randomly sampled from the two variables.

For two category variables, let $P(I, J)$ be the proportion of cases taking value I for the first variable and value J for the second variable. (Suppose I takes values $1, 2, \ldots,$ M and J takes values $1, 2, \ldots, N$.) Let

$$P(I, 0) = \sum \{1 \leq J \leq N\} P(I, J)$$
$$P(0, J) = \sum \{1 \leq I \leq M\} P(I, J).$$

A measure of similarity is

$$\sum \{1 \leq I \leq M, 1 \leq J \leq N\} P(I, J) \log P(I, J)$$
$$- \sum \{1 \leq I \leq M\} P(I, 0) \log P(I, 0)$$
$$- \sum \{1 \leq J \leq N\} P(0, J) \log P(0, J).$$

For a category variable and a real variable, a natural measure of similarity (invariant under permutation of categories and linear transformation of the real variable) is the ratio of between-category mean square to within-category mean square.

What is the relationship between all these measures in the intersecting case of two 0–1 variables?

2.7.2 Distributions of Distances

For many stochastically independent variables, the squared euclidean distance between cases I and K is a sum of independent components and the central limit theorem is applicable. The set of distances $D(I, K)$ is approximately multivariate normal, with the same mean and variance for every $D(I, K)$, with a certain covariance between $D(I, K)$ and $D(I, L)$ not depending on I, K, or L, and with all other covariances zero. It is assumed that the set of cases is fixed and that the number of variables is very large, much larger than the number of cases.

The three parameters mean, variance, and covariance may be estimated in an obvious way from the observed distance matrix.

REFERENCES

BLACKWELDER, R. A. (1966). *Taxonomy*, Wiley, New York. A "traditional" taxonomy book deemphasizing numerical methods. The author argues (p. 150) that some weighting of characters is practically unavoidable, taking place in the first instance in the selection of characters to be used in classifying animals.

CATTELL, R. B. (1949). "R_p and other coefficients of pattern similarity." *Psychometrika* **14**, 279–288. In educational psychology, a pattern or profile is the set of scores obtained by a person on a number of items in a questionnaire. In our terminology, a person is a case, and an item is a variable. Cattell's R_p is a monotone transformation of d, where d is euclidean distance with the variables standardized to have variance 1. Setting k equal to the median of a chi-square variable on n degrees of freedom (approximately $k = n - \frac{2}{3}$),

$$R_p = \frac{2k - d^2}{2k + d^2}.$$

This coefficient takes values between -1 and $+1$, with a negative value corresponding to cases that are further apart than would be expected by chance. Cattell shows empirically that R_p is approximately normally distributed for two randomly selected persons.

CRONBACH, L. J., and GLESER, G. C. (1953). "Assessing similarity between profiles." *Psychol. Bull.* **50**, 456–473. This article contains a good survey of distances used in educational psychology. There is a revealing analysis of the various weighting schemes in terms of eigenvectors. The first principal component is the linear combination of variables (with coefficients whose squares sum to unity), which has maximum variance. The second principal component is that linear combination of variables, which is uncorrelated with the first and has maximum variance among those uncorrelated with the first. And so on. These new variables (it turns out) are obtained by orthogonal transformations from the original ones, and so distances based on them are exactly the same as distances based on the original. As a practical matter, the variances of the later principal components will be small, these components will make only a slight average contribution to distances between cases, and they could perhaps be ignored to reduce storing of data and computation of distances.

Now consider the various weighting schemes. If all variances equal unity, the eigenvectors or principal components will be computed from the correlation matrix of the variables, rather than the covariance matrix. If the variables are uncorrelated, the principal components will all have variance equal to unity (in this case they are not uniquely determined, and a particular set of principal components is the original set of variables). Suppose three variables are nearly identical, but otherwise all variables are nearly uncorrelated. In this case, the first principal component will be proportional to an average of the three identical variables and will have variance 3. Thus weight 3 is "really" being given to these repeated variables.

The Mahalanobis weighting corresponds to standardizing all the principal components to have variance unity. At first sight, this will solve the repeated variables problem described above. Yet all the later principal components (which above were discarded entirely to reduce computation) will have equal weight with the early ones. The feeling is that these later principal components merely represent measurement errors, not the important factors in the data, and so should not be increased in weight.

This argument applies even if the reference covariance matrix is computed within clusters previously given.

FRIEDMAN, H. P., and RUBIN, J. (1967). "On some invariant criteria for grouping data." *J. Am. Stat. Assoc.* **62,** 1159–1178. The Mahalanobis distance requires weights to be the inverse of the covariance matrix computed within clusters, but since the distances are to be used to construct the clusters, the procedure is circular. They suggest a clustering procedure that maximizes the average between cluster distance, with weights computed from the within-cluster covariance matrix.

MAHALANOBIS, P. C. (1936). "On the generalized distance in statistics." *Proc. Indian Nat. Inst. Sci.* (Calcutta) **2,** 49–55. Defines a distance between two populations with mean vectors $\bar{X}_1$ and $\bar{X}_2$ and common covariance matrix W, by $(\bar{X}_1 - \bar{X}_2)'W^{-1}(\bar{X}_1 - \bar{X}_2)$.

PEARSON, K. (1926). "On the coefficient of racial likeness." *Biometrika* **18,** 105. Pearson measures the distance between two groups by

$$\sum_{j=1}^{n} \frac{n_{1j}n_{2j}}{n_{1j} + n_{2j}} \left(\frac{\bar{X}_{1j} - \bar{X}_{2j}}{\sigma_j} \right)^2,$$

where n_{1j} is the number of cases of the jth variable in the first group, n_{2j} is the number of cases of the jth variable in the second group, $\bar{X}_{1j}$ is the average of the jth variable in the first group, $\bar{X}_{2j}$ is the average of the jth variable in the second group, and σ_j is the standard deviation within groups of the jth variable. Actually, $\sigma_j(n_{1j}^{-1} + n_{2j}^{-1})^{1/2}$ is an estimate of measurement error for the jth variable, so this is one of the early weighting schemes where the weight of each variable is based on measurement error. Here the measurement error is estimated from repeated observations. The above quantity is approximately distributed as $2\chi_n^2$ if the variables are independent.

RAO, C. R. (1948). "The utilization of multiple measurements in problems of biological classification." *J. Roy. Stat. Soc.,* B **10,** 159–193. Rao considers the classification of Indian caste groups based on measurement of height, sitting height, nasal depth, nasal height, head length, frontal breadth, bizygometric breadth, head breadth, and nasal breadth. There are 50–200 individuals observed in each caste group, and an initial covariance matrix W is computed within caste groups. Now suppose an individual is known to belong to either the ith or jth caste and is to be classified into one or the other; there is a discriminant function F which is linear in the variables such that, for $F > C$, the individual is classified in the ith caste and, for $F < C$, the individual is classified in the jth caste. The number of individuals misclassified is increasing with $(\bar{X}_i - \bar{X}_j)W^{-1}(\bar{X}_i - \bar{X}_j)$, the Mahalanobis distance between the two castes. This theory assumes a multivariate normal distribution of constant covariance matrix within castes.

SOKAL, R. R., and SNEATH, P. H. (1963). *Principles of Numerical Taxonomy,* Freeman, San Francisco. There is an extensive discussion of distances computed from 0–1 variables (equal to characters in biology). Sokal and Sneath argue that all characters should be given equal weight. If one character is more important than the others, this will show by the presence of several other characters correlated with it, and therefore this "type" of character will have more weight.

One problem that worries them is whether or not two species' agreement in *possessing* a character is equivalent to their agreement in *not* possessing the character. Thus, if the biologist selected for data many characters irrelevant to spiders and, say, one relevant to spiders, the spiders would look unreasonably similar to each other using euclidean distance.

They suggest that measured variables should be standardized to have variance 1 across all cases.

PROGRAMS

DIST   computes Minkowski (including Euclidean) and threshold distances.
WCOV   computes within-cluster covariance matrix.
INVERT inverts a square matrix.
WDIST   computes a euclidean distance with a general weight matrix (such as the inverse of the within-cluster covariance matrix).
MISS   replaces missing values by within-cluster means.
TWO   computes means and covariances of a data matrix.
STAND   standardizes an array so that each column has mean 0 and variance 1.
MOM   computes means and covariances when each row is weighted.

```
      FUNCTION DIST(X,Y,M,N,P)
C.........................................................20 MAY 1973
C..... COMPUTES VARIOUS MEASURES OF DISTANCE BETWEEN VECTORS Z AND Y.
C..... X = FIRST VECTOR
C..... Y = SECOND VECTOR
C..... M = NUMBER OF ELEMENTS
C..... N = SKIP FACTOR, FOR USE IN DISTANCES BETWEEN ROWS OF MATRIX.
C..... TO CALL DIST BETWEEN ITH AND JTH ROW OF A BORDERED M BY N ARRAY,USE
C....      CALL DIST(A(I,2),A(J,2),M1,M2,P) WHERE M1=M*(N-2)+1 , M2=M.)
C... P = PARAMETER
C          P.GT.O PTH POWER DISTANCE
C          P.LE.O THRESHOLD DISTANCE COUNTIN PROPORTION OF DEVIATIONS OVER ABS P
C.........................................................
      DIMENSION X(M),Y(M)
      DD=0
      DP=0
      DO 10 I=1,M,N
      IF(X(I).EQ.99999.OR.Y(I).EQ.99999.) GO TO 10
      DP=DP+1
      DIF=ABS(X(I)-Y(I))
      IF(P.GT.0) DD=DD+DIF**P
      IF(P.LE.0.AND.DIF+P.GT.0) DD=DD+1
   10 CONTINUE
      IF(DP.EQ.0) GO TO 20
      IF(P.GT.0) DD=(DD/DP)**(1./P)
      IF(P.LT.0) DD=DD/DP
   20 DIST=DD
      RETURN
      END
```

```
      SUBROUTINE WCOV(A,M,N,COV,NC)
C.........................................................................20 MAY 1973
C.... COMPUTES COVARIANCE WITHIN CLUSTERS, ASSUMING NO MISSING VALUES
C.... M = NUMBER OF ROWS
C.... N = NUMBER OF COLUMNS
C.... A = M BY N BORDERED ARRAY
C.... NC = M BY 1 ARRAY ASSIGNING CASES TO CLUSTERS
C.... COV = M BY M BORDERED WITHIN CLUSTER COVARIANCE MATRIX
C.........................................................................
      DIMENSION A(M,N),COV(N,N),NC(M)
C.... INITIALIZE COVARIANCES AND COUNT CLUSTERS
      DATA AA/4HWCOV/
      DO 10 I=1,N
      DO 10 J=1,N
   10 COV(I,J)=0
      K=0
      DO 11 I=2,M
   11 IF(NC(I).GT.K) K=NC(I)
C.... COMPUTE MEANS
      Q=0
      DO 20 L=1,K
      P=0
      DO 26 J=2,N
   26 COV(1,J)=0
      DO 21 I=2,M
      IF(NC(I).NE.L) GO TO 21
      P=P+1
      DO 22 J=2,N
   22 COV(1,J)=COV(1,J)+A(I,J)
   21 CONTINUE
      DO 23 J=2,N
   23 IF(P.NE.0) COV(1,J)=COV(1,J)/P
      IF(P.GT.0) Q=Q+P-1
C.... ADD ON TO CROSS PRODUCTS
      DO 24 I=2,M
      IF(NC(I).NE.L) GO TO 24
      DO 25 J=2,N
      DO 25 JJ=2,J
   25 COV(J,JJ)=COV(J,JJ)+(A(I,J)-COV(1,J))*(A(I,JJ)-COV(1,JJ))
   24 CONTINUE
   20 CONTINUE
C.... DIVIDE BY DEGREES OF FREEDOM
      DO 30 J=2,N
      DO 30 JJ=2,J
      IF(Q.GT.0) COV(J,JJ)=COV(J,JJ)/Q
   30 COV(JJ,J)=COV(J ,JJ)
C.... LABEL
      DO 40 J=2,N
      COV(1,J)=A(1,J)
   40 COV(J,1)=A(1,J)
      COV(1,1)=AA
      RETURN
      END
```

```
      SUBROUTINE INVERT(A,D,M)
C.........................................................................20 MAY 1973
C.... INVERTS SQUARE BORDERED ARRAY, PREFERABLY POSITIVE DEFINITE
C.... M = NUMBER OF ROWS
C.... A = M BY N BORDERED ARRAY
C.... D = DETERMINANT
C..............................................................................
      DIMENSION A(M,M)
      D=1.
      TH=10.**(-6)
      DO 10 I=2,M
      D=D*A(I,I)
      IF(A(I,I).LT.TH) WRITE(6,1) A(I,I),I
    1 FORMAT(38H **** ERROR IN INVERT, DIAGONAL VALUE ,F15.10,7H IN ROW
     *,I5)
      IF(A(I,I).LT.TH) D=0
      IF(D.EQ.0) RETURN
      IF(A(I,I).GT.TH) A(I,I)=1./A(I,I)
      DO 13 J=2,M
      IF(J.NE.I) A(I,J)=A(I,J)*A(I,I)
   13 CONTINUE
      DO 11 II=2,M
      IF(II.EQ.I) GO TO 11
      DO 12 J=2,M
      IF(J.EQ.I) GO TO 12
      A(II,J)=A(II,J)-A(I,J)*A(II,I)
   12 CONTINUE
   11 CONTINUE
      DO 14 J=2,M
   14 IF(J.NE.I) A(J,I)=-A(J,I)*A(I,I)
   10 CONTINUE
      RETURN
      END




      FUNCTION WDIST(X,Y,N,WT)
C.........................................................................20 MAY 1973
C.... EUCLIDEAN DISTANCE COMPUTED FROM A WEIGHT MATRIX WT
C.... M = NUMBER OF ELEMENTS
C.... X = M BY 1 VECTOR
C.... Y = M BY 1 VECTOR
C.... WT = M BY M BORDERED WEIGHT MATRIX
C..............................................................................
      DIMENSION X(M),Y(M),WT(M,M)
      DIST=0
      DP=0
      DO 20 I=2,M
      DO 20 J=2,M
      XM=99999.
      IF((X(I)-XM)*(X(J)-XM)*(Y(I)-XM)*(Y(J)-XM).EQ.0.) GO TO 20
      DIST=DIST+(X(I)-Y(I))*(X( J)-Y(J))*WT(I,J)
      DP=DP+WT(I,J)
   20 CONTINUE
      IF(DP.NE.0) DIST=(DIST/DP)**(0.5)
      RETURN
      END
```

```
      SUBROUTINE MISS(A,M,N,NC)
C...............................................................................20 MAY 1973
C.... REPLACES MISSING VALUES BY WITHIN CLUSTER MEANS
C.... M = NUMBER OF ROWS
C.... N = NUMBER OF COLUMNS
C.... A = M BY N BORDERED ARRAY
C.... NC = M BY 1 ARRAY ASSIGNING CASES TO CLUSTERS
C...............................................................................
      DIMENSION A(M,N),NC(M)
C..... FIND NUMBER OF CLUSTERS
      XM=99999.
      K=1
      DO 10 I=2,M
   10 IF(NC(I).GT.K) K=NC(I)
      DO 20 L=1,K
      DO 20 J=2,N
C..... COMPUTE MEANS
      PR=0
      AV=0
      DO 21 I=2,M
      IF(NC(I).NE.L) GO TO 21
      IF(A(I,J).EQ.XM) GO TO 21
      PR=PR+1.
      AV=AV+A(I,J)
   21 CONTINUE
      IF(PR.NE.0) AV=AV/PR
C..... REPLACE MISSING VALUES
      DO 22 I=2,M
      IF(NC(I).EQ.L) GO TO 22
      IF(A(I,J).EQ.XM) A(I,J)=AV
   22 CONTINUE
   20 CONTINUE
      RETURN
      END


      SUBROUTINE TWO(DATA,M,N,COV,AVE)
C...............................................................................20 MAY 1973
C.... COMPUTES MEAN AND COVARIANCE, REPLACING MISSING VALUES BY COLUMN MEANS
C.... M = NUMBER OF ROWS
C.... N = NUMBER OF COLUMNS
C.... DATA = M BY N BORDERED ARRAY
C.... COV = M BY M BORDERED COVARIANCE MATRIX
C.... AVE = 2 BY N BORDERED COLUMN MEANS
C...............................................................................
      DATA CC,AA/3HCOV,3HAVE/
      DIMENSION COV(N,N),AVE(2,N),DATA(M,N)
C.... REPLACE MISSING VALUES BY COLUMN MEANS
      DO 10 J=2,N
      AVE(2,J)=0
      XP=0
      DO 20 I=2,M
      IF(DATA(I,J).EQ.99999.) GO TO 20
      XP=XP+1
      AVE(2,J)=AVE(2,J)+DATA(I,J)
   20 CONTINUE
      IF(XP.NE.0) AVE(2,J)=AVE(2,J)/XP
      DO 30 I=2,M
   30 IF(DATA(I,J).EQ. 99999.) DATA(I,J)=AVE(2,J)
   10 CONTINUE
C.... COMPUTE COVARIANCES
      DO 50 J=2,N
      DO 50 K=2,J
      COV(J,K)=0
      DO 51 I=2,M
   51 COV(J,K)=COV(J,K)+(DATA(I,J)-AVE(2,J))*(DATA(I,K)-AVE(2,K))
      COV(J,K)=COV(J,K)/(M-1)
   50 COV(K,J)=COV(J,K)
C.... LABEL BORDERS
      DO 60 J=2,N
      AVE(1,J)=DATA(1,J)
      COV(1,J)=DATA(1,J)
   60 COV(J,1)=DATA(1,J)
      COV(1,1)=CC
      AVE(1,1)=AA
      RETURN
      END
```

```
      SUBROUTINE STAND(A,M,N)
C..................................................................20 MAY 1973
C.... STANDARDIZES A BORDERED ARRAY A TO HAVE UNIT VARIANCE AND ZERO MEAN
C.... M = NUMBER OF ROWS
C.... A = M BY N ARRAY
C.... N = NUMBER OF COLUMNS
C...................................................................
      DIMENSION A(M,N)
C.... COMPUTE MEANS AND VARIANCES
      DO 20 J=2,N
      S0=0
      S1=0
      S2=0
      DO 30 I=2,M
      IF(A(I,J).EQ.99999.) GO TO 30
      S0=S0+1
      S1=S1+A(I,J)
      S2=S2+A(I,J)**2
   30 CONTINUE
      IF(S0.NE.0) S1=S1/S0
      IF(S0.NE.0) S2=S2/S0-S1**2
      IF(S2.GT.0) S2=S2**0.5
      IF(S2.EQ.0) S2=1
      DO 40 I=2,M
   40 IF(A(I,J).NE.99999.) A(I,J)=(A(I,J)-S1)/S2
   20 CONTINUE
      RETURN
      END




      SUBROUTINE MOM(U,C,P,X,M,N)
C..................................................................20 MAY 1973
C       COMPUTES WEIGHTED MEANS AND COVARIANCES
C.... M = NUMBER OF ROWS
C.... N = NUMBER OF COLUMNS
C.... X = M BY N DATA ARRAY
C.... U = N BY 1 ARRAY OF MEANS
C.... P = M BY 1 ARRAY OF WEIGHTS
C.... C = N BY N ARRAY OF COVARIANCES
C...................................................................
      DIMENSION C(N,N),P(M),U(N),X(M,N)
      SP=0
      DO 10 I=2,M
   10 SP=SP+P(I)
      IF(SP.EQ.0.) SP=10.**(-10)
      DO 20 J=2,N
      SS=0.
      DO 21 I=2,M
   21 SS=SS+X(I,J)*P(I)
   20 U(J)=SS/SP
      DO 30 J=2,N
      DO 30 K=2,J
      SS=0.
      DO 31 I=2,M
   31 SS=SS+(X(I,J)-U(J))*(X(I,K)-U(K))*P(I)
      C(J,K)=SS/SP
   30 C(K,J)=C(J,K)
      RETURN
      END
```

73

CHAPTER 3

Quick Partition Algorithms

3.1 INTRODUCTION

Figure 3.1 consists of the outlines of 20 pieces in a child's jigsaw puzzle. The assembly of jigsaw puzzles is an instructive example of human clustering ability. A standard strategy is to first select the pieces with straight-line edges and construct the border, then to select pieces with some significant color combination and assemble these and to continue this process until most pieces are incorporated. At the end, there is usually a blah background color and these pieces are incorporated on the basis of shape. Characteristically then, some "important" variable is used as a basis for an initial crude partition and other variables are used for more detailed work within the partition.

It is difficult to formalize the patterns of color that are frequently used, but the shapes of the edges are explicitly measurable. In this puzzle (and in many), every piece has four edges and four vertices and two pieces are joined together if and only if they have an exactly matching edge. Three measurements were made for each edge:

(i) the length of the line between the two vertices,

(ii) the maximum deviation of the edge from this line into the piece, and

(iii) the maximum deviation of the edge from this line out from the piece.

There are thus twelve measurements for each piece. These measurements are given in Table 3.1.

Let $A(I, 1)$, $A(I, 2)$, and $A(I, 3)$ denote the three measurements of the first edge of the Ith piece. The first edge of the Ith piece corresponds to the first edge of the Jth piece if $A(I, 1) = A(J, 1)$, $A(I, 2) = A(J, 3)$, and $A(I, 3) = A(J, 2)$. A natural measure of distance between edges taking values $X1, X2, X3$ and $Y1, Y2, Y3$ is thus

$$D = [(X1 - Y1)^2 + (X2 - Y3)^2 + (X3 - Y2)^2]^{1/2}.$$

This is *not* euclidean distance. The distance between two pieces is the minimum distance between any two edges. If any two edges exactly coincide, the distance between the pieces will be zero.

3.2 LEADER ALGORITHM

Preliminaries. It is desired to construct a partition of a set of M cases, a division of the cases into a number of disjoint sets or clusters. It is assumed that a rule for computing the distance D between any pair of objects, and a threshold T are given. The

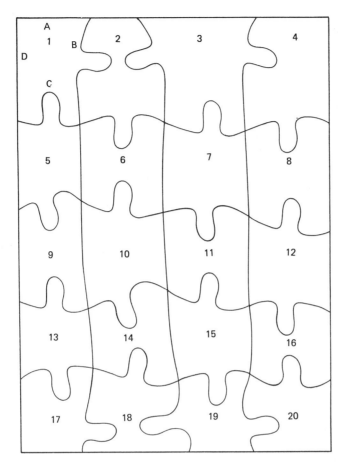

Figure 3.1 Jigsaw puzzle.

algorithm constructs a partition of the cases (a number of clusters of cases) and a leading case for each cluster, such that every case in a cluster is within a distance T of the leading case. The threshold T is thus a measure of the diameter of each cluster. The clusters are numbered $1, 2, 3, \ldots, K$. Case I lies in cluster $P(I)$ $[1 \leq P(I) \leq K]$. The leading case associated with cluster J is denoted by $L(J)$.

The algorithm makes one pass through the cases, assigning each case to the first cluster whose leader is close enough and making a new cluster, and a new leader, for cases that are not close to any existing leaders.

STEP 1. Begin with case $I = 1$. Let the number of clusters be $K = 1$, classify the first case into the first cluster, $P(1) = 1$, and define $L(1) = 1$ to be the leading case of the first cluster.

STEP 2. Increase I by 1. If $I > M$, stop. If $I \leq M$, begin working with the cluster $J = 1$.

STEP 3. If $D(I, J) > T$, go to Step 4. If $D(I, J) \leq T$, case I is assigned to cluster J, $P(I) = J$. Return to Step 2.

Table 3.1   Jigsaw Puzzle Measurements

Piece	Edge A			Edge B			Edge C			Edge D		
1	142	0	0	191	21	48	125	40	12	167	0	0
2	120	0	0	183	59	16	160	16	50	192	48	21
3	186	0	0	208	17	51	152	49	13	183	17	59
4	138	0	0	180	0	0	157	13	42	209	51	18
5	126	13	39	138	0	2	125	20	43	171	0	0
6	159	50	15	163	4	0	152	47	17	139	2	0
7	149	13	50	142	0	6	157	19	49	163	0	3
8	157	42	13	203	0	0	152	50	25	143	3	0
9	125	44	20	190	2	0	138	42	18	159	0	0
10	152	17	46	144	2	0	147	19	56	190	0	2
11	157	48	20	161	0	0	152	49	22	143	0	2
12	152	25	49	139	0	0	153	27	52	160	0	0
13	138	19	42	112	0	1	143	10	41	143	0	0
14	147	55	20	150	6	1	154	36	21	113	1	0
15	151	22	48	126	0	8	137	14	51	160	0	6
16	152	52	27	141	0	0	153	42	21	128	8	0
17	143	42	9	149	7	54	117	0	0	136	0	0
18	154	22	36	130	77	6	192	0	0	150	54	8
19	134	52	13	140	16	51	123	0	0	130	6	78
20	151	21	43	118	0	0	150	0	0	140	52	13

Edges are in clockwise order. There are three measurements for each edge, in hundredths of an inch (the error in each measurement is approximately $\frac{1}{100}$ in.): (i) the length of the line between the vertices, (ii) the maximum deviation of the edge from the line between vertices into the piece, and (iii) the maximum deviation of the edge from the line between vertices out from the piece.

STEP 4.   Increase J to $J + 1$. If $J \leq K$, return to Step 3. If $J > K$, a new cluster is created, with K increased by 1. Set $P(I) = K$, $L(K) = I$, and return to Step 2.

3.3   LEADER ALGORITHM APPLIED TO JIGSAW PUZZLES

The algorithm requires a measure of distance, which will be the one described in Section 3.1. A threshold T is needed also. Since the error in a single measurement is approximately 1 (in hundredths of an inch), a plausible distance between two matching edges is $[2^2 + 2^2 + 2^2]^{1/2} = 3.5$. An initial threshold $T = 4$ will be used.

STEP 1.   Initially $K = 1$, $P(1) = 1$, $L(1) = 1$, and $I = 1$.

STEP 2.   Increase I to $I = 2$. Try the first cluster $J = 1$.

STEP 3.   The distance $D(2, 1) = 1$ by matching edge B of 1 to edge D of 2. Since $D(2, 1) \leq 4$, $P(2) = 1$. Return to Step 2.

STEP 2.   Increase I to $I = 3$. Try $J = 1$.

STEP 3.   The distance $D(3, 1) = 19$ by matching edge D of 1 to edge A of 3. Since $D(3, 1) > 4$, go to Step 4.

STEP 4.   Increase J to 2. Since $J = 2 > K = 1$, a new cluster is created with K increased by 1. Set $P(3) = 2$, $L(2) = 3$, and return to Step 2.

In this way, all cases are assigned to clusters as follows:

Cluster 1:	1	2	5	13*	16*
Cluster 2:	3	4	7		
Cluster 3:	6	10	17*		
Cluster 4:	8	11*	12	15*	
Cluster 5:	9	18*			
Cluster 6:	14				
Cluster 7:	19	20			

The pieces with asterisks do not truly have a matching edge with their cluster leader. The algorithm has identified seven true matches and five false matches. The false matches are mostly due to the algorithm rather than measurement error, because the algorithm assigns each case to the first leader to which its distance is within threshold, rather than to the closest leader. The difficulty is partly solved by reducing the threshold to $T = 2$.

Cluster 1:	1	2	5	13*	16*
Cluster 2:	3	4	7		
Cluster 3:	6	10			
Cluster 4:	8	12			
Cluster 5:	9				
Cluster 6:	11	15			
Cluster 7:	14				
Cluster 8:	17	18			
Cluster 9:	19	20			

Here there are ten true matches, and two false matches. The false matches 13 and 16 to 1 represent edges that are accidentally very close and should not be blamed on the algorithm.

It is sensible to consider clustering edges rather than pieces, since it is the similarities between edges that are used in matching pieces. The results of this clustering are given in Table 3.2. Ideally, there should be 31 clusters of pairs of matched edges, and 18 clusters of single edges corresponding to the borders of the puzzle. The computed partition contains one cluster of four edges, two clusters of three edges, 25 clusters of two edges, and 19 clusters of single edges. There are 23 true matches and 12 false matches, all of which are due to close similarities between edges without indentations. These errors seem unavoidable, whatever the algorithm.

3.4   PROPERTIES OF LEADER ALGORITHM

The positive feature of the leader algorithm is that it is very fast, requiring only one pass through the data. (It is thus not necessary to store the data in core, but it is sufficient to read it once from disk or tape.)

Several negative properties follow from the indecent haste with which objects are assigned to clusters. The first of these is that the partition is not invariant under

Table 3.2

Clusters of edges in a jigsaw puzzle using a leader algorithm with a threshold $T = 2$. False matches are denoted by an asterisk.

CLUSTER	EDGES				CLUSTER	EDGES		
1	1A	13D*	16B*		26	9C	13A	
2	1B	2D			27	9D	11B*	12D*
3	1C	5A			28	10B		
4	1D				29	10C	14A	
5	2A				30	11C	15A	
6	2B	3D			31	12C	16A	
7	2C	6A			32	13B	14D	
8	3A				33	13C	17A	
9	3B	4D			34	14B	15D	
10	3C				35	14C	18A	
11	4A	5B*	17D*	12B*	36	15B	16D	
12	4B				37	15C		
13	4C	8A			38	16C		
14	5C	9A			39	17B	18D	
15	5D				40	17C	20B*	
16	6B	7D			41	18B	9D	
17	6C	10A			42	18C		
18	6D				43	19A		
19	7A				44	19B		
20	7B				45	19C		
21	7C	11A			46	20A		
22	8B				47	20C		
23	8C	12A			48	20D		
24	8D	11D*						
25	9B	10D						

reordering of the cases. For example, the first case is always a cluster leader. A second negative property is that the first clusters are always larger than the later ones, since they get first chance at each case as it is allocated. This property could be changed by allocating each case to the cluster whose leader it is closest to, but this change might require four or five times (or more) the distance calculations.

3.5 SORTING ALGORITHM

Preliminaries. A threshold $T(J)$ is given for the Jth variable ($1 \leq J \leq N$). Cases are partitioned into a set of clusters so that within each cluster the Jth variable has a range less than $T(J)$. The thresholds should be chosen fairly large, especially if there are many variables. The procedure is equivalent to converting each variable to a category variable (using the thresholds to define the categories) and the clusters are then the cells of the multidimensional contingency table between all variables.

STEP 1. Order all cases I $(1 \leq I \leq M)$ according to the integral part of $A(I, 1)/T(1)$.

STEP 2. Reorder all cases I $(1 \leq I \leq M)$ according to the integral part of $A(I, 2)/T(2)$. Then reorder according to the integral part of $A(I, 3)/T(3), \ldots, A(I, N)/T(N)$.

STEP 3. Clusters are now the sequences of cases which have identical values of the integral parts of $A(I, J)/T(J)$ for all J.

HINT. Order by the least important variables first.

3.6 SORTING ALGORITHM APPLIED TO JIGSAW PUZZLES

This algorithm will be used for an initial sorting of edges—the matching of opposite edges would be necessary in a later step. A threshold of 10 will be used for each variable to avoid an excessively large number of clusters.

STEP 1. (Table 3.3). Order by integral part of deviation into piece divided by 10. The minimum value is 0; the maximum is 7.

Table 3.3

Inside deviation/10			Edges												
0	1A	1D	2A	3A	4A	4B	5B	5D	6B	6D	7B	7D	8B	8D	9B	9D
	10B	10D	11B	11D	12B	12D	13B	13D	14B	14D	15B	15D	16B	16D		
	17B	17C	17D	18C	19C	19D	20B	20C								
1	2C	3B	3D	4C	5A	7A	7C	10A	10C	13A	13C	15C	19B			
2	1B	5C	12A	12C	15A	18A	20A									
3	14C															
4	1C	2D	3C	6C	8A	9A	9C	11A	11C	16C	17A					
5	2B	4D	6A	8C	14A	16A	18D	19A	20D							
6																
7	18B															

STEP 2. (Table 3.4). Reorder by integral part of deviation out from piece divided by 10. The minimum value is 0; the maximum is 7.

Table 3.4

Outside deviation/10			Edges												
0	1A	1D	2A	3A	4A	4B	5B	5D	6B	6D	7B	7D	8B	8D	9B	9D
	10B	10D	11B	11D	12B	12D	13B	13D	14B	14D	15B	45D	16B	16D		
	17C	17D	18C	19C	20B	20C	17A	18D	18B							
1	1C	3C	6C	8A	9C	2B	4D	6A	19A	20D						
2	14C	2D	9A	11A	11C	16C	8C	14A	16A							
3	5A	18A														
4	4C	7C	10A	13A	13C	1B	12A	5C	15A	20A						
5	17B	2C	3B	3D	7A	10C	15C	19B	12C							
6																
7	19D															

Next reorder by the integral part of the length divided by 10 (Table 3.5). The minimum value is 11; the maximum is 20.

Table 3.5

Length/10	Edges
11	13B 14D 17C 20B
12	2A 15B 16D 19C 1C 9A 5A 5C
13	4A 5B 6D 12B 17D 18B 9C 19A 13A 15C 19D
14	1A 7B 8D 10B 11D 13D 14B 17A 20D 14A 13C 17B 7A 10C 19B
15	9D 20C 18D 3C 6C 8A 6A 14C 11A 11C 16C 8C 16A 18A 4C 7C 10A 12A 15A 20A 12C
16	1D 6B 7D 11B 12D 14B 15D 2C
17	5D
18	3A 4B 2B 3D
19	9B 10D 18C 2D 1B
20	8B 4D 3B

STEP 3. Clusters are sequences of edges with identical integer parts of $A(I, J)/T(J)$ for all J. Each sequence in a cluster is consistent with the ordering given above.

The clusters are given in Table 3.6. This clustering is not comparable to the ones based on the leader algorithm, because possible candidates for matching are not clustered together. Almost all the nonsingleton clusters are ones with straight edges, including both boundaries and internal straight edges. The other clusters must be matched to find matching edges—for example, the (15, 4, 1) cluster should be matched with the (15, 1, 4) cluster. This yields 6C–10A and 8A–4C as true matches and 3C–7C as a false match.

3.7 PROPERTIES OF SORTING ALGORITHM

The sorting algorithm is fast, requiring as many passes through the data as there are variables. The final clusters are independent of the original order of the cases (unlike clusters from the leader algorithm). They have a simple interpretation in that each variable has a range less than a given threshold $T(J)$ in every cluster.

This algorithm has a peculiar drawback in producing extremely large numbers of clusters when there are many variables. For example, with 10 variables, if every variable has an effect on the clustering (if thresholds are not set larger than the ranges of variables over all cases), there are potentially $2^{10} = 1024$ clusters. Another difficulty is that the clusters are forced to be "rectangular" by the threshold property. Thus, if a cluster is spherical, its cases will be divided among a number of rectangular cells and the cluster will not show clearly.

3.8 THINGS TO DO

3.8.1 Running the Leader Algorithm

The leader algorithm is most appropriate for very large numbers of objects for which a quick initial sorting is required. Suitable data might be the moons and planets, or life expectancies by age and sex, or the dentition of mammals. The threshold must be guessed by using the knowledge that all objects in a cluster are within threshold distance from the cluster leader. For summarizing, the cluster leader represents the cluster.

Table 3.6 Final Clusters from Sorting Algorithm

Length/10	inside deviation/10	outside deviation/10	Edges								
11	0	0	13B	14D	17C	20B					
12	0	0	2A	15B	16D	19C					
12	4	1	1C								
12	4	2	9A								
12	1	3	5A								
12	2	4	5C								
13	0	0	4A	5B	6D	12B	17D				
13	7	0	18B								
13	4	1	9C								
13	5	1	19A								
13	1	4	13A								
13	1	5	15C								
13	0	7	19D								
14	0	0	1A	7B	8D	10B	11D	13D	16B	17A	20D
14	5	2	14A								
14	1	4	13C								
14	0	5	17B								
14	1	5	7A	10C	19B						
15	0	0	9D	20C							
15	5	0	18D								
15	4	1	3C	6D	8A						
15	5	1	6A								
15	3	2	14C								
15	4	2	11A	11C	16C						
15	5	2	8C	16A							
15	2	3	18A								
15	1	4	4C	7C	10A						
15	2	4	12A	15A	20A						
15	2	5	12C								
16	0	0	1D	6B	7D	11B	12D	14B	15D		
16	1	5	2C								
17	0	0	5D								
18	0	0	3A	4B							
18	5	1	2B								
18	1	5	3D								
19	0	0	9B	10D	18C						
19	4	2	2D								
19	2	4	1B								
20	0	0	8B								
20	5	1	4D								
20	1	5	3B								

3.8.2 Improved Leader Algorithm

A one-pass leader algorithm that avoids inflated early clusters measures the distance of a new object to the leaders in reverse order and allocates it to the first leader to which it is close enough. A many-pass algorithm, which is invariant under a change of the input order and does not require the specification of thresholds, begins with an initial central object (say, the mean object on each variable). On the first pass the object furthest from this is discovered. On the second pass, objects are allocated to whichever of the first two objects they are closest, and the object furthest from its leader is discovered to be the new leader in the next pass. In this way, after K passes, K clusters will be discovered. This algorithm is analogous to the K-means algorithm, without updating of cluster leaders.

3.8.3 Number of Clusters

It is always difficult to estimate the threshold and the number of clusters to be obtained for a given threshold. A very small initial threshold will produce a large number of small clusters after expensive computation.

```
      SUBROUTINE QUICK(A,M,N,NC,THRESH,LC,LL,XMISS)
C...............................................................20 MAY 1973
C.... QUICK SUCCESSIVELY ASSIGNS EACH ROW TO THE LAST CLUSTER FOR WHICH THE
C.... DISTANCE BETWEEN THE ROW AND THE CLUSTER LEADER IS LESS THAN THRESHOLD.
C.... IF NO CLUSTER HAS THIS PROPERTY, THE ROW BECOMES A CLUSTER LEADER.
C.... A = M BY N BORDERED ARRAY
C.... M = NUMBER OF ROWS
C.... N = NUMBER OF COLUMNS
C.... NC = NUMBER OF CLUSTERS
C.... THRESH = THRESHOLD FOR ASSIGNING AN OBJECT TO A LEADER.
C.... LL = 1 BY NC ARRAY LISTING LEADERS
C.... LC = 1 BY M ARRAY SPECIFYING LEADER FOR EACH OBJECT
C.... XMISS = VALUE TO BE TREATED AS MISSING
C..................................................................
      DIMENSION A(M,N),LL(NC),LC(M)
      DIMENSION AA(20)
      KC=1
      LL(1)=2
      DO 20 I=2,M
      LC(I)=0
      DO 21 KK=1,KC
      K=KC-KK+1
      L=LL(K)
C.... COMPUTES DISTANCE BETWEEN ROW AND CLUSTER LEADER
      DD=0
      DC=0
      DO 22 J=2,N
      IF (A(L,J).EQ.XMISS.OR.A(I,J).EQ.XMISS) GO TO 22
      DC=DC+1
      DD=DD+(A(L,J)-A(I,J))**2
      IF(DD.GT.THRESH**2*(N-1)) GO TO 21
   22 CONTINUE
      IF(DC.NE.0) DD=(DD/DC)**0.5
      IF (DD.GT.THRESH) GO TO 21
C.... ASSIGN ROW I TO CLUSTER K IF DISTANCE BELOW THRESH
      LC(I)=K
      GO TO 20
   21 CONTINUE
      IF (KC.EQ.NC) GO TO 20
C.... CREATE NEW CLUSTER AND LEADER
      KC=KC+1
      LC(I)=KC
      LL(KC)=I
   20 CONTINUE
C.... OUTPUT CLUSTER LEADERS
      WRITE(6,4)
    4 FORMAT(16H0CLUSTER LEADERS)
      DO 60 K=1,KC
      I=LL(K)
   60 WRITE(6,3) K,(A(I,J),J=1,N)
    3 FORMAT(8H CLUSTER,I4,2X,A4,10F11.4/(18X,10F11.4))
C     OUTPUT CLUSTERS
      KC=KC+1
      DO 50 K=1,KC
      KK=K-1
      J=0
      DO 50 I=2,M
      IF (J.EQ.20) J=0
      IF (LC(I).NE.KK) GO TO 50
      J=J+1
      AA(J)=A(I,1)
   50 IF(J.EQ.20.OR.(I.EQ.M.AND.J.NE.0)) WRITE(6,1) KK,(AA(JJ),JJ=1,J)
    1 FORMAT(' CLUSTER',I5,20(1X,A5))
      RETURN
      END
```

For a metric distance $[D(I, J) \le D(I, K) + D(J, K)$ for each I, J, $K]$ there is a relationship between the threshold and the number of clusters. Suppose that the leader algorithm, for some order of the objects, produces K clusters at threshold T. Then, for any order of the objects, show that it produces no more than K clusters at threshold $2T$.

3.8.4   Size of Clusters

Let the data be real values with each point taken at random from $[0, 1]$. Then the size of the interval corresponding to the Kth cluster is smaller, in probability, than that of the $(K - 1)$th cluster; this means that for every x the Kth cluster is at least as likely as the $(K - 1)$th cluster to be less than x in size. It would be interesting to know by analysis or experiment the distribution of size of the Kth cluster. (This does not depend on the number of points, for large numbers of points.)

PROGRAMS

QUICK   finds quick partition by assigning each object to the first leader object to which its distance is less than a threshold.

CHAPTER 4

The *K*-Means Algorithm

4.1 INTRODUCTION

The principal nutrients in meat, fish, and fowl are listed in Table 4.1. The foods are classified by food type and method of preparation. In the source book, *Yearbook of Agriculture* (1959), there are more details on the mode of preparation, as well as amounts of vitamins and basic amino acids, given for dairy products, meats, fowl, fish, vegetables, fruits, grains, oils, and sugars. There is a ready-made weighting scheme for the nutrients in the estimated daily dietary allowances: food energy (3200 cal), protein (70 g), calcium (0.8 g), and iron (10 mg). In Table 4.2, note that these foods deliver about 7 % of the daily allowances in calories but about 25 % of iron, so that the iron component is rather heavily weighted. An argument could be made that iron is less important than calories or protein and so should be given less weight or ignored entirely.

It is desired to partition the foods so that foods within clusters are close, in some sense, and foods in different clusters are distant. The discordance between the data and a given partition $P(M, K)$ of M objects into K clusters is measured by an error $e[P(M, K)]$. The very large number of possible partitions makes it impractical to search through all for the minimum of e. It is necessary instead to use the technique of local optimization. In this technique, a neighborhood of partitions is defined for each partition. Beginning with an initial partition, search through the set of partitions at each step, moving from a partition to that partition in its neighborhood for which $e[P(M, K)]$ is a minimum. (If the neighborhoods are very large, it is sometimes cheaper computationally to move to the first partition discovered in the neighborhood where $e[P(M, K)]$ is reduced from its present value.) A number of stopping rules are possible. For example, the search stops when $e[P(M, K)]$ is not reduced by movement to the neighborhood. The present partition is then locally optimal in that it is the best partition in its neighborhood.

As an example, consider partitions into three clusters of the beef foods BB, BR, BS, BC, and BH. A plausible neighborhood for a partition is the set of partitions obtained by transferring an object from one cluster to another. Thus, for the partition (BB BR) (BS) (BC BH), the neighborhood consists of the following ten partitions:

> (BR) (BB BS) (BC BH),
> (BR) (BS) (BB BC BH),
> (BB) (BR BS) (BC BH),
> (BB) (BS) (BR BC BH),
> (BB BR BS) () (BC BH),
> (BB BR) () (BS BC BH),
> (BB BR BC) (BS) (BH),
> (BB BR) (BS BC) (BH),
> (BB BR BH) (BS) (BC),
> (BB BR) (BS BH) (BC).

A typical search route beginning with (BB BR) (BS) (BC BH) might be

(BB BR) (BS) (BC BH)	with	$e = 8$
(BB BR) (BS BC) (BH)	with	$e = 6$
(BR) (BS BC) (BB BH)	with	$e = 5$
(BR BC) (BS) (BB BH)	with	$e = 4.$

No neighborhood of this partition reduces e, and the search stops.

4.2 _K_-MEANS ALGORITHM

Preliminaries. The Ith case of the Jth variable has value $A(I, J)$ $(1 \leq I \leq M,$ $1 \leq J \leq N)$. The variables are scaled so that euclidean distance between cases is appropriate. The partition $P(M, K)$ is composed of the clusters $1, 2, \ldots, K$. Each of the M cases lies in just one of the K clusters. The mean of the Jth variable over the cases in the Lth cluster is denoted by $B(L, J)$. The number of cases in L is $N(L)$. The distance between the Ith case and Lth cluster is

$$D(I, L) = (\sum \{1 \leq J \leq N\} \, [A(I, J) - B(L, J)]^2)^{1/2}.$$

The error of the partition is

$$e[P(M, K)] = \sum \{1 \leq I \leq M\} \, D[I, L(I)]^2,$$

where $L(I)$ is the cluster containing the Ith case. The general procedure is to search for a partition with small e by moving cases from one cluster to another. The search ends when no such movement reduces e.

STEP 1. Assume initial clusters $1, 2, \ldots, K$. Compute the cluster means $B(L, J)$ $(1 \leq L \leq K, 1 \leq J \leq N)$ and the initial error

$$e[P(M, K)] = \sum \{1 \leq I \leq M\} \, D[I, L(I)]^2,$$

where $D[I, L(I)]$ denotes the euclidean distance between I and the cluster mean of the cluster containing I.

STEP 2. For the first case, compute for every cluster L

$$\frac{N(L)D(1, L)^2}{N(L) + 1} - \frac{N[L(1)]D[1, L(1)]^2}{N[L(1)] - 1},$$

The increase in error in transferring the first case from cluster $L(1)$, to which it belongs at present, to cluster L. If the minimum of this quantity over all $L \neq L(1)$ is negative, transfer the first case from cluster $L(1)$ to this minimal L, adjust the cluster means of $L(1)$ and the minimal L, and add the increase in error (which is negative) to $e[P(M, K)]$.

STEP 3.   Repeat Step 2 for the Ith case $(2 \leq I \leq M)$.

STEP 4.   If no movement of a case from one cluster to another occurs for any case, stop. Otherwise, return to Step 2.

Table 4.1   Nutrients in Meat, Fish, and Fowl

[*The Yearbook of Agriculture 1959* (The United States Department of Agriculture, Washington, D.C.) p. 244.] The quantity used is always 3 ounces.

		Food Energy (Calories)	Protein (Grams)	Fat (Grams)	Calcium (Milli Grams)	Iron (Milli Grams)
BB	Beef, braised	340	20	28	9	2.6
HR	Hamburger	245	21	17	9	2.7
BR	Beef, roast	420	15	39	7	2.0
BS	Beef, steak	375	19	32	9	2.6
BC	Beef, canned	180	22	10	17	3.7
CB	Chicken, broiled	115	20	3	8	1.4
CC	Chicken, canned	170	25	7	12	1.5
BH	Beef heart	160	26	5	14	5.9
LL	Lamb leg, roast	265	20	20	9	2.6
LS	Lamb shoulder, roast	300	18	25	9	2.3
HS	Smoked ham	340	20	28	9	2.5
PR	Pork roast	340	19	29	9	2.5
PS	Pork simmered	355	19	30	9	2.4
BT	Beef tongue	205	18	14	7	2.5
VC	Veal cutlet	185	23	9	9	2.7
FB	Bluefish, baked	135	22	4	25	.6
AR	Clams, raw	70	11	1	82	6.0
AC	Clams, canned	45	7	1	74	5.4
TC	Crabmeat, canned	90	14	2	38	.8
HF	Haddock, fried	135	16	5	15	.5
MB	Mackerel, broiled	200	19	13	5	1.0
MC	Mackerel, canned	155	16	9	157	1.8
PF	Perch, fried	195	16	11	14	1.3
SC	Salmon, canned	120	17	5	159	0.7
DC	Sardines, canned	180	22	9	367	2.5
UC	Tuna, canned	170	25	7	7	1.2
RC	Shrimp, canned	110	23	1	98	2.6

Table 4.2   Nutrients in Meat, Fish, and Fowl

As a percentage of recommended daily allowances.

	Food Energy	Protein	Fat (Grams)	Calcium	Iron
Beef, braised	11	29	28	1	26
Hamburger	8	30	17	1	27
Beef, roast	13	21	39	1	20
Beef, steak	12	27	32	1	26
Beef, canned	6	31	10	2	37
Chicken, broiled	4	29	3	1	14
Chicken, canned	5	36	7	2	15
Beef, heart	5	37	5	2	59
Lamb leg, roast	8	29	20	1	26
Lamb shoulder, roast	9	26	25	1	25
Ham, smoked	11	29	28	1	25
Pork roast	11	27	29	1	25
Pork simmered	11	27	30	1	25
Beef tongue	6	26	14	1	25
Veal cutlet	6	33	9	1	27
Bluefish, baked	4	31	4	3	6
Clams, raw	2	16	1	10	60
Clams, canned	1	10	1	9	54
Crabmeat, canned	3	20	2	5	8
Haddock, fried	4	23	5	2	5
Mackerel, broiled	6	27	13	1	10
Mackerel, canned	5	23	9	20	18
Perch, fried	6	23	11	2	13
Salmon, canned	4	24	5	20	7
Sardines, canned	6	31	9	46	25
Tuna, canned	5	36	7	1	12
Shrimp, canned	3	33	1	12	26

4.3   *K*-MEANS APPLIED TO FOOD NUTRIENT DATA

To keep the computations manageable, only the first eight foods will be considered for the nutrients' food energy, protein, and calcium as a percentage of recommended daily allowances. The eight foods will be partitioned in three clusters. The calculations appear in Table 4.3.

Table 4.3 Application of *K*-Means Algorithm to Food Data

	Energy	Protein	Calcium
BB	11	29	1
HR	8	30	1
BR	13	21	1
BS	12	27	1
BC	6	31	2
CB	4	29	1
CC	5	36	1
BH	5	37	2

INITIAL CLUSTER MEANS   e = 154.9

	Energy	Protein	Calcium
1. BR, CB	8.5	25	1
2. HR, BS	10	28.5	1
3. BB, BC, CC, BH	6.75	33.25	1.5

FIRST CHANGE   e = 108.2

	Energy	Protein	Calcium
1. BR, CB	8.5	25	1
2. HR, BS, BB	10.33	28.67	1
3. BC, CC, BH	5.33	34.67	1.67

SECOND CHANGE   e = 61.4

	Energy	Protein	Calcium
1. BR	13	21	1
2. HR, BS, BB	10.33	28.67	1
3. BC, CC, BH, CB	5	33.25	1.5

STEP 1.   A quick initial clustering, which often works well, is based on the case sums. Suppose these are denoted by SUM(I), having minimum value MIN and maximum value MAX. To obtain K initial clusters, set case I into the Jth cluster, where J is the integral part of $K[\text{SUM}(I) - \text{MIN}]/(\text{MAX} - \text{MIN}) + 1$. Here the case sums are 41, 39, 35, 40, 41, 34, 42, and 44. The corresponding clusters are 3, 2, 1, 2, 3, 1, 3, and 3. Thus the initial partition is (BR CB) (HR BS) (BB BC CC BH). The values of $B(L, J)$ ($1 \leq L \leq 3$, $1 \leq J \leq 3$) are next computed. For example, $B(1, 1)$, the mean of cases in the first cluster for the first variable, equals $(13 + 4)/2 = 8.5$. (See Table 4.3 for more.) The error for the initial partition is the sum of squared distances of cases from their cluster means,

$$
\begin{aligned}
e[P(8, 3)] = {} & (11 - 6.75)^2 + (29 - 33.25)^2 + (1 - 1.5)^2 + (8 - 10)^2 \\
& + (30 - 28.5)^2 + (1 - 1)^2 + (13 - 8.5)^2 + (21 - 25)^2 \\
& + (1 - 1)^2 + (12 - 10)^2 + (27 - 28.5)^2 + (1 - 1)^2 \\
& + (6 - 6.75)^2 + (31 - 33.25)^2 + (2 - 1.5)^2 + (8.5 - 4)^2 \\
& + (29 - 25)^2 + (1 - 1)^2 + (5 - 6.75)^2 + (36 - 33.25)^2 \\
& + (1 - 1.5)^2 + (5 - 6.75)^2 + (37 - 33.25)^2 + (2 - 1.5)^2 \\
= {} & 154.9.
\end{aligned}
$$

The first three squares are the squared distance of BB from its cluster mean (6·75, 33·25, 1·5), and so on.

STEP 2.   For the first case, the distances to clusters are

$$D(1, 1)^2 = (11 - 8\cdot5)^2 + (29 - 25)^2 + (1 - 1)^2 = 22.25,$$
$$D(1, 2)^2 = 1.25,$$
$$D(1, 3)^2 = 36.4.$$

The increase in error in transferring the first case to cluster 1 is $2 \times 22.5/3 - 4 \times 36.4/3$, and that to cluster 2 is $2 \times 1.25/3 - 4 \times 36.4/3$. The cluster that is best for the first case is thus the second cluster, and the error reduction is 47.7. The new value of $e[P(8, 3)]$ is thus $154.9 - 47.7 = 108.2$.

It is necessary to update the means of clusters 2 and 3, since cluster 2 has gained the first case and cluster 3 has lost it. For example,

$$B(2, 1) = (11 + 2 \times 10)/3 \quad = 10.33,$$
$$B(2, 2) = (29 + 2 \times 28.5)/3 = 28.67,$$
$$B(2, 3) = (1 + 2 \times 1)/3 \quad = 1.00.$$

STEP 3.   Repeating Step 1 on all cases, for case 2, cluster 2 is far closer than any other, and case 2 remains in cluster 2, with no change taking place. Continuing, no change takes place until case 6, which moves to cluster 3. No further changes occur in this pass.

STEP 4.   Since some changes occurred in the last pass, another pass is necessary through all cases. No changes occur on this pass and the algorithm stops with the final cluster (BR) (HR BS BB) (BC CC BH CB). These clusters are characterized by the variables as follows: The first cluster is high in energy and low in protein, the second is high in energy and protein, and the third is low in energy and high in protein Calcium hardly matters. The complete data set is partitioned in Table 4.4.

4.4   ANALYSIS OF VARIANCE

Some distributions which appear frequently in the analysis of variance are the following:

(i)   the normal distribution $N(\mu, \sigma^2)$, which has mean μ and variance σ^2 and density $\exp [-\tfrac{1}{2}(x - \mu)^2\sigma^{-2}]/\sigma\sqrt{2\pi}$. The unit normal is $N(0, 1)$, having mean 0 and variance 1;

(ii)   the chi-square distribution χ_n^2, which is the distribution of a sum of squares of n independent unit normals; and

(iii)   the F distribution $F_{m,n}$, which is the ratio of independent standardized chi squares, $(\chi_m^2/m)(\chi_n^2/n)^{-1}$.

Suppose $P(M, K)$ is a partition of M objects into K clusters, and let $e(M, K, J) = \sum \{1 \le I \le M\} \{A(I, J) - B[L(I), J]\}^2$, where the case I lies in cluster $L(I)$ and $B(L, J)$ is the mean of the Jth variable over cases in cluster L. If the clusters are selected without regard to the Jth variable and if $A(I, J) \sim N\{\mu[L(I)], \sigma^2\}$ independently for each I, $e(M, K, J) \sim \sigma^2\chi_{M-K}^2$. Furthermore, if the partition $P(M, K + 1)$ is obtained

Table 4.4

Clusters and cluster means from *K*-means algorithm applied to food data on energy, protein, and calcium expressed as percentages of daily requirements. The two clusters obtained by splitting cluster 3 are denoted by 31 and 32.

PARTITION	CLUSTERS		ENERGY	PROTEIN	CALCIUM
1	1 : BB HR BR BS BC CB CC BH LL LL HS PR PS BT VC FB AR AC TC HF MB MC PF SC DS UC RC		6.5	27.1	5.5
2	11 : BB HR BR BS BC CB CC BH LL LL HS PR PS BT VC FB AR AC TC HF MB PF UC RC		6.7	27.3	2.6
	12 : MC SC DS		4.7	26.1	28.5
3	12				
	111 : BB HR BR BS BC CB CC BH LL LL HS PR PS BT VC FB HF MB PF UC RC		7.4	29.0	1.8
	112 : AR AC TC		2.1	15.2	8.1
4	112				
	2 : BB HR BR BS BC CB CC BH LL LL HS PR PS BT VC FB HF MB PF UC		7.5	28.8	1.3
	3 : MC SC RC		4.0	26.7	17.3
	4 : DS		5.6	31.4	45.9
5	112, 3, 4				
	21 : HR BC CB CC BH VC FB UC		5.6	32.4	1.5
	22 : BB BR BS LL LL HS PR PS BT HF MB PF		9.1	25.8	1.2
6	112, 21, 22, 4				
	31 : MC SC		4.3	23.6	19.8
	32 : RC		3.4	32.9	12.3
7	112, 21, 31, 32, 4				
	221 : BB BS LL LL HS PR PS BT HF MB PF		8.7	26.5	1.2
	222 : BR		13.1	21.4	.9
8	222, 31, 32, 4				
	5 : BC CC BH VC FB UC		5.2	34.0	1.7
	6 : AR AC		5.6	31.4	45.9
	7 : BB HR BS LL LL HS PR PS MB		9.6	27.8	1.1
	8 : CB BT TC HF PF		4.6	24.0	2.1
9	222, 31, 32, 4, 5, 6,				
	9 : BB HR BS LL LL HS PR PS		10.0	27.9	1.1
	10 : CB BT HF MB PF		5.3	25.4	1.2
	11 : TC		2.8	20.0	4.8

by splitting one of the clusters in $P(M, K)$, then the mean square ratio

$$\left(\frac{e(M, K, J)}{e(M, K + 1, J)} - 1 \right)(M - K - 1) \sim F_{1, M-K-1}.$$

The ratio is a measure of the reduction of within-cluster variance for the Jth variable between the partitions $P(M, K)$ and $P(M, K + 1)$. The F distribution is not correct for evaluating *K*-means partitions because each variable influences the partition. The partition $P(M, K + 1)$ is chosen to minimize

$$\sum \{1 \leq J \leq N\}\, e(M, K + 1, J),$$

and this will tend to increase all the mean square ratios. Also, the partition $P(M, K + 1)$ is not necessarily obtained by splitting one of the clusters in $P(M, K)$, so the mean square ratio is conceivably negative. Nevertheless, as a crude rule of thumb, large values of the ratio (say, >10) justify increasing the number of clusters from K to $K + 1$.

Suppose again that $P(M, K)$ is a given partition into K clusters, that $P(M, K + 1)$ is obtained from it by splitting one of the clusters, and that

$$A(I, J) \sim N\{\mu[L(I), J], \sigma^2\}$$

independently over all I and J.

Then the overall mean square ratio

$$R = \left(\frac{e[P(M, K)]}{e[P(M, K + 1)]} - 1 \right)(M - K + 1) \approx F_{N, (M-K-1)N}.$$

Again this F distribution is not applicable in the K-means case, because the partition $P(M, K + 1)$ is chosen to maximize the overall mean square ratio. Again, as a crude rule of thumb, overall mean square ratios greater than 10 justify increasing partition size.

Some ratio measures are given for the food data in Table 4.5. For the variables, notice that calcium is very much reduced at the second, fourth, and sixth partitions,

Table 4.5 Ratio Due to K-Partition

Decrease in the sum of squares from the $(K - 1)$th to the Kth partition, divided by the mean sum of squares within the Kth partition.

MAXIMUM CLUSTER SIZE	PARTITION SIZE	OVERALL	ENERGY	PROTEIN	CALCIUM
24	2	23.6	1.0	0.1	64.1
21	3	13.0	9.6	25.8	4.1
20	4	18.4	2.4	0	200.8
12	5	16.1	15.4	20.3	0.3
12	6	6.7	0.1	6.6	35.0
11	7	3.2	4.8	3.0	0.1
9	8	12.5	18.4	10.4	11.6
8	9	6.5	11.9	3.5	31.9

while energy and protein are reduced for the third and fifth. The larger ratios for calcium follow from the large initial variance for calcium [the initial variances are 10 (energy), 37 (protein), and 95 (calcium)].

Plausible stopping points in the clustering are $K = 2$, $K = 5$, and $K = 8$, where the ratios are unusually large.

4.5 WEIGHTS

The weights will depend on considerations outside the data. If persons eating the food were known to have a diet abundant in calcium, then the calcium component would be down-weighted. If protein were scarce in other foods, then it would be given more weight. It is clear that in the initial analysis calcium was much the most important variable in the first few partitions. This is partly due to the scaling and partly due to the good clustering qualities of calcium, which is extremely high in a few sea foods and low elsewhere.

Another weighting scheme scales all variables to equal variance. As previously explained, this may be self-defeating in reducing the weight of variables that cluster well. Another weighting scheme repeats the partition a number of times, with one of the variables given weight 0 each time. The effect of the omitted variable on the clustering may be evaluated in this way. The 8-partitions corresponding to these weighting schemes are given in Table 4.6. Note there that calcium and protein are the best

Table 4.6 Effect of Changing Weights on 8-Partitions of Food Data by *K*-Means Algorithm

MEAN SQUARE ERROR WITHIN CLUSTERS

WEIGHTING	Energy (Calories^2)	Protein (Grams^2)	Fat (Grams^2)	Calcium (Mgms^2)	Iron (Mgms^2)
% DAILY ALLOWANCE	4665	4.1	60	11	.99
EQUAL VARIANCE	924	3.0	10	1500	.42
OMITTING ENERGY	1392	3.8	17	1199	.45
OMITTING PROTEIN	1151	19.5	11	320	.20
OMITTING FAT	1632	3.6	20	382	.41
OMITTING CALCIUM	816	2.6	9	6787	.36
OMITTING IRON	791	4.1	9	167	1.25

clustering variables in that their omission from the sum of squares to be minimized vastly increases their mean square error within the clusters obtained. Iron is the worst in that omitting it does not much increase the iron mean square and does much reduce the other mean squares. Since iron is subjectively less important anyway, a good final scheme would be to weight so that all variables would have variance 1 except iron, which would have variance 0.

4.6 OTHER DISTANCES

The *K*-means algorithm searches for partitions $P(M, K)$ with a small error

$$e[P(M, K)] = \sum \{1 \leq I \leq M\} \, D^2[I, L(I)],$$

where D is the euclidean distance from I to the average object in $L(I)$, the cluster to which I belongs. The essential characteristics of the *K*-means method are the search method, changing partitions by moving objects from one cluster to another, and the measure of distance. Euclidean distance leads naturally to cluster means and an analysis of variance for each of the variables.

To consider more general measures of distance, denote the *I*th case $\{A(I, J), 1 \leq J \leq N\}$ by $\mathbf{A}(I)$, and let $\mathbf{B}(L) = \{B(L, J), J = 1, \ldots, N\}$ denote a set of values corresponding to the *L*th cluster.

A distance $F[\mathbf{A}(I), \mathbf{A}(K)]$ is defined between the *I*th and *K*th cases. The central case of the *L*th cluster is $\mathbf{B}(L)$ minimizing $\sum \{I \in L\} F[\mathbf{A}(I), \mathbf{B}(L)]$.

The error of a particular partition is

$$e[P(M, K)] = \sum \{1 \leq I \leq M\} \, F\{\mathbf{A}(I), \mathbf{B}[L(I)]\}$$

where $L(I)$ is the cluster containing I.

Locally optimal clusters may be obtained by moving cases from one cluster to another, if this decreases e, and updating the central cases after each movement. For example, if the distance between two cases is the sum of the absolute deviations between the cases over variables, then the central case of a cluster is the median for each variable. The contribution of a cluster to the error is the sum of absolute deviations from the median, over all objects in the cluster and over all variables.

Approaching the error function from a statistical point of view, the cases $\{A(I, J), 1 \leq J \leq N\}$ will be denoted by $\mathbf{A}(I)$, and the parameters $\boldsymbol{\theta}(1), \boldsymbol{\theta}(2), \ldots, \boldsymbol{\theta}(K)$ will be associated with each of the clusters. The cases in cluster L are a random sample from a probability distribution determined by $\boldsymbol{\theta}(L)$; the probability of observing these cases is

$$\prod P[\mathbf{A}(I) \mid \boldsymbol{\theta}(L)],$$

where the product is over the cases in the Lth cluster. The probability of observing all cases is

$$\prod \{1 \leq I \leq M\} \, P\{\mathbf{A}(I) \mid \boldsymbol{\theta}[L(I)]\}.$$

The error function associated with the partition $P(M, K)$ is minus the log likelihood:

$$e = -\sum \{1 \leq I \leq M\} \log P\{\mathbf{A}(I) \mid \boldsymbol{\theta}[L(I)]\}.$$

To minimize this error function for a particular partition, the parameters $\boldsymbol{\theta}(L)$ are chosen by maximum likelihood. Searching over all possible partitions may now take place by using the K-means procedure. Choosing the values $\boldsymbol{\theta}(L)$ corresponds to selecting the cluster "centers" $B(L, J)$ in the K-means procedure. The probability distribution $P[\mathbf{A}(I) \mid \boldsymbol{\theta}(L)]$ specifies the joint distribution of all variables, given the cluster center $\boldsymbol{\theta}(L)$. If the variables are independent normal with equal variance and mean vector $\boldsymbol{\theta}(L)$, the K-means error function is minus the log likelihood as indicated above. Note that this requires that the variables have equal variance *within* clusters.

Independent Laplace distributions for each variable within clusters implies a distance function summing absolute deviations. Uniform distributions within clusters implies a distance between cases equal to the maximum deviations between the cases, over variables.

A further refinement is to use Bayes techniques, with some prior distribution of the parameters, $\boldsymbol{\theta}(1), \ldots, \boldsymbol{\theta}(K)$, which will be assumed to be independent and identical random variables. The random variables $\mathbf{A}(I)$ are marginally independent between clusters but dependent within, so that the probability of the observed cases is $\prod \{1 \leq L \leq K\} \, P(L)$, where $P(L)$ is the probability of the cases lying in cluster L. The value $-\log P(L)$ is a reasonable measure of cluster diameter, $e = \sum [-\log P(L)]$ is a measure of partition error, and the same search procedure as before is used to find good partitions.

4.7 THE SHAPE OF *K*-MEANS CLUSTERS

Some properties of the K-means algorithm, including the convexity of the clusters, are discussed in Fisher and Van Ness (1971). Consider a partition that is locally optimal in

$$e[P(M, K)] = \sum \{1 \leq I \leq M\} \, D^2[I, L(I)]$$

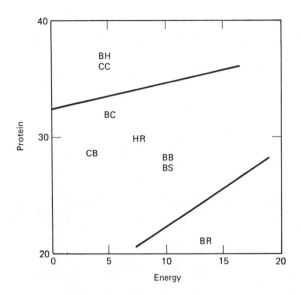

Figure 4.1   Clusters separated by hyperplanes based on food data in Table 4.3.

using the *K*-means search pattern. Here $L(I)$ is the cluster containing case *I*, and $D(I, L)$ is the distance between case *I* and the mean of cases in cluster *L*.

Let $L1$ and $L2$ be two different clusters. For each *I* in $L1$, $D(I, L1) < D(I, L2)$, for otherwise *I* would be removed from $L1$ during step 2 of the algorithm. Therefore, each case *I* in $L1$ satisfies

$$\sum \{1 \leq J \leq N\} \, [B(L1, J) - B(L2, J)]A(I, J) > c,$$

where

$$c = \sum \{1 \leq J \leq N\} \tfrac{1}{2}[B(L1, J)^2 - B(L2, J)^2],$$

and each case *I* in $L2$ satisfies

$$\sum \{1 \leq J \leq N\} \, [B(L1, J) - B(L2, J)]A(I, J) < c.$$

Geometrically, the cases in $L1$ and $L2$ are separated by a hyperplane normal to $\{B(L1, J) - B(L2, J)\}$ (see Figure 4.1). (This hyperplane is a linear discriminant function for separating cases into the clusters $L1$ and $L2$.)

Each cluster is convex, which means that a case lies in a cluster if and only if it is a weighted average of cases in the cluster. To show this, suppose that case *II* lies in cluster $L2$, but it is a weighted average of cases in cluster $L1$. For all cases in $L1$

$$\sum \{1 \leq J \leq N\} \, [B(L1, J) - B(L2, J)]A(I, J) > c,$$

and the reverse holds for cases in $L2$.

Since $A(II, J) = \sum W(I)A(I, J)$, where the summation is over cases in cluster $L1$ and $W(I) \geq 0$,

$$\sum \{1 \leq J \leq N\} \, [B(L1, J) - B(L2, J)]A(II, J) > c.$$

Therefore case *II* does not lie in $L2$. Thus each cluster is convex, as required.

The partition that is optimal over all partitions is also, of course, locally optimal in the neighborhood of the *K*-means search procedure. The globally optimal clusters are therefore also convex. In searching for a globally optimal partition, it is necessary to consider only convex clusters, and in certain special cases this constraint makes it feasible to insist on the global optimum.

In univariate problems, the convexity requirement means that clusters consist of all the cases lying in an interval. The clustering proceeds first by ordering all *M* cases and then by applying the Fisher algorithm (Chapter 6) to the ordered points. For example, the optimal 2-partition consists of clusters of the form $\{I \mid A(I, 1) < c\}$ and $\{I \mid A(I, 1) > c\}$, where there are only *M* choices of *c*. For the protein data in Table 4.3, the case values are BB = 29, HR = 30, BR = 21, BS = 27, BC = 31, CB = 29, CC = 36, and BH = 37. The clusters must be intervals in the sequence BR BS CB BB HR BC CC BH. The 2-partition is settled by trying the eight possible cuts giving (BR) (BS CB BB HR BC CC BH). The 3-partition is (BR) (BS CB BB HR BC) (CC BH), and so on.

For 2-partitions with two variables, convexity requires that each partition be determined by a line. The first cluster lies on one side of the line and the second cluster lies on the other. The number of such partitions is $M(M - 1)/2 + 1$. It is thus feasible for just two variables to find exactly optimal 2-partitions. The number of 2-partitions for *N* variables is $\sum \{0 \leq J \leq N\} \binom{M - 1}{J}$, where $\binom{M - 1}{J}$ is the number of ways of choosing *J* objects from $M - 1$. For reasonably large *N*—say, $N > 4$—it quickly becomes impractical to obtain the optimal 2-partition by searching over hyperplanes.

To find the optimal 2-partition, consider a projection $V(I) = \sum \{1 \leq J \leq N\}$ $E(J)A(I, J)$, where the coefficients $E(J)$ sum square to unity. Find the optimal 2-partition of the variable *V* and compute the mean square error between clusters—say, $B(V)$. The optimal 2-partition of all the data is the optimal 2-partition of *V* for that *V* maximizing $B(V)$. Searching over all coefficients $E(J)$ corresponds to searching over all hyperplanes. The error $B(V)$ is continuous but not differentiable everywhere as a function of the $E(J)$; it has many local maxima, and so setting derivatives equal to zero is useless. Fix $J = J1$ and consider coefficients $E(J1)$, $\{\alpha E(J), J \neq J1\}$, where $E(J1)$ and α vary but $E(J)$ for $J \neq J1$ are fixed. The variable *V* is determined by a single parameter, and the optimal $E(J1)$ may be found by using a similar procedure to the two-variable problem. In this way, each of the $E(J)$'s may be optimized one at a time. The final partition is not guaranteed globally optimal.

Now, asymptotic arguments will be used to speculate about cluster shape. For *M* infinite, with one variable, normally distributed, the clusters for 2, 3, 4, . . . , partitions have been computed by Cox (1957) and Dalenius (1951). For example, the 2-partition contains clusters $(- \infty, 0)$ $(0, \infty)$, each with 50% of the cases. The 3-partition contains clusters $(- \infty, -0.612)$ $(-0.612, 0.612)$ $(0.612, \infty)$, containing 27%, 46%, 27%, and so on (see Table 4.8). Note that the length of intervals increases towards the tails and the proportion of cases contained decreases. The cut points must be equidistant from the cluster means on either side, and the cluster means are determined by integration between the cutpoints. Beginning with an initial set of cutpoints, the cluster means are computed, then the cut points are taken anew as the averages of the neighboring cluster means, and so on until convergence. This is very similar to the *K*-means algorithm in concept and will not necessarily

Table 4.7   Relation between Weights and Mean Square Error within Clusters

For 27 observations from five-dimensional spherical normal. The weights used are 1, 2, 3, 4, 5. Table entries are weights multiplied by the mean square error within clusters.

PARTITION SIZE	VARIABLE				
	1	2	3	4	5
1	1.2	1.9	2.7	3.7	4.5
2	1.1	1.8	2.5	3.3	1.9
3	1.0	1.7	1.9	2.4	1.9
4	1.0	1.7	1.5	1.8	1.7
5	.9	1.6	1.3	1.3	1.5
10	.7	.8	1.1	1.0	.8
15	.5	.6	.5	.6	.6

converge to the optimum partition for any distribution (one with many modes, for example).

For N variables, M infinite, assume some joint distribution with continuous density on the variables such that each variable has finite variance. As $K \to \infty$, the mean square error within clusters approaches zero. If the joint density is positive everywhere, for each case there must be a cluster mean arbitrarily close to the case for K large enough. There will be some asymptotic N-dimensional distribution of cluster means

Table 4.8   Optimal Clusters of Normal Distribution

PARTITION SIZE		PROPORTION IN CLUSTERS					
2	0	.50	.50				
3	$\pm$ .612	.27	.46	.27			
4	0, $\pm$ .980	.16	.34	.34	.16		
5	$\pm$ 1.230, $\pm$ .395	.11	.24	.31	.24	.11	
6	0, $\pm$ 1.449, $\pm$ .660	.07	.18	.25	.25	.18	.07

that will depend on the original N-dimensional distribution of cases. In the neighborhood of a point, the density of cases is nearly constant. A certain large number of cluster means will be located in the neighborhood, and an approximately optimal location would occur if the density were exactly constant in the neighborhood. Therefore, there is no reason for the cluster shapes to be oriented in any one direction (the shapes will not be spheres, but polyhedra) and the within-cluster covariance matrix will be proportional to the identity matrix.

This is a heuristic argument. It has an important practical consequence. For large M and K, the within-cluster covariance matrix will be proportional to the identity matrix. Thus a hoped-for iterative weighting procedure may not work. It is desired to obtain weights from the within-cluster variances. The clusters must first be computed by using equal weights. The above argument shows, at least for large K, that the final within-cluster variances will be nearly equal and no change in weights will be suggested.

To test this empirically, 27 observations from a five-dimensional normal with mean zero and unit covariance matrix were clustered, using a distance function,

$$D(I, L) = \sum \{1 \leq J \leq 5\} \, W(J)[A(I, J) - A(L, J)]^2,$$

where the weights $W(J)$ take the values 1, 2, 3, 4, 5. In a later step, the inverses of the within-cluster variances might be used as weights. In Table 4.7, it will be seen that the inverses of the within-cluster variances approach the original weights as the number of clusters increases.

Thus in specifying the weights you are specifying the within-cluster variances. In specifying between-variable weights, you are specifying the within-cluster covariance matrix to be the inverse of the weight matrix. These consequences occur only for a large number of clusters, so that if the clustering is stopped soon enough there may be some value to iteration. For example, if there are very distinct clusters separated well by every variable, the partitioning might stop when these clusters are discovered and the weights might be based on these within-cluster variances. But, of course, if there are such distinct clusters, they will appear under any weighting scheme and careful choice of the weights is not important.

4.8 SIGNIFICANCE TESTS

Consider the division of M observations from a single variable into two clusters minimizing the within-cluster sum of squares. Since this is the maximum likelihood division, under the model that observations in the first cluster are $N(\mu_1, \sigma^2)$ and observations in the second cluster are $N(\mu_2, \sigma^2)$, it will be plausible to test $\mu_1 = \mu_2$ versus $\mu_1 \neq \mu_2$ on this normal model.

Let $L(I) = 1$ or 2, according as the observation $X(I)$ lies in the first or second cluster. Define

$$N(J) = \sum \{L(I) = J\}1 \qquad \text{(the number of observations in cluster } J\text{)},$$

$$Y(J) = \sum \{L(I) = J\} \frac{X(J)}{N(J)} \qquad \text{(the average in cluster } J\text{)},$$

$$\text{SSW} = \sum \{L(I) = J\}[X(I) - Y(J)]^2,$$

$$\text{SSB} = \frac{[Y(1) - Y(2)]^2}{1/N(1) + 1/N(2)}.$$

The likelihood ratio criterion is monotone in SSB/SSW, rejecting $\mu_1 = \mu_2$ if this quantity is large enough. The empirical distribution of this quantity for less than 50 observations is tabulated in Engelman and Hartigan (1969).

Suppose $\mu_1 = \mu_2 = 0$. It is sufficient to consider partitions where the first cluster consists of observations less than some split point c, and the second cluster consists of observations greater than c. Asymptotically, SSB and SSW vary negligibly over splits in the neighborhood of μ_1, so the split may be assumed to occur at $\mu_1 = 0$ and the second cluster of observations will be a sample from the half-normal. The half-normal density $f(x) = \exp\left(-\frac{1}{2}x^2\right)\sqrt{2/\pi}$, $(x > 0)$ has mean $\sqrt{2/\pi}$, variance $1 - 2/\pi$, third moment $\sqrt{2/\pi}(4/\pi - 1)$, and fourth moment $3 - 4/\pi - 12/\pi^2$. From this it follows, by using standard asymptotic normal theory on the sums SSB and SSW,

$$\text{SSB} \approx N\left(\frac{2M}{\pi}, \frac{8(\pi - 2)M}{\pi^2}\right),$$

$$\text{SSW} \approx N\left[M\left(1 - \frac{2}{\pi}\right), 2M\left(1 - \frac{8}{\pi^2}\right)\right],$$

and the covariance between them is $M(16/\pi^2 - 4/\pi)$. Note that SSB + SSW $\approx$ $N(M, 2M)$, which is correct because SSB + SSW is just the sum of squared deviations from the overall mean.

More generally, for an arbitrary symmetric parent distribution X, for which the optimal split point converges asymptotically to zero, define

$$\mu(1) = E|X|$$

and

$$\mu(I) = E[|X| - \mu(1)]^I.$$

Then

$$\text{SSB} \approx N[M\mu(1)^2, 4M\mu(1)^2\mu(2)]$$

and

$$\text{SSW} \approx N\{M\mu(2), [\mu(4) - \mu(2)^2]M\},$$

with covariance $2\mu(3)\mu(1)$. It follows that SSB/SSW is approximately normal with mean $\mu(1)^2/\mu(2)$ and variance

$$\left(\frac{4\mu(1)^2}{\mu(2)} + \frac{[\mu(4) - \mu(2)^2]\mu(1)^4}{\mu(2)^4} - \frac{4\mu(3)\mu(1)^3}{\mu(2)^3}\right)M^{-1}$$

(you should forgive the expression).

In the normal case,

$$\frac{\text{SSB}}{\text{SSW}} \approx N\left(\frac{2}{\pi - 2}, \left(\frac{8}{\pi}\right)\left(1 - \frac{3}{\pi}\right)\left(1 - \frac{2}{\pi}\right)^{-4}M^{-1}\right)$$

or

$$\frac{\text{SSB}}{\text{SSW}} \approx N\left(1.75, \frac{6.58}{M}\right).$$

This asymptotic distribution is not applicable except for very large M because SSB/SSW is extremely skew. Empirical investigation shows that log (SSB/SSW) is nearly nonskew for small M (say, $M > 8$) and that the actual distribution is much closer to the asymptotic one,

$$\log\left(\frac{\text{SSB}}{\text{SSW}}\right) \approx N\left(0.561, \frac{2.145}{M}\right).$$

A comparison between SSB/SSW and log (SSB/SSW) for small sample sizes is given in Table 4.9 (see also Engelman and Hartigan, 1969). There remains a substantial bias in log (SSB/SSW) that is incorporated in the formula

$$\log\left(\frac{\text{SSB}}{\text{SSW}}\right) \approx N\left(0.561 + \frac{0.5}{M - 1}, \frac{2.145}{M}\right).$$

Table 4.9 Empirical Distribution of SSB/SSW

Using random samples from a normal distribution, with 100 repetitions for sample size $n = 5$ and 10, with 200 repetitions for sample sizes $n = 20$ and 50.

SAMPLE SIZE		SSB/SSW		LOG (SSB/SSW)	
		MEAN	VARIANCE	MEAN	VARIANCE
5	OBSERVED	8.794	382.007	1.708	.582
	ASYMPTOTIC	1.752	1.316	.561	.429
10	OBSERVED	3.286	2.613	1.093	.180
	ASYMPTOTIC	1.752	.658	.561	.214
20	OBSERVED	2.316	.565	.792	.089
	ASYMPTOTIC	1.752	.329	.561	.107
50	OBSERVED	2.004	.207	.668	.052
	ASYMPTOTIC	1.752	.132	.561	.042

The quantity $\mu(1)^2/\mu(2)$ is a measure of the degree of bimodality of the distribution. It will be a maximum when the distribution (assumed symmetric) is concentrated at two points and a minimum (zero) for long-tailed distributions with infinite variance but finite first moment. Since SSB/SSW estimates this quantity, for symmetric distributions it might be better to estimate it directly by

$$\mu_1 = \sum \{1 \leq I \leq M\} \frac{|X(I) - \mathrm{XBAR}|}{M}$$

$$\mu_2 = \sum \frac{[X(I) - \mathrm{XBAR}]^2}{M} - \mu_1^2,$$

where $\mathrm{XBAR} = \sum \{1 \leq I \leq M\} X(I)/M$. This method of estimation is faster than the splitting method, which requires ordering the M observations and then checking M possible splits. In a similar vein, a quick prior estimate of within-cluster variance for weighting purposes (the estimate works well if the two clusters are of approximately equal size) is

$$\sum \{1 \leq I \leq M\} \frac{[X(I) - \mathrm{XBAR}]^2}{M} - \left(\sum \{1 \leq I \leq M\} \frac{|X(I) - \mathrm{XBAR}|}{M} \right)^2.$$

In N dimensions, assume that the null distribution is multivariate normal and that the covariance matrix has eigenvalues $E(1) > E(2) > \cdots > E(N)$. The asymptotic argument reduces to the one-dimensional case by orthogonal transformation to the independent normal variables with variances $E(1), \ldots, E(N)$. The split will be essentially based on the first variable, and the remaining variables will contribute standard chi-square-like terms to the sums of squares:

$$\text{SSB} \approx N\left(\frac{2ME(1)}{\pi}, 8(\pi - 2)M\frac{E(1)^2}{\pi^2}\right),$$

$$E(\text{SSW}) = M \sum \{1 \leq I \leq N\}E(I) - 2M\frac{E(1)}{\pi},$$

$$\text{VAR}(\text{SSW}) = 2M \sum \{1 \leq I \leq N\}E(I)^2 - 16M\frac{E(1)^2}{\pi^2},$$

$$\text{COV}(\text{SSB}, \text{SSW}) = ME(1)^2\left(\frac{16}{\pi^2} - \frac{4}{\pi}\right).$$

From this, asymptotically,

$$E\left(\frac{\text{SSB}}{\text{SSW}}\right) = \frac{2E(1)}{\pi}\left(\sum \{1 \leq I \leq N\}E(I) - \frac{2E(1)}{\pi}\right)^{-1},$$

$$\text{VAR}\left(\frac{\text{SSB}}{\text{SSW}}\right) = \left[\left(\frac{8}{\pi^2}\right)E(1)^2(\sum \{1 \leq I \leq N\}E(I)^2 + (\pi - 2)[\sum \{1 \leq I \leq N\}E(I)]^2)\right.$$

$$\left. - \left(\frac{16}{\pi^2}\right)E(1)^3 \sum \{1 \leq I \leq N\}E(I)\right]\left[M\left(\frac{E(\text{SSW})}{M}\right)^4\right]^{-1}.$$

For N large, with $E(1)$ making a relatively small contribution to $\sum \{1 \leq I \leq N\} E(I)$,

$$E\left(\frac{\text{SSB}}{\text{SSW}}\right) = \frac{2E(1)}{\pi}(\sum \{1 \leq I \leq N\}E(I))^{-1}$$

and

$$\text{VAR}\left(\frac{\text{SSB}}{\text{SSW}}\right) = \left(\frac{8}{\pi^2}\right)E(1)^2[\sum \{1 \leq I \leq N\}E(I)]^{-2}\frac{\pi - 2}{M}.$$

This reveals the obvious—that SSB/SSW has larger expectation if $E(1)$ is larger relative to the other eigenvalues.

In the important case where the null distribution is spherical normal (all eigenvalues equal), the asymptotic distribution of SSB and SSW is not joint normal and the complicated calculations will not be given here.

4.9 THINGS TO DO

4.9.1 Running *K*-Means

A good trial data set is expectations of life by country, age, and sex, in Table 4.10. Try running the *K*-means algorithm on these data. An initial decision is the question of rescaling the variables. The variances and covariances of the variables should be examined. In a problem like this, where the variables are on the same scale (here, years), no change should be made except for a compelling reason.

The number of clusters K should not be decided in advance, but the algorithm should be run with several different values of K. In this problem, try $K = 1, 2, \ldots, 6$. Analysis of variance, on each of the variables, for each clustering will help decide which number of clusters is best. It is also desirable to compute covariances within each cluster, the overall covariance matrix within clusters, and the overall covariance

Table 4.10 Expectations of Life by Country, Age, and Sex

Keyfitz, N., and Flieger, W. (1971). *Population*, Freeman.

COUNTRY (YEAR)	AGE	MALE 0	25	50	75	AGE	FEMALE 0	25	50	75
1. ALGERIA 65		63	51	30	13		67	54	34	15
2. CAMEROON 64		34	29	13	5		38	32	17	6
3. MADAGASCAR 66		38	30	17	7		38	34	20	7
4. MAURITIUS 66		59	42	20	6		64	46	25	8
5. REUNION 63		56	38	18	7		62	46	25	10
6. SEYCHELLES 60		62	44	24	7		69	50	28	14
7. SOUTH AFRICA (COL) 61		50	39	20	7		55	43	23	8
8. SOUTH AFRICA (WH) 61		65	44	22	7		72	50	27	9
9. TUNISIA 60		56	46	24	11		63	54	33	19
10. CANADA 66		69	47	24	8		75	53	29	10
11. COSTA RICA 66		65	48	26	9		68	50	27	10
12. DOMINICAN REP. 66		64	50	28	11		66	51	29	11
13. EL SALVADOR 61		56	44	25	10		61	48	27	12
14. GREENLAND 60		60	44	22	6		65	45	25	9
15. GRENADA 61		61	45	22	8		65	49	27	10
16. GUATEMALA 64		49	40	22	9		51	41	23	8
17. HONDURAS 66		59	42	22	6		61	43	22	7
18. JAMAICA 63		63	44	23	8		67	48	26	9
19. MEXICO 66		59	44	24	8		63	46	25	8
20. NICARAGUA 65		65	48	28	14		68	51	29	13
21. PANAMA 66		65	48	26	9		67	49	27	10
22. TRINIDAD 62		64	43	21	7		68	47	25	9
23. TRINIDAD 67		64	43	21	6		68	47	24	8
24. UNITED STATES 66		67	45	23	8		74	51	28	10
25. UNITED STATES (NON-W) 66		61	40	21	10		67	46	25	11
26. UNITED STATES (W) 66		68	46	23	8		75	52	29	10
27. UNITED STATES 67		67	45	23	8		74	51	28	10
28. ARGENTINA 64		65	46	24	9		71	51	28	10
29. CHILE 67		59	43	23	10		66	49	27	12
30. COLUMBIA 65		58	44	24	9		62	47	25	10
31. ECUADOR 65		57	46	25	9		60	49	28	11

101

matrix between clusters. If the overall within covariance matrix differs significantly from a multiple of the unit matrix, transformation of the data to make it proportional to a unit matrix is suggested.

4.9.2 Varieties of the *K*-Means Algorithm

There are a number of versions of the *K*-means algorithm that need to be compared by sampling experiments or asymptotic analysis. The changeable components are (i) the starting clusters, (ii) the movement rule, and (iii) the updating rule. The criteria for evaluation are (i) the expected time of calculation and (ii) the expected difference between the local optimum and the global optimum. It is often required that an algorithm produce clusters that are independent of the input order of the cases; this requirement is not necessarily satisfied by the *K*-means algorithm but can always be met by some initial reordering of the cases. For example, reorder all cases by the first, second, . . . , *N*th variables, in succession. (Note that this reordering will usually reduce the number of iterations in the algorithm.)

The following are some starting options:

(i) Choose the initial clusters at random. The algorithm is repeated several times from different random starting clusters with the hope that the spread of the local optima will give a hint about the likely value of the true global optimum. To justify this procedure, it is necessary to have a distribution theory, finite or asymptotic, connecting the local and global optima.

(ii) Choose a single variable, divide it into *K* intervals of equal length, and let each cluster consist of the cases in a single interval. The single variable might be the average of all variables or the weighted combination of variables that maximizes variance, the first row eigenvector.

(iii) Let the starting clusters for *K* be the final clusters for $K - 1$, with that case furthest from its cluster mean split off to form a new cluster.

The following are some movement options:

(i) Run through the cases in order, assigning each case according to the cluster mean it is closest to.

(ii) Find the case whose reassignment most decreases the within-cluster sum of squares and reassign it.

(iii) For each cluster, begin with zero cases and assign every case to the cluster, at each step finding the case whose assignment to the cluster most decreases (or least increases) the within-cluster sum of squares. Then take the cluster to consist of those cases at the step where the criterion is a minimum. This procedure makes it possible to move from a partition which is locally optimal under the movement of single cases.

The following are some updating options:

(i) Recompute the cluster means after no further reassignment of cases decreases the criterion.

(ii) Recompute the cluster means after each reassignment.

4.9.3 Bounds on the Global Optimum

It would be good to have empirical or analytic results connecting the local optima and the global optimum. How far is the local optimum likely to be from the global optimum? How different are the two partitions?

For some data configurations, the local optimum is more likely global than for others. Some bad things happen. Let $\{A(I, J), 1 \leq I \leq M, 1 \leq J \leq N\}$ be divisible into K clusters such that the euclidean distance between any pair of cases inside the same clusters is less than ρ and the euclidean distance between any pair of cases in different clusters is greater than ρ. Then this partition is a local optimum but not necessarily global.

Yet, if the clusters are widely separated, it should be possible to prove that there is only one local optimum. Let there be K clusters and fix the distances inside each of the clusters, but let the distances between cluster means all approach infinity. Then eventually there is a single local optimum.

The interesting problem is to make the relation precise between the within-cluster distances and the between-cluster distances, so that there is a unique local optimum. For example, suppose there are K clusters, $M(I)$ points in the Ith cluster, $D(I)$ is the maximum distance within the Ith cluster, and $E(I, J)$ is the minimum distance between the Ith and Jth clusters. For what values of $M(I)$, $D(I)$, and $E(I, J)$ is there a unique local optimum at this partition?

Some asymptotic results suggest that for large sample sizes there will be only a few local optima, differing only a little from the global optimum. The assumption required is that the points are drawn from some parent distribution which itself has a unique local optimum. If these results are expected, it means that a crude algorithm arriving quickly at some local optimum will be most efficient. It would be useful to check the asymptotics in small samples by empirical sampling.

4.9.4 Other Criteria

The points in cluster J may come from a population with parameters, possibly vector valued, $\boldsymbol{\theta}(J)$, φ. The log likelihood of the whole data set is then

$$-\sum \{1 \leq I \leq M\}\, F\{\mathbf{A}(I), \boldsymbol{\theta}[L(I)], \varphi\},$$

where $\mathbf{A}(I)$ denotes the Ith case, $L(I)$ denotes the cluster to which it belongs, and φ denotes a general parameter applying across clusters. A generalized K-means algorithm is obtained by first assigning $\mathbf{A}(I)$ to minimize the criteria and then changing the parameter values $\boldsymbol{\theta}[L(I)]$ and φ to minimize the criteria.

A first generalization is to allow an arbitrary within-cluster covariance matrix. The algorithm will first assign each case according to its distances from cluster means relative to the covariance matrix. It will then recompute the covariance matrix according to the redefined clusters. Both steps increase the log likelihood. The final clusters are invariant under arbitrary linear transformations of the variables, provided the initial clusters are invariant.

4.9.5* Asympototics

Consider the simplest case of division of real observations into two clusters. (The following results generalize to arbitrary numbers of dimensions and clusters and to more general optimization criteria.) The cluster means are θ and φ, and a typical point is x. Define the 2-vector

$$W(x, \theta, \varphi) = \begin{pmatrix} \theta - x \\ 0 \end{pmatrix}, \qquad \text{if} \quad |\theta - x| \leq |\varphi - x|$$

$$= \begin{pmatrix} 0 \\ \varphi - x \end{pmatrix}, \qquad \text{if} \quad |\varphi - x| < |\theta - x|.$$

For data $X(1), \ldots, X(M)$, the criterion

$$\sum \{X(I) \in C(1)\}[X(I) - \theta]^2 + \sum \{X(I) \in C(2)\} [X(I) - \varphi]^2$$

has a local minimum if no reallocation of an $X(I)$ between clusters $C(1)$ and $C(2)$ reduces it and if no change of θ or φ reduces it. The criterion has a local minimum if and only if $\sum \{1 \leq I \leq M\} W[X(I), \theta, \varphi] = 0$.

Asymptotic distributions for θ and φ follow from asymptotic distributions for $\sum W$, which for each fixed θ, φ is a sum of identically distributed independent random variables. Let $X(I)$ be sampled from a population with three finite moments. Let

$$E(W) = \begin{bmatrix} \displaystyle\int_{X \leq 1/2(\theta+\varphi)} (\theta - X) \, dP \\[2em] \displaystyle\int_{X > 1/2(\theta+\varphi)} (\varphi - X) \, dp \end{bmatrix} \qquad \text{for} \qquad \theta < \varphi,$$

$$V(W) = \begin{bmatrix} \displaystyle\int_{X \leq 1/2(\theta+\varphi)} (\theta - X)^2 \, dP & 0 \\[2em] 0 & \displaystyle\int_{X > 1/2(\theta+\varphi)} (\varphi - X)^2 \, dP \end{bmatrix} - E(W)E(W').$$

Suppose $E(W) = 0$ for a unique $\theta, \varphi, \theta < \varphi$—say, θ_0, φ_0. Suppose the population has a density $f > 0$ at $x_0 = \frac{1}{2}(\theta_0 + \varphi_0)$, and that $V(W)$ is evaluated at θ_0, ϕ_0.

Let $\hat{\theta}_n$ and $\hat{\varphi}_n$ denote solutions to the equation $\sum W[X(I), \theta, \varphi] = 0$. Asymptotically,

$$\sqrt{nE}[W(X, \hat{\theta}_n, \hat{\varphi}_n)] \approx N[0, V(W)].$$

This means

$$\sqrt{n} \begin{bmatrix} \dfrac{\partial EW}{\partial \theta_0} & \dfrac{\partial EW}{\partial \varphi_0} \end{bmatrix} \begin{bmatrix} \hat{\theta}_n - \theta_0 \\ \hat{\varphi}_n - \varphi_0 \end{bmatrix} \approx N[0, V(W)]$$

and

$$\sqrt{n} \begin{bmatrix} \hat{\theta}_n - \theta_0 \\ \hat{\phi}_n - \varphi_0 \end{bmatrix} \approx N[0, U],$$

where

$$U = \mathbf{\Sigma}^{-1} V \mathbf{\Sigma}^{-1},$$

$$\mathbf{\Sigma} = \begin{bmatrix} P(X \leq x_0) + \delta & \delta \\ \delta & P(X > x_0) + \delta \end{bmatrix},$$

and

$$\delta = \tfrac{1}{4}(\theta_0 - \varphi_0)f(x_0).$$

It turns out that different locally optimal solutions $\hat{\theta}_n$ and $\hat{\varphi}_n$ differ from one another by terms of $O(n^{-1})$.

For symmetric parent distributions,

$$\sqrt{n} (\hat{\phi}_n - \hat{\theta}_n) \approx N(-\sqrt{n}2\theta_0, V),$$

$$\theta_0 = 2 \int_{X < 0} X \, dP,$$

and

$$V = 8 \int_{X < 0} (X - \theta_0)^2 \, dP.$$

For a unit normal variable

$$\sqrt{n}(\hat{\varphi}_n - \hat{\theta}_n) \approx N[2\sqrt{2n/\pi}, 4(1 - 2/\pi)].$$

In general, the cluster centers are asymptotically normal with the covariance matrix computed in a similar way to the above. Each cluster center is the average of a number of observations lying closest to it. Its covariance matrix is just the covariance matrix of a mean with a δ term added due to the boundary of the region varying.

Since usually the quantities $\partial EW/\partial\theta$ and $V(W)$ are not known, they must be estimated from the data—$V(W)$ from the observed quantities $W[X(I), \theta, \varphi]$ at $\theta = \hat{\theta}_n$ and $\varphi = \hat{\varphi}_n$, and the derivatives from

$$\sum \{W[X(I), \theta, \varphi] - W[X(I), \hat{\theta}_n, \hat{\varphi}_n]\}$$

for θ, φ near $\hat{\theta}_n$, $\hat{\varphi}_n$.

It would be useful to check the asymptotics by empirical sampling, at least for the normal distribution above. It would be useful to check that the different locally optimal solutions vary by $O(n^{-1})$. It would be useful to check the formulas of Section 4.8 empirically.

4.9.6 Subsampling

To avoid thought and asymptotic formulas, distributions for the cluster means that agree with the asymptotic ones may be obtained empirically as follows. A subset of cases is formed by randomly including each case with probability 0.5. The algorithm produces cluster means. A new subset is formed, and the algorithm produces a new set of cluster means. Repeating this procedure a few times, a sample of cluster means is obtained that agrees asymptotically with a sample from the posterior distribution of true cluster means.

4.9.7 Symmetric Paradox

In the univariate case, find a set of values, symmetric about zero, for which an optimal division into two clusters does not occur at zero.

4.9.8 Large Data Sets

For large numbers of cases (say, $M = 5000$) it is wasteful to do many runs on all cases. No matter what the eventual analysis, there will usually be no great loss in reducing the data to 100 cases using the leader algorithm. Each data point will be within a threshold distance d of one of these 100 cases. Suppose that the leading case is replaced by the average case in each cluster. A K-means algorithm is run on these 100 cases, with each case weighted by the number of original cases in the corresponding cluster. This produces a partition of the average cases with a weighted within-cluster sum of squares w. Show that the within-cluster sum of squares of the corresponding partition of the original cases lies between w and $w + Md^2$.

REFERENCES

COX, D. R. (1957). "Note on grouping." *J. Am. Stat. Assoc.* **52,** 543–547. It is desired to classify an individual into one of K groups using an observation X. Define a grouping function $\xi(X)$ that takes at most K different values. Measure the loss associated with the grouping by $E[X - \xi(X)]^2$. Then $\xi_i P[\xi(X) = \xi_i] = \int_{\xi(X)=\xi_i} X \, dP$. This equation determines the value of $\xi(X)$ in any group, but it does not specify the groups. The groups are specified, without proof, for $K = 1, 2, \ldots, 6$ from a normal distribution.

DALENIUS, TORE (1951). "The problem of optimum stratification." *Skandinavisk Aktuarietidskrift* **34**, 133–148. In dividing a continuous population into K clusters to minimize within-cluster variance, the cut point between neighboring clusters is the average of the means in the clusters. An iterative method is proposed for attaining this condition from an initial trial division.

ENGELMAN, L., and HARTIGAN, J. A. (1969). "Percentage points of a test for clusters." *J. Am. Stat. Assoc.* **64**, 1647–1648. This paper gives empirical distributions of SSB/SSW for sample sizes 3–10, 15, 20, 25, 50. From examination of the empirical distributions, it is suggested that

$$\log \left(1 + \frac{\text{SSB}}{\text{SSW}} \right) \approx N \left[- \log \left(1 - \frac{2}{\pi} \right) + \frac{2.4}{M - 2}, \frac{1}{M - 2} \right].$$

This formula agrees approximately with the asymptotic theory of Section 4.8.

FISHER, L., and VAN NESS, J. W. (1971). "Admissible clustering procedures." *Biometrika* **58**, 91–104. The authors compare a number of "joining" algorithms and the *K*-means algorithm, which they call "hill climbing least squares." Their technique is to test whether or not the procedures produce clusters satisfying certain admissibility conditions for every possible set of data.

The *K*-means algorithm shows up rather badly failing every test but one—the final clusters are convex. That is, if a case is a weighted average of cases in a cluster, the case also lies in the cluster. The tests it fails include the following:

(i) *Perfect data.* If there exists a clustering such that all distances within clusters are less than all distances between clusters, K means might not discover this clustering.

(ii) *Duplication of cases.* If a case is repeated exactly, the final clustering might change.

They also note that it is practicable for very large numbers of observations, when a joining algorithm requires too much storage and computer time.

MACQUEEN, J. (1967). "Some methods for classification and analysis of multivariate observations." *5th Berkeley Symposium on Mathematics, Statistics, and Probability*. Vol. 1, pp. 281–298. The *K*-means procedure starts with K groups, each of which contains a single random point. A sequence of points is sampled from some distribution, and each point is added in turn to the group whose mean it is closest to. After a point is added, the group mean is adjusted.

Suppose the population has density $p(z)$ and that the cluster means, after n points are sampled, are $x_1, \ldots, x_K$.

Set

$$W(x_1, \ldots, x_K) = \sum_{i=1}^{K} \int_{S_i} |z - x_i|^2 \, p(z) \, dz,$$

where

$$S_i = \left\{ z \mid |z - x_i| = \min_j |z - x_j| \right\}.$$

Then the principal theorem is that $W(x_1, \ldots, x_k)$ converges to $W(u_1, \ldots, u_k)$, where $u_i = (\int_{S_i} z \, p(z) \, dz)/\int_{S_i} p(z) \, dz$. In words, the population variance within the sample clusters converges to the population variance within a locally optimal clustering of the population.

SEBESTYEN, GEORGE S. (1962). *Decision Making Processes in Pattern Recognition*, Macmillan, New York. "Pattern detection is the process of learning the characterization of a class of inputs by detecting the common pattern of attributes of inputs of the same class. Pattern recognition is a process of decision making in which a new input is recognized as a member of a given class by a comparison of its attributes with the already known pattern of common attributes of members of that class." A principal part of the book is concerned with the second problem, discrimination between known classes. A *K*-means-type of algorithm is considered on p. 47, although it is described only in general terms. A number of inputs are introduced in sequence. Each input is assigned to one of a number of classes, according to its distance to the mean input of each class. The input is left unassigned if it is not close enough to any mean input. The mean inputs are updated after each assignment.

PROGRAMS

There are a series of routines that construct partitions of various sizes and print summary statistics about the clusters obtained.

BUILD   calls programs constructing optimal partition of given size and increases partition size by splitting one of the clusters.
KMEANS assigns each case optimally.
SINGLE   computes summary statistics.
OUTPUT prints information about clusters.

The following programs perform the basic operations of *K* means.

RELOC   moves cluster center to cluster means.
ASSIGN   reassigns each case to closest cluster center.

```
      SUBROUTINE BUILD(A,M,N,K,SUM,XMISS,NCLUS,DCLUS,X,ITER)
C...................................................................20 MAY 1973
C     BUILDS K CLUSTERS BY K-MEANS METHOD.  A CLUSTER IS ADDED AT EACH STEP, THE
C     WORST OBJECT FROM THE PREVIOUS STEP.
C     FOR VERY LARGE MATRICES, THIS PROGRAM MUST BE MODIFIED TO AVOID STORING
C     THE COMPLETE DATA MATRIX A IN CORE.   THERE WILL BE K PASSES THROUGH THE
C     DATA MATRIX FOR K CLUSTERS.  THE DIMENSION STATEMENT MUST BE MODIFIED, AND
C     STATEMENT NUMBERED 12.
C.... A = M BY N BORDERED ARRAY
C.... M = NUMBER OF ROWS
C.... N = NUMBER OF COLUMNS
C.... K = NUMBER OF CLUSTERS
C.... XMISS = MISSING VALUE
C.... NCLUS = M BY 1 ARRAY SPECIFYING A CLUSTER NUMBER FOR EACH CASE
C.... DCLUS = 1 BY M ARRAY SPECIFYING DISTANCE OF EACH ROW TO CLOSEST CLUSTER.
C.... X = N BY 1 SCRATCH VECTOR
C.... ITER = NUMBER OF ITERATIONS AT EACH PARTITION SIZE
C...................................................................
      DIMENSION SUM(8,N,K),A(M,N),X(N),NCLUS(M),DCLUS(M)
      DO 20 I=1,8
      DO 20 J=2,N
      DO 20 KK=1,K
   20 SUM(I,J,KK)=0
      KL=K-1
      DO 10 KK=1,KL
      DO 14 NC=1,ITER
      DMAX=0.
      ERR=0.
      DO 13 KKK=1,KK
      DO 13 J=2,N
      IF(NC.EQ.1.OR.SUM(1,J,KKK).NE.SUM(3,J,KKK)) ERR=1.
   13 CONTINUE
      IF(ERR.EQ.0) GO TO 15
      DO 16 KKK=1,KK
      DO 16 J=2,N
      SUM(8,J,KKK)=SUM(2,J,KKK)
      IF(NC.EQ.1) SUM(8,J,KKK)=1.
      SUM(2,J,KKK)=0
   16 SUM(1,J,KKK)=SUM(3,J,KKK)
      DO 11 I=2,M
      DO 12 J=2,N
   12 X(J)=A(I,J)
      NCLUS(I)=NC
      CALL KMEANS(N,KK,SUM,X,NCLUS(I),DCLUS(I),XMISS)
   11 CONTINUE
   14 CONTINUE
   15 CONTINUE
      CALL OUTPUT(M,N,KK,SUM,A,NCLUS,DCLUS)
C.... CREATE A NEW CLUSTER BY SPLITTING VARIBLE WITH LARGE WITHIN VARIANCE
      SM=0
      DO 30 J=2,N
      DO 30 KKK=1,KK
      IF(SUM(4,J,KKK).LT.SM) GO TO 30
      SM=SUM(4,J,KKK)
      JM=J
      KM=KKK
   30 CONTINUE
      KN=KK+1
      DO 31 JJ=2,N
      SUM(2,JJ,KN)=0
      SUM(2,JJ,KM)=0
      SUM(3,JJ,KM)=0
   31 SUM(3,JJ,KN)=0
      DO 32 I=2,M
      IF(NCLUS(I).NE.KM) GO TO 32
      DO 33 JJ=2,N
      IF(A(I,JJ).EQ.XMISS) GO TO 33
      IF(A(I,JJ).LT.SUM(1,JJ,KM)) GO TO 34
      SUM(2,JJ,KN)=SUM(2,JJ,KN)+1
      SUM(3,JJ,KN)=SUM(3,JJ,KN)+A(I,JJ)
      GO TO 33
   34 SUM(2,JJ,KM)=SUM(2,JJ,KM)+1
      SUM(3,JJ,KM)=SUM(3,JJ,KM)+A(I,JJ)
   33 CONTINUE
   32 CONTINUE
      DO 35 JJ=2,N
      IF(SUM(2,JJ,KN).NE.0) SUM(3,JJ,KN)=SUM(3,JJ,KN)/SUM(2,JJ,KN)
      IF(SUM(2,JJ,KM).NE.0) SUM(3,JJ,KM)=SUM(3,JJ,KM)/SUM(2,JJ,KM)
   35 CONTINUE
   10 CONTINUE
      RETURN
      END
```

108

```
      SUBROUTINE KMEANS(N,K,SUM,X,JMIN,DMIN,XMISS)
C.............................................................20 MAY 1973
C      ASSIGNS THE VECTOR X TO THAT CLUSTER WHOSE CLUSTER CENTRE IT IS CLOSEST TO
C      UPDATES FOR THIS CLUSTER, VARIOUS SUMMARY STATISTICS SUCH AS MEAN,SD,MIN,
C      MAX,SSQ.   NOTE THAT CLUSTER CENTERS ARE NOT CHANGED BY THE ADDITION OF X.
C.... N = LENGTH OF VECTOR X
C.... K = TOTAL NUMBER OF CLUSTERS
C.... SUM = 7 BY N BY K ARRAY, CHANGED DURING SUBROUTINE
C      SUM(1,J,I) = VALUE OF JTH VARIABLE AT CLUSTER CENTER
C      SUM(2,J,I) = NUMBER OF NON MISSING OBSERVATIONS,JTH VARIABLE, ITH CLUSTE
C      SUM(3,J,I) = AVERAGE, JTH VARIABLE, ITH CLUSTER
C      SUM(4,J,I) = STANDARD DEVIATION
C      SUM(5,J,I) = MINIMUM
C      SUM(6,J,I)= MAXIMUM
C      SUM(7,J,I) = SUM OF SQUARED DEVIATIONS FROM CLUSTER MEAN
C.... X = N BY 1 VECTOR TO BE ALLOCATED AMONG THE K CLUSTERS
C.... JMIN = NUMBER OF CLUSTER WHOSE CENTRE X IS CLOSEST TO.
C.... DMIN = EUCLIDEAN DISTANCE BETWEEN X AND CENTER OF JMIN CLUSTER
C.... XMISS = MISSING VALUE
C.........................................................................
      DIMENSION SUM(8,N,K),X(N)
      JMIN=1
      DMIN=10.**20
      DO 20 J=1,K
      XP=10.**(-10)
      DD=0
      DO 21 I=2,N
      IF (X(I).EQ.XMISS) GO TO 21
      DD=DD+(X(I)-SUM(1,I,J))**2
      XP=XP+1.
   21 CONTINUE
      DD=(DD/XP)**0.5
      IF(DD.GT.DMIN) GO TO 20
      DMIN=DD
      JMIN=J
   20 CONTINUE
      XM=N
      DO 31 I=2,N
      IF(X(I).EQ.XMISS) GO TO 31
   30 CALL SINGLE(X(I),SUM(2,I,JMIN),SUM(3,I,JMIN),SUM(4,I,JMIN)
     1,SUM(5,I,JMIN),SUM(6,I,JMIN),SUM(7,I,JMIN))
   31 CONTINUE
      RETURN
      END




      SUBROUTINE SINGLE(X,COUNT,AVE,SD,XMIN,XMAX,SSQ)
C.............................................................20 MAY 1973
C      INCORPORATES NEW VALUE X INTO SUMMARY STATISTICS
C      THE MEANING OF EACH VARIABLE IS GIVEN IN KMEANS.
C.........................................................................
      IF(COUNT.NE.0.)GO TO 10
      AVE=0
      SD=0
      XMIN=10.**20
      XMAX=-10.**20
      SSQ=0
   10 COUNT=COUNT+1.
      AVE=AVE+(X-AVE)/COUNT
      IF(COUNT.NE.1) SSQ=SSQ+COUNT*(X-AVE)**2/(COUNT-1.)
      SD=(SSQ/COUNT)**0.5
      IF(XMIN.GT.X) XMIN=X
      IF(XMAX.LT.X) XMAX=X
      RETURN
      END
```

```
          SUBROUTINE OUTPUT(M,N,KK,SUM,A,NCLUS,DCLUS)
C...........................................................................20 MAY 1973
C       OUTPUT ROUTINE FOR KMEANS ALGORITHM
C       PUTS OUT SUMMARY STATISTICS FOR EACH VARIABLE FOR EACH CLUSTER
C       ALSO PUTS OUT OVERALL ANALYSIS OF VARIANCE FOR EACH VARIABLE
C.... A = M BY N BORDERED DATA MATRIX
C.... M = NUMBER OF CASES
C.... N = NUMBER OF VARIABLES
C.... KK = NUMBER OF CLUSTERS
C.... SUM = 7 BY N BY KK MATRIX OF SUMMARY STATISTICS(SEE KMEANS ROUTINE)
C.... NCLUS = M BY 1 ARRAY IDENTIFYING CLUSTER TO WHICH EACH ROW BELONGS
C.... DCLUS = EUCLIDEAN DISTANCE OF EACH ROW TO CLOSEST CLUSTER
C...........................................................................
          DIMENSION SUM(8,N,KK),NCLUS(M),DCLUS(M),A(M,N)
          DIMENSION AA(10),DD(10)
          DIMENSION R(50)
          DATA NPAGE/0/
          DATA LC/0/
C.... MEAN SQUARE CALCULATION OVER ALL CLUSTERS
          NPAGE=NPAGE+1
          WRITE(6,7) NPAGE
    7 FORMAT(1H1,110X,I5)
          WRITE(6,9) KK
    9 FORMAT(' OVERALL MEAN SQUARE CALCULATIONS, FOR EACH VARIABLE, ',
     1' WITH',I5,'  CLUSTERS'.)
          ASSW=0
          DO 40 J=2,N
          SD=0.
          SC=0.
          SSB=0.
          SSW=0.
          DO 41 K=1,KK
          SD=SD+SUM(3,J,K)*SUM(2,J,K)
          SSB=SSB+SUM(3,J,K)**2*SUM(2,J,K)
          SSW=SSW+SUM(7,J,K)
   41 SC=SC+SUM(2,J,K)
          DFB=KK-1
          DFW=SC-DFB-1.
          TH=10.**(-10)
          IF(SC.EQ.0) SC=TH
          IF(DFW.EQ.0) DFW=TH
          IF(DFB.EQ.0) DFB=TH
          ASSW=ASSW+SSW
          SSB=SSB-SD**2/SC
          SSB=SSB/DFB
          SSW=SSW/DFW
          IF(SSW.EQ.0) SSW=TH
          RATIO=0
          IF(LC.NE.0)RATIO=(R(J)/SSW-1)*(1+DFW)+1
          R(J)=SSW
          WRITE(6,8)A(1,J),SSW,DFW,SSB,DFB,RATIO
    8 FORMAT(' VARIABLE',A8,F20.6,'(WITHIN MEAN SQ.)',F4.0,'(WITHIN DF)'
     1,F20.6,'(BETWEEN MSQ)',F4.0,'(BETWEEN DF)',F6.1,'(FRATIO)')
   40 CONTINUE
          WRITE(6,10) ASSW
   10 FORMAT(' OVERALL WITHIN SUM OF SQUARES',F20.6)
          LC=LC+1
          DO 20 K=1,KK
          WRITE(6,11)
   11 FORMAT(1X,131(1H-))
          WRITE(6,1) K,KK
    1 FORMAT(I5,'   TH CLUSTER OF',I5)
          WRITE(6,2)(I,I=1,10)
    2 FORMAT('OCLUSTER MEMBERS WITH THEIR DISTANCES TO THE CLUSTER CENTRE'/(13X,
     1E'/13X,10I11)
          L=0
          DO 21 I=2,M
          IF(NCLUS(I).NE.K) GO TO 22
          L=L+1
          AA(L)=A(I,1)
          DD(L)=DCLUS(I)
   22 IF (L.LT.10.AND.I.LT.M) GO TO 21
          IF(L.EQ.0) GO TO 21
          WRITE(6,3)(AA(LL),LL=1,L)
    3 FORMAT(15X,10(7X,A4))
          WRITE(6,12)(DD(LL),LL=1,L)
   12 FORMAT(15X,10F11.4)
          L=0
```

110

```
   21 CONTINUE
      WRITE(6,4)
    4 FORMAT('0SUMMARY STATISTICS FOR THE CLUSTER')
      WRITE(6,5)
    5 FORMAT(' LABEL',5X,'CENTRE',8X,
      1        'COUNT',12X,'AVE',13X,'SD',11X,'XMIN',11X,'XMAX',12X,'SSQ''
      1SQ')
      DO 30 J=2,N
   30 WRITE(6,6)A(1,J),(SUM(I,J,K),I=1,7)
    6 FORMAT(1X,A4,7F15.6)
   20 CONTINUE
   81 CONTINUE
      RETURN
      END




      SUBROUTINE RELOC(M,N,K,A,X,NC)
C...................................................................20 MAY 1973
C.... RELOCATES EACH CLUSTER CENTRE TO BE A CLUSTER MEAN
C.... M = NUMBER OF ROWS
C.... N = NUMBER OF COLUMNS
C.... K = NUMBER OF CLUSTERS
C.... A = M BY N BORDERED ARRAY
C.... X = N BY K BORDERED ARRAY OF CLUSTER CENTRES
C.... NC = M BY 1 ARRAY ASSIGNING EACH ROW TO A CLUSTER
C...................................................................
      DIMENSION A(M,N),X(N,K),NC(M)
      DIMENSION CC(10)
      DATA CC/4HCLUS,2HC1,2HC2,2HC3,2HC4,2HC5,2HC6,2HC7,2HC8,2HC9/
      XM=99999
C.... COMPUTE MEANS
      DO 10 L=2,K
      DO 10 J=2,N
      X(J,L)=0
      IF(NC(I).NE.L) GO TO 20
      IF(A(I,J).EQ.XM) GO TO 20
      P=0
      DO 20 I=2,M
      P=P+1
      X(J,L)=X(J,L)+A(I,J)
   20 CONTINUE
      IF(P.NE.0) X(J,L)=X(J,L)/P
      IF(P.EQ.0) X(J,L)=XM
   10 CONTINUE
C.... LABEL CLUSTER MEANS
      DO 40 J=2,N
   40 X(J,1)=A(1,J)
      DO 50 L=1,K
      IF(L.GT.10) RETURN
   50 X(1,L)=CC(L)
      RETURN
      END
```

```
      SUBROUTINE ASSIGN(M,N,K,A,X,NC)
C.............................................................20 MAY 1973
C.... ASSIGNS EACH ROW OF BORDERED ARRAY TO CLOSEST OF CLUSTER CENTRES X
C.... M = NUMBER OF ROWS
C.... N = NUMBER OF COLUMNS
C.... K = NUMBER OF CLUSTERS
C.... A = M BY N BORDERED ARRAY
C.... X = BORDERED ARRAY OF CLUSTER CENTRES
C.... X = N BY K BORDERED ARRAY OF CLUSTER CENTRES
C.... NC = M BY 1 ARRAY ASSIGNING EACH ROW TO A CLUSTER
C          NC(1) = ROW FURTHEST FROM ITS CLUSTER CENTRE
C .......................................................................
      DIMENSION A(M,N),X(N,K),NC(M)
C.... INITIALISE
      D=0
      NC(1)=0
      XM=99999.
      DO 10 I=2,M
      DC=10.**10
      DO 20 J=2,K
      DJ=0
      DP=0
C.... FIND DISTANCE TO CLUSTER CENTRE
      DO 30 L=2,N
      IF(X(L,J).EQ.XM.OR.A(I,L).EQ.XM) GO TO 20
      DJ=DJ+(X(L,J)-A(I,L))**2
   30 CONTINUE
      IF(DP.GT.0) DJ=(DJ/DP)**0.5
C.... FIND CLOSEST CLUSTER CENTRE
      WRITE(6,1) I,J,NC(I),DP,DC,DJ
    1 FORMAT(3I5,3F20.6)
      IF(DJ.GT.DC) GO TO 20
      DC=DJ
      NC(I)=J
   20 CONTINUE
C.... FIND ROW FURTHEST FROM ITS CENTRE
      IF(DC.LT.D) GO TO 10
      D=DC
      NC(1)=I
   10 CONTINUE
      RETURN
      END
```

CHAPTER 5

Mixtures

5.1 INTRODUCTION

A statistical model for partitions supposes that each of M observations is selected at random from one of K populations. If the probability density for an observation $\mathbf{x}$ given a parameter value $\boldsymbol{\theta}$ is written $f(\mathbf{x} \mid \boldsymbol{\theta})$, the probability density of the mixture will be $\sum \{1 \leq J \leq K\} P(J) f[\mathbf{x} \mid \boldsymbol{\theta}(J)]$. Here $P(J)$ is the probability that the observation $\mathbf{x}$ belongs to the Jth population, which is determined by parameter $\boldsymbol{\theta}(J)$.

The log likelihood of observations $\mathbf{x}(1), \mathbf{x}(2), \ldots, \mathbf{x}(M)$ is

$$\sum \{1 \leq I \leq M\} \log \left(\sum \{1 \leq J \leq K\} P(J) f[\mathbf{x}(I) \mid \boldsymbol{\theta}(J)] \right),$$

and this may be maximized to estimate the parameters $P(J)$ and $\boldsymbol{\theta}(J)$. The quantity

$$P(I, J) = P(J) f[\mathbf{x}(I) \mid \boldsymbol{\theta}(J)] \left(\sum \{1 \leq J \leq K\} P(J) f[\mathbf{x}(I) \mid \boldsymbol{\theta}(J)] \right)^{-1}$$

is the probability that the Ith observations belongs to the Jth population (or Jth cluster), given its value, $\mathbf{x}(I)$. These conditional probabilities are useful in assigning objects to populations after the parameters have been estimated. They also play an important role in the iterative process to maximize the likelihood with $P(I, J)$ being the weight that the Ith observation has in determining the Jth parameter.

Stationary values of the likelihood occur when the following equations are satisfied:

$$\sum \{1 \leq I \leq M\} P(I, J) \frac{\partial \log f[\mathbf{x}(I) \mid \boldsymbol{\theta}(J)]}{\partial \boldsymbol{\theta}(J)} = 0, \tag{1a}$$

$$\sum (1 \leq I \leq M) \frac{P(I, J)}{M} = P(J). \tag{2a}$$

and

$$P(I, J) = P(J) \frac{f[\mathbf{x}(I) \mid \boldsymbol{\theta}(J)]}{\left(\sum \{1 \leq J \leq K\} P(J) f[\mathbf{x}(I) \mid \boldsymbol{\theta}(J)] \right)}. \tag{3a}$$

Note that eq. $1a$ corresponds to the maximum likelihood estimate of $\boldsymbol{\theta}$ when the Ith observation is observed $P(I, J)$ times. In this way, the $P(I, J)$ are the weights of the Ith observation in the Jth parameter estimate.

There are no guarantees that the above equations determine the maximum likelihood estimate, that a solution exists, or that a solution is unique. These questions are answered differently for different densities $f(\mathbf{x} \mid \boldsymbol{\theta})$. A plausible iterative technique first guesses $P(J)$ and $\boldsymbol{\theta}(J)$ [the $\boldsymbol{\theta}(J)$'s should be unequal], computes $P(I, J)$ from eq.

3a, and then estimates $\theta(J)$, $P(J)$, and $P(I, J)$ in succession, until the procedure converges. This approach is analogous to the K-means approach, in which cluster means are first guessed [corresponding to $P(J)$ and $\theta(J)$] and then objects are assigned to the various clusters [corresponding to estimating $P(I, J)$], after which cluster means are recomputed, objects are reassigned, and so on. The same sort of local optimization problems are to be expected. A solution to the equations may not be unique, and other solutions may have larger likelihoods.

For the multivariate normal $\mathbf{x} \sim N(\boldsymbol{\mu}, \boldsymbol{\Sigma})$, the equations will be

$$\mu(J) = \frac{\sum \{1 \le I \le M\} \, \mathbf{x}(I) \, P(I, J)}{\sum \{1 \le I \le M\} \, P(I, J)}, \tag{1b}$$

$$\boldsymbol{\Sigma}(J) = \frac{\sum (1 \le I \le M) \, [\mathbf{x}(I) - \mu(J)][x(I) - \mu(J)]' \, P(I, J)}{\sum (1 \le I \le M) \, P(I, J)} \tag{1b'}$$

$$P(J) = \sum \{1 \le I \le M\} \frac{P(I, J)}{M}, \tag{2b}$$

and

$$P(I, J) = |\boldsymbol{\Sigma}(J)|^{-1/2} \, P(J) B(I) \exp \{-\tfrac{1}{2}[\mathbf{x}(I) - \mu(J)]'\boldsymbol{\Sigma}^{-1}(J)[\mathbf{x}(I) - \mu(J)]\} \tag{3b}$$

with $B(I)$ determined by $\sum \{1 \le J \le K\} P(I, J) = 1$. This model is difficult to fit if there are many clusters and variables and just a few observations. In any case, there are solutions for which the likelihood is infinite, in which some of the $\boldsymbol{\Sigma}(J)$'s are zero, and these must be forbidden somehow.

In order to reduce the many parameters to be estimated in the covariance matrices, a number of simplifying assumptions are suggested. First, suppose that the $\boldsymbol{\Sigma}(J)$ are all equal. The equations remain as above, except that Eq. 1b' becomes

$$\boldsymbol{\Sigma}(J) = \sum \{1 \le I \le M\}\{1 \le J \le K\} \, [\mathbf{x}(I) - \mu(J)][\mathbf{x}(I) - \mu(J)]' \frac{P(I, J)}{M}.$$

Second, suppose that $\boldsymbol{\Sigma}(J) = \Delta$, where Δ is diagonal. The equations remain the same, except that the off-diagonal terms in the estimate of $\boldsymbol{\Sigma}(J)$ are always set zero.

Finally, if all variables are assumed uncorrelated with equal variance σ^2, the equations become

$$\mu(J) = \frac{\sum \{1 \le I \le M\} \, \mathbf{x}(I) P(I, J)}{\sum \{1 \le I \le M\} \, P(I, J)}, \tag{1c}$$

$$\sigma^2 = \frac{\sum \{1 \le I \le M\}\{1 \le J \le K\} \, [\mathbf{x}(I) - \mu(J)]'[\mathbf{x}(I) - \mu(J)] \, P(I, J)}{MN} \tag{1c'}$$

$$P(J) = \sum \{1 \le I \le M\} \frac{P(I, J)}{M}, \tag{2c}$$

and

$$P(I, J) = P(J) B(I) \exp \left(\frac{-\tfrac{1}{2}[\mathbf{x}(I) - \mu(J)]' \, [\mathbf{x}(I) - \mu(J)]}{\sigma^2} \right), \tag{3c}$$

where $B(I)$ is determined by $\sum \{1 \le J \le K\} P(I, J) = 1$.

The mixture model is connected to K-means models as follows. Return to the general model, where M observations are assumed each to come from one of K populations and where the probability density for the Ith observation from the Lth population is $f[\mathbf{x}(I) \mid \boldsymbol{\theta}(L)]$. Let $E(I, L)$ be a parameter that is unity if I comes from the Lth population and zero otherwise. Then the log likelihood of the observations is

$$\sum \{1 \leq I \leq M\}\{1 \leq L \leq K\} \, E(I, L) \log f[x(\mathbf{J}) \mid \boldsymbol{\theta}(L)].$$

For fixed $\{E(I, L), 1 \leq I \leq M, 1 \leq L \leq K\}$, the maximum log likelihood is

$$LL(E) = \sum \{1 \leq I \leq M\} \log f\{\mathbf{x}(I) \mid \hat{\theta}[L(I)]\},$$

where $\hat{\theta}(L)$ is the value of $\boldsymbol{\theta}(L)$ maximizing

$$\sum \{1 \leq I \leq M, E(I, L) = 1\} \log f[\mathbf{x}(I) \mid \boldsymbol{\theta}(L)],$$

and $L(I)$ is the unique value of L for which $E(I, L) = 1$.

The vector $\{E(I, L), 1 \leq I \leq M, 1 \leq L \leq K\}$ determines a partition of the M observations into K clusters, which is evaluated by the error function $LL(E)$ given above. In the multivariate normal case, with covariance matrix spherical normal within all clusters, the quantity $LL(E)$ is just the sum of squared euclidean distances from the cluster means. Other assumptions on the covariance matrix will give different error functions appropriate for, say, constant covariance matrices within clusters or covariance matrices that are diagonal, etc.

Now suppose that the parameters $E(I, L)$ are random variables. A reasonable assumption is that $\{E(I, L), 1 \leq L \leq K\}$ are independent and identically distributed over the M observations, so that $E[I, L] = 1$ with probability $P(L)$,

$$\sum \{1 \leq L \leq K\} P(L) = 1.$$

Then the log likelihood is

$$\sum \{1 \leq I \leq M\} \log \left(\sum \{1 \leq L \leq K\} P(L) f[\mathbf{x}(I) \mid \boldsymbol{\theta}(L)] \right)$$

and the mixture model is obtained. An important feature of the model is that the conditional distribution of $\{E(I, L), 1 \leq L \leq K\}$, given the observations $\mathbf{x}(I)$, gives probabilities that each observation belongs to each of the K clusters. These probabilities are the quantities $P(I, K)$ which played a pivotal role in the iterative maximum likelihood solution. Unfortunately, the $P(I, K)$ involve the unknown parameters $\boldsymbol{\theta}(L)$, which must be estimated, and the estimated inclusion probabilities are not quite trustworthy as an indication of definition of the various clusters.

The mixture model follows from the straight partition model by assuming a prior probability distribution of partitions. An important outcome of this approach is the posterior distribution of partitions given the data. A fully Bayesian approach would also require prior distributions on $\boldsymbol{\theta}(L)$ and the parameters $P(L)$. Usually $\boldsymbol{\theta}(1)$, $\boldsymbol{\theta}(2), \ldots, \boldsymbol{\theta}(K)$ and $P(1), P(2), \ldots, P(K)$ should be exchangeable, since there is no reason to distinguish between clusters. [Exchangeability might offer a control on wildly different estimates of $\boldsymbol{\theta}(1), \boldsymbol{\theta}(2), \ldots, \boldsymbol{\theta}(K)$ obtained when the data are few.]

There are considerable difficulties in generalizing this approach from partitions to trees.

5.2 NORMAL MIXTURE ALGORITHM

Preliminaries. The data $\{A(I, J), 1 \leq I \leq M, 1 \leq J \leq N\}$ are assumed to be a random sample of size M from a mixture of K multivariate normal distributions in N dimensions. The Lth multivariate normal is determined by its mean $\{U(J, L), 1 \leq J \leq N\}$ and covariance $\{C(J1, J2, L), 1 \leq J1, J2 \leq N\}$. The probability density for the Ith observation from the Lth normal is

$$F(I, L) = (2\pi)^{-N/2} D(L)^{-1/2}$$
$$\times \exp\left(-\tfrac{1}{2}\sum \{1 \leq J1, J2 \leq N\} [A(I, J1) - U(J1, L)]\right.$$
$$\left. \times [A(I, J2) - U(J2, L)]CC(J1, J2, L)\right),$$

where $D(L)$ is the determinant, and $\{CC(J1, J2, L), 1 \leq J1, J2 \leq N\}$ is the inverse of the covariance $\{C(J1, J2, L), 1 \leq J1, J2 \leq N\}$. In the mixture model, each observation is obtained by selecting from the K multivariate normals, drawing the Lth with probability $W(L)$ and then sampling the observation from the normal distribution selected. The probability density of the Ith observation is thus

$$G(I) = \sum \{1 \leq L \leq K\} F(I, L)W(L).$$

The quantity

$$P(I, L) = \frac{F(I, L)}{G(I)}$$

is the probability that the Ith observation was drawn from the Lth normal. These quantities play an important role in estimating W, U, C, and also in interpreting the fit. Essentially the $P(I, L)$ determine to which of the K clusters the Ith observation probably belongs.

The criterion to be maximized is the log likelihood $\sum \{1 \leq I \leq M\} \log G(I)$. The procedure iterates toward a stationary (locally optimal under small changes in W, U, C) likelihood, but there is no guarantee that this local optimum is global. In general, there will be more than one local optimum.

The technique first guesses at the probabilities $\{P(I, L), 1 \leq I \leq M, 1 \leq L \leq K\}$, like guessing an initial partition in the K-means algorithm, then estimates means U and covariance C as weighted averages over all observations with the weights $\{P(I, L), 1 \leq I \leq M, 1 \leq L \leq K\}$; it then estimates the weights W as averages of the P, and finally estimates the probabilities $P(I, L)$ using $P(I, L) = F(I, L)/G(I)$. This cycle is repeated until the values of W, U, C converge. It is not known whether the log likelihood surely increases after each cycle. A reasonable stopping rule is: Stop when the log likelihood increases by less than 0.01.

The very many parameters present in the covariance matrices $\{C(J1, J2, L), 1 \leq J1, J2 \leq N, 1 \leq L \leq K\}$ require a lot of data for their estimation. A rule of thumb is that M should be greater than $\tfrac{1}{2}(N + 1)(N + 2)K$. Even with many observations, the procedure is vulnerable to nonnormality, or linear dependence among the variables. A way to reduce this sensitivity is by assumption on the covariance matrices. There are four plausible stages from very general to very specific:

(i)   Covariance matrices are arbitrary;

(ii)   Covariance matrices in different normals are equal;

(iii)   Covariance matrices are equal and diagonal;

(iv)   All variables have the same variance and are pairwise independent.

STEP 1.   Initialize $P(I, L)$ by setting

$$P(1, 1) = P(2, 1) = \cdots = P[I(1), 1] = 1,$$
$$P[I(1) + 1, 2] = P[I(1) + 2, 2] = \cdots = P[I(2), 2] = 1,$$

.

.

.

$$P[I(K - 1) + 1, K] = \cdots = P[I(K), K] = 1,$$

where $I(J) = JM/K$. The remainder of the $P(I, L)$ are set equal to zero. This initialization is equivalent to partitioning the observations into equal size groups. A different initialization is obtained by reordering the observations.

STEP 2.   Update the weights W by

$$W(L) = \sum \{1 \leq I \leq M\} P(I, L).$$

The quantity $W(L)$ is the estimated number of observations from the Lth distribution.

STEP 3.   Update the means U by

$$U(J, L) = \sum \{1 \leq I \leq M\} A(I, J) \frac{P(I, L)}{W(L)}.$$

STEP 4.   Update the covariances C by

$$C(J1, J2, L) = \sum \{1 \leq I \leq M\} [A(I, J1) - U(J1, L)][A(I, J2) - U(J2, L)] \frac{P(I, L)}{W(L)}.$$

Under option **(i)**, go to Step 5.
Under options **(ii)–(iv)**, replace $C(J1, J2, L)$ for each L $(1 \leq L \leq K)$ by

$$\sum \{1 \leq L \leq K\} C(J1, J2, L) \frac{W(L)}{M}.$$

Under option **(iii)**, replace $C(J1, J2, L)$, for each L and for each $J1 \neq J2$ by zero.
Under option **(iv)**, replace $C(J, J, L)$ for each J, L by

$$\sum \{1 \leq J \leq N\} \frac{C(J, J, 1)}{N}.$$

STEP 5.   Compute the determinants and inverses of the covariance matrices $\{C(J1, J2, L), 1 \leq J1, J2 \leq N\}$ for each L and then the probability densities $F(I, L)$, the average densities $G(I)$, and the log likelihood $\sum \{1 \leq I \leq M\} \log G(I)$. Update $P(I, L)$ by $P(I, L) = F(I, L)/G(I)$. If the log likelihood does not exceed its previous value by 0.01, stop. Otherwise go to Step 2.

NOTE.   For covariance structure options **(iii)** and **(iv)**, independence between variables is assumed, and considerable simplification in the computations in Steps 4 and 5 is possible.

5.3 NORMAL MIXTURE ALGORITHM APPLIED
TO NEW HAVEN SCHOOL SCORES

Table 5.1 contains the scores in fourth grade reading, fourth grade arithmetic, sixth grade reading, and sixth grade arithmetic in 25 New Haven primary schools. A reasonable standardization would divide the fourth grade scores by the national average, four, and the sixth grade scores by the national average, six. Since the general mixture model is invariant under linear transformation, this scaling is not necessary.

For two clusters, the number of parameters to be estimated in the most general mixture model is $\frac{1}{2}(N + 1)(N + 2)K - 1 = 29$, which exceeds the number of observations. The algorithm produced infinite likelihoods for several starting configurations

Table 5.1   Achievement Test Scores, New Haven Schools

The measurements are in years and months, in terms of national averages. There are ten months in a school year. At the beginning of fourth grade, the national average score is 4.0.

SCHOOL	FOURTH GRADE		SIXTH GRADE	
	READING	ARITHMETIC	READING	ARITHMETIC
Baldwin	2.7	3.2	4.5	4.8
Barnard	3.9	3.8	5.9	6.2
Beecher	4.8	4.1	6.8	5.5
Brennan	3.1	3.5	4.3	4.6
Clinton	3.4	3.7	5.1	5.6
Conte	3.1	3.4	4.1	4.7
Davis	4.6	4.4	6.6	6.1
Day	3.1	3.3	4.0	4.9
Dwight	3.8	3.7	4.7	4.9
Edgewood	5.2	4.9	8.2	6.9
Edwards	3.9	3.8	5.2	5.4
Hale	4.1	4.0	5.6	5.6
Hooker	5.7	5.1	7.0	6.3
Ivy	3.0	3.2	4.5	5.0
Kimberly	2.9	3.3	4.5	5.1
Lincoln Bassett	3.4	3.3	4.4	5.0
Lovell	4.0	4.2	5.2	5.4
Prince	3.0	3.0	4.6	5.0
Ross	4.0	4.1	5.9	5.8
Scranton	3.0	3.2	4.4	5.1
Sherman	3.6	3.6	5.3	5.4
Truman	3.1	3.2	4.6	5.0
West Hills	3.2	3.3	5.4	5.3
Winchester	3.0	3.4	4.2	4.7
Woodward	3.8	4.0	6.9	6.7

by moving toward a cluster containing only four objects. It was decided to reduce the number of parameters to be estimated by constraining the covariance matrices within each cluster to be equal.

STEP 1.   Initially,
$$P(1, 1) = P(2, 1) = \cdots = P(12, 1) = 1,$$
$$P(13, 2) = P(14, 2) = \cdots = P(25, 2) = 1,$$
and all other $P(I, J)$ are zero.

STEP 2.   Compute $W(1) = 12$ and $W(2) = 13$. These are the number of observations initially in the two clusters.

STEP 3.   The $U(J, 1)$ are the means of the four variables over the first twelve observations: $U(1, 1) = 3.81$, $U(2, 1) = 3.82$, $U(3, 1) = 5.42$, and $U(4, 1) = 5.43$. Similarly, the $U(J, 2)$ are the means of the four variables over the last thirteen observations.

STEP 4.   The covariance matrix $C(J1, J2, 1)$ $(1 \leq J1, J2 \leq 4)$ is computed on the first twelve points, and the covariance matrix $C(J1, J2, 2)$ is computed on the next thirteen points. These covariances are averaged with weights 12 and 13, and the weighted covariance matrix is used in later steps.

STEP 5.   The determinant and inverse of the within-cluster covariance matrix is used to compute the probability density of each observation under each of the two multivariate normal distributions. Return to Step 2.

The values of various parameters at the first and 11th steps are given in Tables 5.2 and 5.3. The parameters take almost the same values at the 21st step as at the 11th,

Table 5.2   Normal Mixture Model with Equal Covariances, Applied to New Haven School Data, First Cycle

INITIAL PROBABILITIES OF BELONGING TO FIRST NORMAL

Baldwin	1.000	Conte	1.000	Edwards	1.000	Lincoln	0.000	Sherman	0.000
Barnard	1.000	Davis	1.000	Hale	1.000	Lovell	0.000	Truman	0.000
Beecher	1.000	Day	1.000	Hooker	1.000	Prince	0.000	West Hills	0.000
Brennan	1.000	Dwight	1.000	Ivy	0.000	Ross	0.000	Winchester	0.000
Clinton	1.000	Edgewood	1.000	Kimberly	0.000	Scranton	0.000	Woodward	0.000

WEIGHTS = EXPECTED NUMBER OF OBSERVATIONS IN EACH NORMAL
        = (12, 13)

MEANS	REA4	ARI4	REA6	ARI6
FIRST NORMAL	3.81	3.82	5.42	5.43
SECOND NORMAL	3.52	3.61	5.15	5.37

VARIANCES AND CORRELATIONS

	REA4	ARI4	REA6	ARI6
REA4	.544	.955	.875	.774
ARI4	.955	.270	.865	.806
REA6	.875	.865	1.137	.931
ARI6	.774	.806	.931	.377

LOG LIKELIHOOD = 60.032

Table 5.3   Normal Mixture Model with Equal Covariances, Applied to New Haven School Data, 11th Cycle

Baldwin	.942	Conte	.924	Edwards	.222	Lincoln	.000	Sherman	.071
Barnard	.000	Davis	.988	Hale	.763	Lovell	1.000	Truman	.004
Beecher	1.000	Day	.004	Hooker	1.000	Prince	.000	West Hills	.031
Brennan	1.000	Dwight	.970	Ivy	.006	Ross	.971	Winchester	.989
Clinton	.040	Edgewood	1.000	Kimberly	.044	Scranton	.000	Woodward	.000

WEIGHTS = EXPECTED NUMBER OF OBSERVATIONS IN EACH NORMAL

= (12, 13)

MEANS	REA4	ARI4	REA6	ARI6
FIRST NORMAL	4.01	4.00	5.59	5.44
SECOND NORMAL	3.33	3.44	4.99	5.36

VARIANCES AND CORRELATIONS

	REA4	ARI4	REA6	ARI6
REA4	.454	.948	.874	.826
ARI4	.948	.205	.878	.897
REA6	.874	.878	1.064	.950
ARI6	.826	.897	.950	.376

LOG LIKELIHOOD = 62.210

so these values are a particular set of maximum likelihood solutions. It will be noted that the belonging probabilities in Table 5.3 are all nearly unity or zero, so that, except for Edwards and Hale, membership in the clusters is clear cut. This is not a sign of stability or statistical significance of the clusters.

The iterative algorithm converges rather slowly and needs to be speeded up. The acceleration technique, after five iterations, sets each probability equal to unity or zero, whichever the probability was closer to. This recognizes the fact that final belonging probabilities tend to be close to unity or zero.

The two clusters of schools are well distinguished by the difference in arithmetic performance between fourth and sixth grades. Schools in the first cluster advance 1.44 years, whereas schools in the second cluster advance 1.92 years, with a within-cluster standard deviation of 0.3 years.

5.4 THINGS TO DO

5.4.1 Running the Mixture Algorithm

The mixture algorithm is a fancy version of the *K*-means algorithm, with the extra twist that each case is assigned only a probability of belonging to the various clusters. Therefore, the results of the mixture algorithm should be compared with the results of a *K*-means algorithm. Usually the probabilities that each case belongs to various clusters are very close to unity or zero, but this should not be taken as an indication of sharply defined clusters. Setting about half the data missing will disclose different clusters with equally well-defined "belonging" probabilities.

Table 5.4 Civil War Battles in Chronological Order

[From T. L. Livermore (1957). *Numbers and Losses in the Civil War*, Indiana University Press, Bloomington.] The variables are union forces, union shot, confederate forces, and confederate shot.

Bull Run	28452	1492	33232	1969
Wilson's Creek	5400	944	11600	1157
Ft. Donelson	27000	2608	21000	2000
Pea Ridge	11250	1183	14000	600
Shiloh	62682	10162	40335	9735
Williamsburg	40768	1866	31823	1570
Fair Oaks	41797	4384	41816	5729
Mechanicsville	15631	256	16356	1484
Gaines' Mill	34214	4001	57018	8751
Malvern Hill	83345	4969	86748	8602
Seven Days	91169	9796	95481	19739
Cedar Mt.	8030	1759	16868	1338
Manassas	75696	10096	48527	9108
Richmond, Ky.	6500	1050	6850	450
South Mt.	28480	1728	18714	1885
Antietam	75316	11657	51844	11724
Corinth	21147	2196	22000	2470
Perryville	36940	3696	16000	3145
Fredericksburg	100007	10884	72497	4656
Chickasaw Bayou	30720	1213	13792	197
Stone's River	41400	9220	34732	9239
Arkansas Post	28944	1032	4564	109
Chancellorsville	97382	11116	57352	10746
Champions Hill	29373	2254	20000	2181
Pt. Hudson I	13000	1838	4192	235
Pt. Hudson II	6000	1604	3487	47
Gettysburg	83289	17684	75054	22638
Ft. Wagner	5264	1126	1785	169
Chicamanga	58222	11413	66326	16986
Chatanooga	56359	5475	46165	2521
Mine Run	69643	1272	44426	680
Olustee	5115	1355	5200	934
Pleasant Hill	12647	994	14300	1000
Drewry's Bluff	15800	2770	18025	2860
Atlanta I	110123	10528	66089	9187
Weldon RR	20289	1303	14787	1200
Kenesaw Mt.	16225	1999	17733	270
Tupelo	14000	636	6600	1326
Peach Tree Ck.	20139	1600	18832	2500
Atlanta II	30477	1989	36934	7000
Atlanta III	13226	559	18450	4100
Jonesborough	14170	179	23811	1725
Winchester	37711	4680	17103	2103
Cedar Ck.	30829	4074	18410	1860
Franklin	27939	1222	26897	5550
Bentonville	16127	933	16895	1508

[Note: Confederate casualties not known in Grant's Richmond campaign.]

Table 5.5 Planets and Moons

[From P. Moore (1970) *The Atlas of the Universe* Rand McNally, New York.] Distance: in thousands of miles, of satellite from planet; diameter: in miles, of satellite; period: in days, of satellite about planet.

PLANET	DIST	DIAM	PERIOD	PLANET	DIST	DIAM	PERIOD
EARTH	239	2160	655	SATURN	116	300	23
MARS	5.8	10	7.7		148	400	33
	14.6	10	30		183	600	45
JUPITER	112	100	12.0		235	600	66
	262	2020	42		327	810	108
	417	1790	85		759	2980	383
	665	3120	172		920	100	511
	1171	2770	401		2213	500	1904
	7133	50	6014		8053	100	13211
	7295	20	6232	URANUS	77	200	34
	7369	10	6325		119	500	60
	13200	10	15146		166	300	100
	14000	10	16620		272	600	209
	14600	10	17734		365	500	323
	14700	10	18792	NEPTUNE	220	2300	141
					3461	200	8626

SUN	DIST	DIAM	PERIOD	Mass (relative to earth)
MERCURY	35950	3100	2112	.054
VENUS	67180	7700	5393	.81
EARTH	92900	7927	8766	1.00
MARS	141500	4190	16488	.11
JUPITER	483200	88640	103911	316.94
SATURN	886000	74100	258420	95.20
URANUS	1782000	32000	735840	14.70
NEPTUNE	2793000	31000	1445400	17.20
PLUTO	3670000	?	2172480	?

Various degrees of vagueness are possible in specifying the error distribution within clusters, with the most general being arbitrary covariance matrices within the various clusters. A suitable trial data set is Table 5.4, the numbers and losses in Civil War battles, which should be logged before analysis. Also, follow Kepler in analysis of satellite data (Table 5.5).

5.4.2 Singularities

Let the data be a single variable $\{X(I), 1 \leq I \leq M\}$. Let the model for a single observation be

$$f[x \mid \mu(1), \mu(2), \sigma(1), \sigma(2), p]$$

$$= p \frac{\exp\{-\frac{1}{2}[x - \mu(1)]^2/\sigma(1)^2\}}{\sigma(1)\sqrt{2\pi}} + (1-p) \frac{\exp\{-\frac{1}{2}[x - \mu(2)]^2/\sigma(2)^2\}}{\sigma(2)\sqrt{2\pi}}.$$

Show that the log likelihood is infinite at $\mu(1) = X(1)$, $\sigma(1) = 0$. More generally, for N-dimensional data with arbitrary covariances within clusters, show that there exists an infinite log likelihood whenever there is a cluster containing less than $N + 1$ cases.

5.4.3* A More Probabilistic Approach

Consider again the general two-cluster normal model for a single variable. Maximum likelihood is not available since the likelihood is infinite at $\sigma(1) = 0$, $\mu_1 = x(1)$. Suppose that it is assumed that $\mu(1)$ and $\mu(2)$ are randomly sampled from $N(\mu, \sigma^2)$ and that $\sigma(1)^2$ and $\sigma(2)^2$ are random sampled from ds^2/χ_d^2. [These assumptions will rule out $\sigma(1) = 0$ or $\sigma(2) = 0$.]

The posterior density for $\mu(1)$, $\mu(2)$, $\sigma(1)$, and $\sigma(2)$, given the observations X and the parameters μ, σ^2, s^2, and d, is maximized when

$$\mu(J) = \frac{P(I, J)X(I)/\sigma^2(J) + \mu/\sigma^2}{M/\sigma^2(J) + 1/\sigma^2}, \tag{1}$$

$$\sigma^2(J) = \frac{P(I, J)[X(I) - \mu(J)]^2 + ds^2}{M + d}, \tag{2}$$

$$P(I, J) = C(I)P(J)\exp\left(-\frac{1}{2}\frac{[X(I) - \mu(J)]^2}{\sigma(J)^2}\right)[\sigma(J)]^{-1} \tag{3}$$

and

$$P(I, 1) + P(I, 2) = 1,$$

$$P(J) = \sum\{1 \leq I \leq M\}\frac{P(I, J)}{M}. \tag{4}$$

This still leaves the parameters μ, σ^2, s^2, and d, but these may be determined, to maximize the posterior density, by

$$\mu = \tfrac{1}{2}[\mu(1) + \mu(2)], \tag{5}$$

$$\sigma^2 = \tfrac{1}{4}[\mu(1) - \mu(2)]^2, \tag{6}$$

$$\frac{1}{s^2} = \frac{1}{2}\left(\frac{1}{\sigma(1)^2} + \frac{1}{\sigma(2)^2}\right), \tag{7}$$

and

$$\frac{2\Gamma'(d/2)}{\Gamma(d/2)} = \log\left(\frac{ds^2}{2\sigma(1)^2}\right) + \log\left(\frac{ds^2}{2\sigma(2)^2}\right). \tag{8}$$

Alternatively, these parameters might be fixed in advance from considerations outside the data. For large M, the effects of the parameters will be minor except they prevent the infinite likelihood at $\sigma_1 = 0$ or at $\sigma_2 = 0$.

This technique generalizes in an obvious way to larger numbers of clusters and with some complications to a larger number of variables.

5.4.4 Trees

Consider a single variable $\{X(I), 1 \leq I \leq M\}$. For K clusters, one model for a single case is

$$f(x) = \sum p(J)f[x \mid \mu(J)],$$

where

$$f[x \mid \mu(J)] = \frac{\exp\{-\frac{1}{2}[x - \mu(J)]^2/\sigma_1^2\}}{\sigma_1\sqrt{2\pi}}.$$

The $\mu(J)$'s here are cluster means, which must be grouped in a tree approach.
Therefore treat the $\mu(J)$ as observations from the mixture

$$g(x) = \sum r(K)f[x \mid \theta(K)],$$

where

$$f[x \mid \theta(K)] = \frac{\exp[-\frac{1}{2}[x - \theta(K)]^2/\sigma_2^2]}{\sigma_2\sqrt{2\pi}}.$$

Finally let the $\theta(K)$ be distributed as $N(\varphi, \sigma_3^2)$, where this last "mixture" of just one distribution corresponds to the root of the tree. This model corresponds to a tree with four levels: the cases, the parameters μ, the parameters θ, and the parameter φ.

Theory for fitting the parameters awaits development.

5.4.5 Initialization

There may be many different solutions of the maximum likelihood equations. Using the iterative technique, the final solution will depend on the initial choice of "belonging probabilities" $P(I, L)$. One method of making the initial choice is to run the K-means algorithm first; the initial "belonging probabilities" will then be a zero–one array determined by the membership of observations in clusters. If the data consist of observations on a single variable, the Fisher exact algorithm is suggested.

5.4.6 Coincidence with K Means

In one dimension, let $X(1), X(2), \ldots, X(J)$ and $X(J + 1), \ldots, X(M)$ denote two clusters of observations. Assume that $X(1) \leq X(2) \leq \cdots \leq X(M)$. Show that, if $[X(J) - X(1)]/[X(J + 1) - X(J)]$ and $[X(M) - X(J + 1)]/[X(J + 1) - X(J)]$ are both sufficiently small, then K means and the mixture model with equal variances within clusters coincide in finding the two clusters $\{X(1), \ldots, X(J)\}$ and $\{X(J + 1), \ldots, X(M)\}$.

PROGRAMS

MIX   computes maximum likelihood fit to mixture model under various constraints on within-cluster covariances. Needs INVERT, COVOUT, MOM.

COVOUT prints output from mixture model after every five iterations.

MIXIND mixture model with variables having constant variances and zero covariances within clusters.

MIXOUT prints output for MIXIND.

```
      SUBROUTINE MIX(P,W,M,N,K,C,U,X,ITER,NCOV,Q)
C...................................................................20 MAY 1973
C.... ITERATES TO MAXIMUM LIKELIHOOD ESTIMATES IN MIXTURE MODEL, ARBITRARY MEANS
C     AND COVARIANCES WITHIN CLUSTERS.
C.... NEEDS INVERT,MOM,COVOUT
C.... M = NUMBER OF ROWS
C.... N = NUMBER OF COLUMNS
C.... K = NUMBER OF CLUSTERS
C.... C = N BY N BY K COVARIANCE MATRICES, ONE FOR EACH CLUSTER
C.... U = N BY K ARRAY OF CLUSTER MEANS
C.... X = M BY N BORDERED ARRAY, DATA MATRIX
C.... ITER = NUMBER OF ITERATIONS , TRY 10
C.... P = M BY K BORDERED ARRAY, P(I,J)= PROBABILITY THAT ITH CASE BELONGS TO
C           THE JTH CLUSTER.
C.... Q = M BY K SCRATCH ARRAY
C.... W = K BY 1 ARRAY OF MIXTURE PROBABILITIES
C.... NCOV = INTEGER DETERMINING STRUCTURE OF COVARIANCE MATRIX
C           NCOV = 1    GENERAL COVARIANCES
C           NCOV = 2    COVARIANCES EQUAL BETWEEN CLUSTERS
C           NCOV = 3    COVARIANCES EQUAL AND DIAGONAL
C           NCOV = 4    COVARIANCES SPHERICAL
C...................................................................................
      DIMENSION P(M,K),W(K),C(N,N,K),U(N,K),X(M,N)
      DIMENSION Q(M,K)
C.... INITIALIZE P
      DO 20 I=2,M
      DO 19 J=1,K
   19 P(I,J)=0
      J=((I-1)*K)/M+1
   20 P(I,J)=1.
C.... UPDATE MEANS AND COVARIANCES
      DO 70 IT=1,ITER
      DO 30 J=1,K
   30 CALL MOM(U(1,J),C(1,1,J),P(1,J),X,M,N)
C.... UPDATE WEIGHTS
      WW=0
      DO 42 J=1,K
      W(J)=0.
      DO 41 I=2,M
   41 W(J)=W(J)+P(I,J)
   42 WW=WW+W(J)
      DO 43 J=1,K
   43 IF(WW.NE.0) W(J)=W(J)/WW
C.... ADJUST FOR COVARIANCE STRUCTURE
      IF(NCOV.EQ.1) GO TO 39
      DO 32 I=2,N
      DO 32 II=2,N
      C(I,II,1)=W(1)*C(I,II,1)
      DO 31 J=2,K
   31 C(I,II,1)=C(I,II,1)+C(I,II,J)*W(J)
      IF(NCOV.GE.3.AND.I.NE.II) C(I,II,1)=0
      DO 33 J=1,K
   33 C(I,II,J)=C(I,II,1)
   32 CONTINUE
      CC=0
      DO 35 I=2,N
   35 CC=CC+C(I,I,1)
      CC=CC/(M-1)
      DO 36 I=2,N
      DO 36 J=1,K
   36 IF(NCOV.EQ.4) C(I,I,J)=CC
   39 CONTINUE
      II=IT-1
      IF((II/5)*5.EQ.II) CALL COVOUT(X,M,N,U,C,P,W,K)
C.... UPDATE BELONGING PROBABILITIES
      DO 50 J=1,K
      DET=1
      IF(NCOV.GT.2) C(1,1,J)=1./C(1,1,J)
      IF(NCOV.LE.2) CALL INVERT(C(1,1 ,J),DET,N)
      IF(DET.EQ.0) RETURN
      DET=DET**0.5
      C(1,1,J)=DET
      DO 50 I=2,M
      S=0.
      DO 60 L=2,N
      DO 60 LL=2,N
   60 S=S+C(L,LL,J)*(X(I,L)-U(L,J))*(X(I,LL)-U(LL,J))
      IF(S.GT.100) S=100
```

126

```
      Q(I,J)=P(I,J)
   50 P(I,J)=EXP(-S/2.)*W(J)/DET
C.... COMPUTES LOG LIKELIHOOD
      XLL=0
      DO 61 I=2,M
      S=0.
      DO 63 J=1,K
   63 S=S+P(I,J)
      IF(S.EQ.0.) S=10.**(-10)
      XLL=XLL+ALOG(S)
      DO 61 J=1,K
   61 P(I,J)=P(I,J)/S
    . WRITE(6,1) IT,XLL
    1 FORMAT(13H ITERATION = ,I5,11HLOG LIKE = ,F10.6)
      DO 62 I=2,M
      DO 62 J=1,K
      XIT=ITER
      A=1.+.7*IT/XIT
      P(I,J)=A*P(I,J)-(A-1.)*Q(I,J)
      IF(IT.EQ.5.AND.P(I,J).GT.0.5) P(I,J)=1.
      IF(IT.EQ.5.AND.P(I,J).LE.0.5) P(I,J)=0.
      IF(P(I,J).GT.1) P(I,J)=1
      IF(P(I,J).LT.0) P(I,J)=0
   62 CONTINUE
   70 CONTINUE
      RETURN
      END




      SUBROUTINE COVOUT(A,M,N,XA,C,XMP,XP,K)
C.......................................................................20 MAY 1973
C.... PRINTS OUT PARAMETERS OF MIXTURE MODEL
C.... SEE SUBROUTINE MIXIND FOR ARGUMENT LIST
C.... C = N BY N BY K ARRAY, COVARIANCES WITHIN EACH CLUSTER
C.......................................................................
      DIMENSION A(M,N),XA(N,K),C(N,N,K),XMP(M,K),XP(K)
      DIMENSION DD(100)
      WRITE(6,9) A(1,1),K
    9 FORMAT(18H1MIXTURE MODEL FOR,A5,5H WITH,I5,9H CLUSTERS )
      WRITE(6,2)(KK,KK=1,K)
    2 FORMAT(8H CLUSTER,2X,10(5X,I4,3X))
C.... PRINT CLUSTER PROBABILITIES
      WRITE(6,3)(XP(KK),KK=1,K)
    3 FORMAT(22H MIXTURE PROBABILITIES/(10X,10F12.6))
C.... PRINT MEANS
      WRITE(6,6)
    6 FORMAT(14H CLUSTER MEANS )
      DO 10 J=2,N
   10 WRITE(6,4) A(1,J),(XA(J,KK),KK=1,K)
    4 FORMAT(A7,3X,10F12.4)
      WRITE(6,12)(C(1,1,J),J=1,K)
   12 FORMAT(13H DETERMINANTS/(10X,10G12.4))
      WRITE(6,1)
    1 FORMAT(42H WITHIN CLUSTER VARIANCES AND CORRELATIONS   )
      DO 50 I=2,N
      DO 50 J=I,N
      DO 51 KK=1,K
      Z=C(I,I,KK)*C(J,J,KK)
      DD(KK)=C(I,J,KK)
      IF(I.EQ.J) Z=0
   51 IF(Z.NE.0) DD(KK)=C(I,J,KK)*Z**(-0.5)
      IF(I.EQ.J) WRITE(6,11)
   11 FORMAT(1H )
      WRITE(6,8) A(1,I),A(1,J),(DD(KK),KK=1,K)
    8 FORMAT(2A5,10F12.4)
   50 CONTINUE
C.... PRINT PROBABILITIES
      WRITE(6,7)
    7 FORMAT(24H BELONGING PROBABILITIES )
      DO 20 I=2,M
   20 WRITE(6,5) A(I,1),(XMP(I,KK),KK=1,K)
    5 FORMAT(A7,3X,10F12.6)
      RETURN
      END
```

```
      SUBROUTINE MIXIND(A,M,N,XA,XV,XMP,XP,K)
C.................................................................20 MAY 1973
C.... FITS MIXTURE MODEL, K MULTIVARIATE NORMALS, EACH VARIABLE HAVING VARIANCE
C.... CONSTANT OVER DIFFERENT CLUSTERS.
C.... M = NUMBER OF ROWS
C.... N = NUMBER OF COLUMNS
C.... A = M BY N BORDERED ARRAY
C.... XA = N BY K ARRAY OF CLUSTER MEANS
C.... XV = N BY 1 ARRAY OF VARIANCES WITHIN CLUSTERS
C.... XMP = M BY K ARRAY, XMP(I,K) = PROBABILITY THAT CLUSTER K CONTAINS CASE I.
C.... XP = K BY 1 ARRAY, MIXTURE PROBABILITIES
C.... K = NUMBER OF CLUSTERS
C.................................................................
      DIMENSION A(M,N),XA(N,K),XV(N),XP(K),XMP(M,K)
C.... INITIALIZE PROBABILITIES
      XM=99999.
      DO 30 KK=1,K
      KB=KK-1
C.... FIND FURTHEST CASE FROM PRESENT MEANS
      DM=0
      IM=2
      IF(KB.EQ.0) GO TO 35
      DO 31 I=2,M
      DI=10.**10
      DO 32 KL=1,KB
      DD=0
      XC=0
      DO 33 J=2,N
      IF(A(I,J).EQ.XM) GO TO 33
      XC=XC+1
      DD=DD+(A(I,J)-XA(J,KL))**2 /XV(J)
      IF(DD.GT.DI*(N-1)) GO TO 32
   33 CONTINUE
      IF(XC.EQ.0) GO TO 31
      DD=DD/XC
   32 IF(DD.LT.DI) DI=DD
      IF(DI.LT.DM) GO TO 31
      DM=DI
      IM=I
   31 CONTINUE
   35 CONTINUE
C.... BEGIN A NEW CLUSTER WITH THIS CASE
      DO 40 J=2,N
   40 XA(J,KK)=A(IM,J)
C.... ITERATE A FEW TIMES
      XP(KK)=EXP(0.5*N)
      ITER=5
      DO 50 IT=1,ITER
C.... UPDATE PROBABILITIES OF BELONGING
      DO 51 I=2,M
      PP=0
      DO 52 KL=1,KK
      DD=0
      DO 53 J=2,N
      IF(A(I,J).EQ.XM) GO TO 53
      IF(KB.EQ.0) GO TO 53
      DD=DD+(A(I,J)-XA(J,KL))**2/(XV(J)*2.)
   53 CONTINUE
      IF(DD.GT.100) DD=100
      XMP(I,KL)=XP(KL)*EXP(-DD)
   52 PP=PP+XMP(I,KL)
      IF(PP.EQ.0) GO TO 51
      PN=0
      DO 54 KL=1,KK
      XMP(I,KL)=XMP(I,KL)/PP
      TH=.0001
      IF(XMP(I,KL).LT.TH) XMP(I,KL)=0
      PN =PN+XMP(I,KL)
   54 CONTINUE
      DO 91 KL=1,KK
   91 XMP(I,KL)=XMP(I,KL)/PN
   51 CONTINUE
C.... UPDATE MIXTURE PROBABILITIES
      DO 60 KL=1,KK
      XP(KL)=0
      DO 60 I=2,M
   60 XP(KL)=XP(KL)+XMP(I,KL)/(M-1)
C.... UPDATE CLUSTER ESTIMATES, EACH ONE A WEIGHTED MEAN
```

128

```
      DO 70 KL=1,KK
      DO 70 J=2,N
      XA(J,KL)=0
      DO 71 I=2,M
   71 XA(J,KL)=XA(J,KL)+A(I,J)*XMP(I,KL)
   70 IF(XP(KL).NE.0) XA(J,KL)=XA(J,KL)/(XP(KL)*(M-1))
      DO 81 J=2,N
      XV(J)=0
      DO 80 I=2,M
      DO 80 KL=1,KK
   80 XV(J)=XV(J)+(A(I,J)-XA(J,KL))**2*XMP(I,KL)
   81 XV(J)=XV(J)/(M-1)
   50 CONTINUE
      CALL MIXOUT(A,M,N,XA,XV,XMP,XP,KK)
   30 CONTINUE
      RETURN
      END




      SUBROUTINE MIXOUT(A,M,N,XA,XS,XMP,XP,K)
C.....................................................................20 MAY 1973
C.... PRINTS OUT PARAMETERS OF MIXTURE MODEL
C.... SEE SUBROUTINE MIXIND FOR ARGUMENT LIST
C....................................................................
      DIMENSION A(M,N),XA(N,K),XS(N),XMP(M,K),XP(K)
      WRITE(6,9) A(1,1),K
    9 FORMAT(18H1MIXTURE MODEL FOR,A5,5H WITH,I5,9H CLUSTERS )
C.... PRINT VARIANCES
      WRITE(6,1)(XS(J),A(1,J),J=2,N)
    1 FORMAT(25H WITHIN CLUSTER VARIANCES/5(F15.6,1H(,A4,1H)))
C.... PRINT CLUSTER PROBABILITIES
      WRITE(6,2)(KK,KK=1,K)
    2 FORMAT(10X,8H CLUSTER,9(I3,1X,8H CLUSTER),I3)
      WRITE(6,3)(XP(KK),KK=1,K)
    3 FORMAT(22H MIXTURE PROBABILITIES/(10X,10F12.6))
C.... PRINT MEANS
      WRITE(6,6)
    6 FORMAT(14H CLUSTER MEANS )
      DO 10 J=2,N
   10 WRITE(6,4) A(1,J),(XA(J,KK),KK=1,K)
    4 FORMAT(A7,3X,10F12.4)
C.... PRINT PROBABILITIES
      WRITE(6,7)
    7 FORMAT(24H BELONGING PROBABILITIES )
      DO 20 I=2,M
   20 WRITE(6,5) A(I,1),(XMP(I,KK),KK=1,K)
    5 FORMAT(A7,3X,10F12.6)
      RETURN
      END
```

CHAPTER 6

Partition by Exact Optimization

6.1 INTRODUCTION

A partition of a set of objects is a family of subsets such that each object lies in exactly one member of the family. If $R(M, K)$ is the number of partitions of M objects into K sets, it can be seen by induction on M that

$$R(M, K) = KR(M - 1, K) + R(M - 1, K - 1).$$

Also $R(M, 1) = R(M, M) = 1$. As a result, the number of partitions increases approximately as K^M; for example, $R(10, 3) = 9330$.

Optimization techniques in clustering associate an error function $e[P(M, K)]$ with each partition $P(M, K)$ of M objects into K clusters and seek that partition which minimizes $e[P(M, K)]$. Because of the very large number of possible partitions it is usually impractical to do a complete search, so it is necessary to be satisfied with a "good" partition rather than a best one. Sometimes prior constraints or mathematical consequences of the error function e reduce the number of possible partitions, and a complete search to find the exact optimum is possible.

Such a case occurs when the data points are ordered, for example, by time, as in Table 6.1, which reports the winning times in the 100-m run in the Olympics from 1896 to 1964. The data points are ordered by the year of the Olympics. It is sensible to require each cluster to correspond to an interval of years. The partition will reveal whether or not there were periods of years when the winning times remained fairly constant. The number of possible partitions into K clusters is now $O(M^K)$ rather than $O(K^M)$, and even further reduction is possible when an error criterion is used that is additive over clusters.

6.2 FISHER ALGORITHM

Preliminaries. This algorithm is due to Fisher (1958). Objects are labeled $1, 2, \ldots, M$, and clusters are constrained to consist of intervals of objects $(I, I + 1, I + 2, \ldots, J - 1, J)$. There are only $\frac{1}{2}M(M + 1)$ possible clusters. There is a diameter $D(I, J)$ associated with the cluster $(I, I + 1, \ldots, J)$ such that the error of a partition $P(M, K)$ into K clusters $(I_1 = 1, I_1 + 1, \ldots, I_2 - 1), (I_2, I_2 + 1, \ldots, I_3 - 1), \ldots, (I_{K-1}, \ldots, I_K - 1), (I_K, I_K + 1, \ldots, M)$ is

$$e[P(M, K)] = \sum \{1 \leq J \leq K\} D(I_J, I_{J+1} - 1).$$

The error of a partition is thus the sum of cluster diameters over the clusters it contains.

Table 6.1   Olympic Track 1896–1964

From *The World Almanac* (1966), New York World-Telegram, New York.

In tenths of seconds (- denotes missing)

	100M	200M	400M	800M	1500M	5000M	10000M
1896	120	-	542	1310	2732	-	-
1900	108	222	494	1214	2460	-	-
1904	110	216	492	1160	2454	-	-
1906	112	-	532	1212	2520	-	-
1908	108	224	500	1128	2434	-	-
1912	108	217	482	1119	2368	8766	18808
1920	108	220	496	1134	2418	8956	19058
1924	106	216	476	1124	2336	8712	18232
1928	108	218	478	1118	2332	8780	18188
1932	103	212	462	1098	2312	8700	18114
1936	103	207	465	1129	2278	8622	18154
1948	103	211	462	1092	2298	8576	17996
1952	104	207	459	1092	2252	8460	17570
1956	105	206	467	1077	2212	8196	17256
1960	102	205	449	1063	2156	8234	17122
1964	100	203	451	1051	2181	8288	17044

The spring of this algorithm is the relation between optimum partitions into K clusters and optimum partitions into $K - 1$ clusters. Let $\mathbf{P}(I, L)$ denote the optimum partition of objects $1, 2, \ldots, I$ into L clusters for $I \leq M$, $L \leq K$. Suppose that $\mathbf{P}(M, K) = (I_1, \ldots, I_2 - 1), (I_2, \ldots, I_3 - 1), \ldots, (I_K, \ldots, M)$. Then necessarily $\mathbf{P}(I_K - 1, K - 1) = (I_1, \ldots, I_2 - 1)(I_2, \ldots, I_3 - 1) \cdots (I_{K-1}, \ldots, I_K - 1)$. Since error is additive, if this were not true, $e[\mathbf{P}(M, K)]$ could be reduced by varying

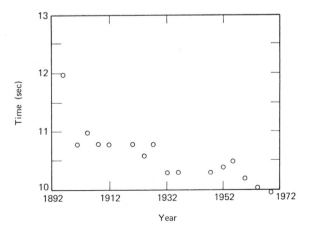

Figure 6.1   Times in the 100-m run, Olympic games.

$I_2, I_3, \ldots, I_{K-1}$. Using this relationship, the algorithm proceeds by successively computing optimal partitions into 2, 3, 4, . . . , K clusters, building on the $(K-1)$-partitions to find the K-partitions. This is a "dynamic programming" procedure (Bellman and Dreyfus, 1962).

STEP 1.   Compute the diameter $D(I, J)$ for the cluster $(I, I+1, \ldots, J)$, for all I, J such that $1 \leq I < J \leq M$.

STEP 2.   Compute the errors of the optimal partitions, $2 \leq I \leq M$, by $e[P(I, 2)] =$ min $[D(1, J-1) + D(J, I)]$ over the range $2 \leq J \leq I$.

STEP 3.   For each L $(3 \leq L \leq K)$ compute the errors of the optimal partitions $e[P(I, L)]$ $(L \leq I \leq M)$ by

$$e[P(I, L)] = \min \{e[P(J-1, L-1)] + D(J, I)\}$$

over the range $L \leq J \leq I$.

STEP 4.   The optimal partition $P(M, K)$ is discovered from the table of errors $e[P(I, L)]$ $(1 \leq L \leq K, 1 \leq I \leq M)$ by first finding J so that

$$e[P(M, K)] = e[P(J-1, K-1)] + D(J, M).$$

The last cluster is then $(J, J+1, \ldots, M)$. Now find J^* so that $e[P(J-1, K)] = e[P(J-1, K-1)] + D(J^*, J-1)$. The second-to-last cluster of $P(M, K)$ is $(J^*, J^*+1, \ldots, J-1)$, and so on.

NOTE.   The partition is guaranteed optimal, but it is not necessarily unique.

6.3 FISHER ALGORITHM APPLIED TO OLYMPIC TIMES

It is first necessary to define cluster diameter. Let $X(I)$ be the value associated with the Ith object. A standard measure of diameter of the cluster $I, I+1, \ldots, J$ is

$$D(I, J) = \sum \{I \leq L \leq J\} [X(L) - \bar{X}]^2,$$

where

$$\bar{X} = \sum \{I \leq L \leq J\} \frac{X(L)}{J - I + 1}$$

is the mean of the values in the cluster.

Another measure, more convenient for hand calculation, is

$$D(I, J) = \sum \{I \leq L \leq J\} |X(L) - \tilde{X}|,$$

where the median $\tilde{X}$ is both no greater than and no less than half the values $X(L)$ $(I \leq L \leq J)$.

STEP 1.   Compute the diameter of all clusters. With 16 objects there are $16 \times \frac{17}{2} = 126$ diameters. For example, $D(7, 12)$ is computed from the times 10.6, 10.8, 10.3, 10.3, and 10.4, which have median 10.4. The deviations are 0.2, 0.4, 0.1, 0.1, 0 which have a sum of 0.8, so $D(7, 12) = 0.8$. All diameters are given in Table 6.2.

Table 6.2 Diameters of Clusters of Olympic Times (in Tenths of Seconds)

Diameter in the interval (I, J) is the sum of absolute deviations from the median of observations $X(I), \ldots, X(J)$.

J	1	2	3	4	5	6	7	8	9	10	11	12	13	14	15	16
I 1	0	12	12	14	14	14	16	16	21	26	31	35	38	44	50	57
2		0	2	2	2	2	4	4	9	14	19	23	26	30	36	42
3			0	2	2	2	4	4	9	14	19	23	24	28	33	39
4				0	0	0	2	2	7	12	17	19	20	23	28	33
5					0	0	2	2	7	12	15	17	17	20	24	29
6						0	2	2	7	10	13	13	14	16	20	24
7							0	2	5	8	8	9	10	12	15	16
8								0	5	5	5	6	8	9	12	16
9									0	0	0	1	3	4	7	11
10										0	0	1	3	4	7	11
11											0	1	2	4	7	11
12												0	1	3	7	11
13													0	3	5	8
14														0	2	3
15															0	1
16																0

STEP 2. All optimal 2-partitions $\mathbf{P}(I, 2)$ are to be computed. It is necessary to remember only $e[\mathbf{P}(I, 2)]$ for later steps. As an example, $e[\mathbf{P}(4, 2)]$ is the minimum of

$$D(1, 3) + D(4, 4) = 12,$$

$$D(1, 2) + D(3, 4) = 12 + 2 = 14,$$

and

$$D(1, 1) + D(2, 4) = 2.$$

Thus $e[\mathbf{P}(4, 2)] = 2$. All the errors for optimal partitions are given in Table 6.3.

STEP 3. The optimal 3-partitions are developed from the optimum 2-partitions. For example, $e[\mathbf{P}(6.3)]$ is the minimum of

$$e[\mathbf{P}(5, 2)] + D(6, 6) = 2 + 0 = 2,$$

$$e[\mathbf{P}(4, 2)] + D(5, 6) = 2 + 0 = 2,$$

$$e[\mathbf{P}(3, 2)] + D(4, 6) = 2 + 0 = 2,$$

and

$$e[\mathbf{P}(2, 2)] + D(3, 6) = 0 + 2 = 2,$$

so $e[\mathbf{P}(6, 3)] = 2$.

Similarly, the optimal 4-partitions are developed from the optimum 3-partitions.

Table 6.3   Errors of Optimal Partitions

$e[\mathbf{P}(I, L)]$ is the error of optimal partition of objects $1, 2, \ldots, I$ into L clusters.

I	1	2	3	4	5	6	7	8	9	10	11	12	13	14	15	16
	0	12	12	14	14	14	16	16	21	26	31	35	38	44	50	57
		0	2	2	2	2	4	4	9	14	16	17	19	20	23	27
			0	2	2	2	2	4	4	4	4	5	7	8	11	15
				0	0	0	2	2	4	4	4	4	5	7	8	9
					0	0	0	2	2	2	2	3	4	5	7	8
						0	0	0	2	2	2	2	3	4	5	6
							0	0	0	0	0	1	2	3	4	5
								0	0	0	0	0	1	2	3	4
									0	0	0	0	0	1	2	3
										0	0	0	0	0	1	2
											0	0	0	0	0	1
												0	0	0	0	0
													0	0	0	0
														0	0	0
															0	0
																0

STEP 4.   To find the optimal partition of 16 into, say four clusters, first find J such that

$$e[\mathbf{P}(16, 4)] = e[\mathbf{P}(J - 1, 3)] + D(J, 16).$$

Such a J is $J = 15$. The last cluster is $(15, 16)$. Now find J so that

$$e[\mathbf{P}(14, 3)] = e[\mathbf{P}(J - 1, 2)] + D(J, 14).$$

Thus $J = 9$, and the second last cluster is $(9, 14)$.

Since $e[\mathbf{P}(8, 2)] = D(1, 1) + D(2, 8)$, the first two clusters are (1) and $(2, 8)$.

Thus, $\mathbf{P}(16, 4) = (1), (2, \ldots, 8), (9, \ldots, 14), (15, 16)$. In terms of the observations,

$\mathbf{P}(16, 4)$

$= (12) \ (10.8 \ 11 \ 10.8 \ 10.8 \ 10.8 \ 10.6 \ 10.8) \ (10.3 \ 10.3 \ 10.3 \ 10.4 \ 10.5 \ 10.2) \ (10.0 \ 9.9).$

This seems a reasonable partition of the data. Some idea of the best number of clusters may be obtained by plotting $e[\mathbf{P}(16, K)]$ against K, as in Figure 6.2. There are sharp decreases in error at $K = 2$ and $K = 3$, a noticeable decrease at $K = 4$, and trivial decreases for larger K. The correct number of clusters is 3 or 4.

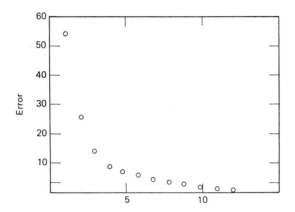

Figure 6.2 Errors of optimal K-partitions of Olympic data, using absolute deviations to measure diameter.

6.4 SIGNIFICANCE TESTING AND STOPPING RULES

Let $X(1), X(2), \ldots, X(M)$ denote the M observations in order, and consider a statistical model for a partition into K clusters:

$$(I_1 = 1, I_1 + 1, \ldots, I_2 - 1)(I_2, I_2 + 1, \ldots, I_3), \ldots, (I_K, I_K + 1, \ldots, M),$$

for which $X(1), X(2), \ldots, X(I_2 - 1)$ are independent observations from the density $f(X \mid \theta_1)$, $X(I_2), X(I_2 + 1), \ldots, X(I_3 - 1)$ are independent observations from the density $f(X \mid \theta_2)$, and so on, up to $X(I_K), \ldots, X(M)$ are independent observations from the density $f(X \mid \theta_K)$. (Note that both X and θ may have dimension greater than 1.) The likelihood of the joint parameters $\theta_1, \theta_2, \ldots, \theta_K$ is

$$\prod \{1 \leq L \leq K\} \prod \{I_L \leq J < I_{L+1}\} f[X(J) \mid \theta_L].$$

The maximum log likelihood of the observations is

$$LL = \sum \{1 \leq L \leq K\} \sum \{I_L \leq J < I_{L+1}\} \log f[X(J) \mid \hat{\theta}_L],$$

where $\hat{\theta}_L$ is the maximum likelihood estimate of θ_L, based on observations in the Lth clusters. Note that, if $D(I, J)$, the cluster diameter, is defined by

$$D(I, J) = -\max_{\theta} \sum \{I \leq L \leq J\} f[X(L) \mid \theta],$$

then $-LL$ is the sum of $D(I, J)$'s corresponding to the clusters. Minimizing $-LL$ means that clusters are found, and parameter values within clusters are estimated, to make the given observations most probable.

In this way, the density f, which relates the observations X to the cluster parameter θ, generates a cluster diameter and an additive error measure for partitions. For example, with the double exponential $f(X \mid \theta) = 0.5 \exp(-|X - \theta|)$, the measure of cluster diameter is

$$D(I, J) = \sum \{I \leq L \leq J\} |X(L) - \tilde{X}|,$$

where $\tilde{X}$ is the cluster median. The more common normal density, $f(X, \theta) = \exp(-\frac{1}{2}(X - \theta)^2)/\sqrt{2\pi}$ leads to the familiar sum of squared deviations diameter $D(I, J) = \sum \{I \leq L \leq J\} [X(L) - \bar{X}]^2$, where $\bar{X}$ is the cluster mean.

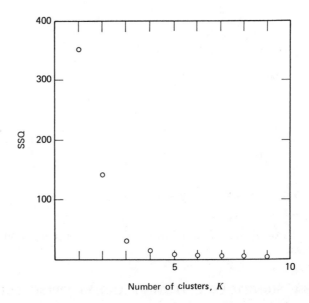

Figure 6.3   Sum of squared errors within optimal partitions of Olympic data.

The optimal partitions of the Olympic data with this criterion are shown in Table 6.4 and Figure 6.3. It will be seen that the principal decreases in the sum of squares come in the second and third partitions, and later decreases are relatively minor. This agrees with the intuition that about three clusters seem right to describe this data.

When are $K + 1$ rather than K clusters necessary? A naive approach supposes that K distinct clusters are present in the data, and that $K + 1$ clusters are obtained by splitting one of these clusters in two arbitrarily. [In reality, one of the clusters would be split optimally if the clusters were really distinct, and in general there will be no simple relation between the optimal K-partition and the optimal $(K + 1)$-partition.] Then the mean square ratio

$$(N - K - 1)\left(\frac{e[P(N, K)]}{e[P(N, K + 1)]} - 1\right)$$

is distributed as $F_{1, N-K-1}$ if the observations are normal. It therefore seems worthwhile to study this ratio of mean square reduction to mean square within clusters, and to take a large value of this quantity to indicate that $(K + 1)$-clusters are necessary. For the Olympic data, the 3-cluster stands out in the mean square ratio table. (The values for $K = 10, 11$ are large but should be ignored; in these cases the mean square error within clusters is grossly underestimated because the data are rounded to tenths of seconds.)

When is the mean square error ratio larger than would be expected from random normal observations? One approach to this is through asymptotic theory as the number of observations becomes large. It is possible to show (by an argument too long to fit in the margin) that the number of objects in the clusters alternates between very large and very small; that is, if $n(L)$ denotes the number of objects in the Lth cluster, $n(L)/n(L + 1)$ is near zero or unity as $M \to \infty$. For example, an optimal partition

Table 6.4 Optimal Partitions of Olympic Data Using Sum of Squared Deviations From Mean as Cluster Diameter

PARTITION SIZE	MSQ RATIO	SUM OF SQUARES	PARTITION (OBSERVATION IN TENTHS OF SECONDS)
1		352.00	120 108 110 108 108 108 106 108 103 103 103 104 105 102 100 99
2	20.7	142.93	120/108 110 108 108 108 106 108 103 103 103 104 105 102 100 99
3	49.0	30.00	120/108 110 108 108 108 106 108/103 103 103 104 105 102 100 99
4	14.0	13.83	120/108 110 108 108 108 106 108/103 103 103 104 105 102/100 99
5	2.8	11.03	120/108 110 108 108 108/106 108/103 103 103 104 105 102/100 99
6	2.4	9.03	120/108 110 108 108 108/106 108/103 103 103/104 105/102 100 99
7	4.1	6.20	120/108 110 108 108 108/106 108/103 103 103/104 105/102/100 99
8	3.8	4.20	120/108 110 108 108 108/106/108/103 103 103/104 105/102/100 99
9	2.8	3.00	120/108 110/108 108 108/106/108/103 103 103/104 105/102/100 99
10	12.0	1.00	120/108/110/108 108 108/106/108/103 103 103/104 105/102/100 99
11	5.0	0.50	120/108/110/108 108 108/106/108/103 103 103/104/105/102/100 99
12	∞	0.00	120/108/110/108 108 108/106/108/103 103 103/104/105/102/100/99

of 500 random normals into 10 clusters yielded cluster sizes 183, 3, 63, 1, 100, 2, 10, 39, 98, 1. Also the mean square error ratio slowly approaches infinity as $M \to \infty$.

Empirical sampling for moderate M shows surprising uniformity in the distributions of the root mean square ratio for various K (Table 6.5). The root mean square ratio has an expectation very close to 2 for a wide range of M and K. The variances depend on M, the total number of objects, but not much on K. The expected values increase slightly with M but are still near 2, even for $M = 500$. The convergence to infinity demanded by asymptotic theory is slow. The various ratios for different K are approximately independent normal variables with the same mean and variance (Figure 6.4). Therefore, under the null hypothesis that no clusters exist, the largest ratio has approximately the distribution of the largest of a number of independent normals. This reference distribution thus provides an approximate significance level for the largest ratio.

6.5 TIME AND SPACE

The Fisher algorithm requires $O(M^2 K)$ additions, where M is the number of objects and K is the number of clusters. It is thus feasible for 100–1000 objects. It may be a high price to pay to get exact optimization. During the algorithm it is necessary to store $M \times K$ errors corresponding to optimal partitions $\mathbf{P}(I, L)$ $(1 \leq I \leq M, 1 \leq L \leq K)$.

In Table 6.4, the optimal partitions for various K almost have hierarchical structure; that is, for any two clusters in any two partitions, one cluster includes the other or they are disjoint. The only exception to this rule is the cluster (103 103 103 104 105 102). If it is known that the final partitions have hierarchical structure, a shorter algorithm proceeds as follows.

STEP 1. Split the sequence $1, 2, \ldots, M$ optimally into two clusters by choosing one of M possible split points.

Table 6.5 Empirical Distributions of Root Mean Square Ratio, Based on Random Normal, 500 Trials.

N = 6	K = 2	K = 3
Expectation	2.154	2.271
Variance	1.731	1.755

N = 11	K = 2	K = 3	K = 4	K = 5	K = 6	K = 7
Expectation	2.030	2.191	1.961	2.010	2.013	1.947
Variance	.592	.586	.476	.528	.660	.658

N = 16	K = 2	K = 3	K = 4	K = 5	K = 6	K = 7	K = 8	K = 9	K = 10
Expectation	2.020	2.260	1.992	1.968	1.963	1.893	1.932	1.871	1.898
Variance	.529	.487	.359	.287	.268	.243	.361	.256	.374

N = 50	K = 2	K = 3	K = 4	K = 5	K = 6	K = 7	K = 8	K = 9	K = 10
Expectation	2.090	2.433	2.144	2.134	2.115	2.003	2.004	1.971	1.969
Variance	.360	.295	.193	.189	.148	.113	.094	.080	.080
(200 TRIALS)									

N = 100	K = 2	K = 3	K = 4	K = 5	K = 6	K = 7	K = 8	K = 9	K = 10
Expectation	2.184	2.528	2.246	2.285	2.211	2.179	2.216	2.105	2.108
Variance	.294	.242	.192	.120	.100	.094	.091	.082	.062
(50 TRIALS)									

N = 500	K = 2	K = 3	K = 4	K = 5	K = 6	K = 7	K = 8	K = 9	K = 10
Expectation	2.3	2.9	2.6	2.4	2.5	2.3	2.5	2.4	2.4
(2 TRIALS)									

(N = Number of observations in sequence; K = Number of clusters in partition)

STEP 2. Split optimally into three clusters by choosing one of M possible split points—some in the first cluster from Step 1, the remainder in the second cluster.

STEP 3. Split optimally into four clusters by choosing one of M possible split points, and so on. This procedure requires only $O(MK)$ additions. This algorithm could be used to obtain good (not guaranteed best) partitions of long sequences.

6.6 THINGS TO DO

6.6.1 Running the Fisher Algorithm

This algorithm is appropriate when objects are already ordered by some overriding variable such as time. There is no particular requirement that the data be one-dimensional. Vietnam combat deaths over time (Table 6.6) might be analyzed to detect different phases of U.S. involvement.

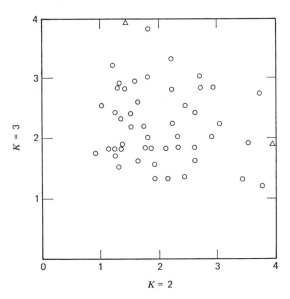

Figure 6.4 Independence of root mean square ratios, computed from 50 normal samples. For 11 observations, ratios for 2- and 3-partitions.

Table 6.6 Combat Deaths in Indochina

		US	SVN	THIRD	ENEMY		US	SVN	THIRD	ENEMY		US	SVN	THIRD	ENEMY
'66	JAN	282	903	74	2648	'68	1202	2905	111	15217	'70	343	1768	69	9187
	FEB	435	1359	58	4727		2124	5025	147	39867		386	1417	36	8828
	MAR	507	1145	59	5685		1543	2570	88	17371		449	1674	75	10335
	APR	316	945	30	2818		1410	1922	85	12215		526	2642	79	13063
	MAY	464	961	19	4239		2169	3467	85	24086		754	2851	58	17256
	JUN	507	1185	41	4815		1146	1974	92	10319		418	2873	63	7861
	JUL	435	1006	32	5297		813	1409	65	6653		332	1711	71	7183
	AUG	396	914	44	5860		1080	2393	73	15478		319	1720	63	6446
	SEP	419	803	30	4459		1053	2164	58	12543		219	1734	46	6138
	OCT	340	844	63	5665		600	1169	70	8168		170	1491	57	5549
	NOV	475	907	87	5447		703	1408	38	9632		167	1619	48	5607
	DEC	432	981	29	3864		749	1509	67	9600		138	1846	39	6185
'67	JAN	520	914	77	6064	'69	795	1664	76	10955	'71	140	1616	30	6155
	FEB	662	885	95	7341		1073	2072	85	14086		221	2435	48	11704
	MAR	944	1297	54	9351		1316	2186	90	19805		272	3676	104	19858
	APR	710	1057	56	6227		847	1710	52	14539		226	2198	86	10457
	MAY	1233	1184	112	9808		1209	2251	92	17443		138	2091	50	9094
	JUN	830	981	74	7354		1100	1867	75	16825		108	1846	44	7648
	JUL	781	676	102	7923		638	1455	64	10237		65	1389	44	6247
	AUG	535	1068	90	5810		795	1625	74	12373		67	1488	32	6165
	SEP	775	1090	149	6354		477	1543	60	10369		78	1607	27	6300
	OCT	733	1066	96	6272		377	1597	80	8747		29	1574	20	5744
	NOV	881	1299	98	7662		446	2105	62	11639		19	1161	14	4283
	DEC	774	1199	102	7938		341	1758	56	9936		17	988	26	4439

From *Unclassified Statistics on Southeast Asia* (1972), Department of Defense, OASD (Comptroller), Directorate for Information Operations.

6.6.2   Real-Valued Data

For clustering a single real variable to minimize the within-cluster sum of squares, the clusters must be convex, which means they must consist of points lying in an interval. The Fisher algorithm may be applied to the ordered points to find the exactly optimal partition into K clusters. This exact partition may be compared with locally optimal partitions obtained by approximate techniques, such as the K-means algorithm.

6.6.3*   Estimating Densities

Given that a density is unimodal in an interval, there is a well-known maximum likelihood technique for estimating it (each point in the interval is tried for the mode, and for each modal point the density is first estimated as the reciprocal of the intervals between points and then neighboring intervals are averaged if they violate the monotonicity required by unimodality). Using the Fisher algorithm, maximum likelihood densities with K modes may be computed.

6.6.4   Sequential Splitting

As justification for the hierarchical algorithm in Section 6.5, suppose the data consist of M real values in time and that the time interval is divided into K intervals within each of which the values are constant. The error function is within-cluster sum of squares. There is a K-partition for which the error is zero, and this will be discovered by the hierarchical algorithm.

6.6.5   Updating Sums of Absolute Deviations

The median of a set of numbers $X(1), \ldots, X(M)$ is any number $\tilde{X}$ such that $\tilde{X} \leq X(I)$ occurs at least $M/2$ times and $\tilde{X} \geq X(I)$ occurs at least $M/2$ times. Let $\tilde{X}^*$ be the median (there may be more than one) closest to X. Then the minimum sum of absolute deviations for $X(1), \ldots, X(M), X$ is the minimum sum of absolute deviations for $X(1), \ldots, X(M)$ plus $|X - \tilde{X}^*|$.

REFERENCES

BELLMAN, R. E., and DREYFUS, S. E. (1962). *Applied Dynamic Programming*, Princeton U. P., Princeton N.J. On p. 15, the principle of optimality is stated. An optimal policy has the property that, whatever the initial state and initial decision are, the remaining decisions must constitute an optimal policy with regard to the state resulting from the first decision. A typical formulation of a dynamic programming problem is as follows: Maximize $R(x_1, x_2, \ldots, x_N) = \Sigma g_i(x_i)$ over the region $x_i \geq 0$, $\Sigma x_i = x$. Define $f_N(x)$ to be this maximum. Then

$$f_N(x) = \max_{0 \leq x_N \leq x} [g_N(x_N) + f_{N-1}(x - x_N)].$$

As an example on p. 104, consider the problem of meeting a series of demands in time when it is expensive to have excess capacity or to change the capacity. Thus, if the demands are r_k, the capacities are x_k, the loss due to excess capacity is $\varphi_k(x_k - r_k)$, and the loss due to a change in capacity is $\psi_k(x_k - x_{k-1})$, the problem is to minimize $\sum [\varphi_k(x_k - r_k) + \psi_k(x_k - x_{k-1})]$ subject to $x_k \geq r_k$.

FISHER, W. D. (1958). On grouping for maximum homogeneity. *J. Am. Stat. Assoc.* **53,** 789–798. Given a set of K objects, such that each object has a weight w_i and a numerical measure a_i, Fisher discusses techniques of assigning the objects to G groups so as to minimize $\sum w_i(a_i - \bar{a}_i)^2$, where $\bar{a}_i$ is the weighted mean of a's in the group to which a_i is assigned. He shows that each group must be convex in the a's; that is, each group consists of all the a's in an interval, if the optimization takes place without prior constraints. If the objects are ordered *a priori*, so that clusters are sequences of objects in the original order, he remarks that the additive property of the sum of squares criterion makes it possible to reduce the computation by relating optimum partitions into G clusters to optimum partitions into fewer clusters. He also notes a nonlinear programming formulation of the problem: Let x_{hi} denote the fractional part of a_i that is assigned to group h ($h = 1, 2, \ldots, G$). Define $\bar{a}_h = (\sum_i x_{hi} w_i a_i)(\sum_i x_{hi} w_i)^{-1}$, $S = \sum_i \sum_h w_i x_{hi}(a_i - \bar{a}_h)^2$, and minimize S subject to $x_{hi} \geq 0, \sum_h x_{hi} = 1$. The solutions will always have just one of $x_{hi} = 1, h = 1, \ldots, G$, and so will solve the grouping problem. It is not clear whether or not this formulation simplifies the solution.

PROGRAMS

FISH   partitions data, consistently with input order, to maximize between cluster sum of squares.

PFISH prints output from Fisher algorithm.

```
      SUBROUTINE FISH(X,SG,MG,N,K)
C.... X = M BY 1 VECTOR
C     PROGRAM GROUPS REAL VALUED OBSERVATIONS X(1), . . X(N).
C     THE OBSERVATIONS NEED NOT BE ORDERED, BUT THE GROUPS WILL ALL CONSIST OF
C     SEQUENCES OF OBSERVATIONS X(I), . . X(J).
C     IF A PARTITION INTO K CLUSTERS IS REQUESTED, PARTITIONS LE K ARE AUTOMATIC
C     PRINT OUT INFORMATION ABOUT CLUSTERS BY CALLING PFISH
C.... X = N BY 1 ARRAY OF OBSERVATIONS TO BE FITTED
C.... MG = N BY K ARRAY,WHERE MG(I,J) IS LOWER BOUNDARY OF JTH GROUP, IN OPTIMAL
C         SPLIT OF X(1),... X(I) INTO J GROUPS.
C.... SG = N BY K ARRAY, SG(I,J) IS SUM OF SQUARES WITHIN GROUPS FOR
C         X(1),.... X(I) SPLIT OPTIMALLY INTO J GROUPS.
C......................................................................
      DIMENSION X(N),SG(N,K),MG(N,K)
C.... INITIALIZE SG,MG
      DO 20 J=1,K
      MG(1,J)=1
      SG(1,J)=0.
      DO 20 I=2,N
   20 SG(I,J)=10.**10
C.... COMPUTE SG,MG ITERATIVELY
      DO 30 I=2,N
      SS=0.
      S=0.
      DO 31 II=1,I
      III=I-II+1
      SS=SS+X(III)**2
      S=S+X(III)
      SN=II
      VAR=SS-S**2/SN
      IK=III-1
      IF (IK.EQ.0) GO TO 31
      DO 32 J=1,K
      IF (J.EQ.1) GO TO 32
      IF (SG(I,J).LT.VAR+SG(IK,J-1)) GO TO 32
      MG(I,J)=III
      SG(I,J)=VAR+SG(IK,J-1)
   32 CONTINUE
   31 CONTINUE
      SG(I,1)=VAR
   30 MG(I,1)=1
      RETURN
      END
```

```
      SUBROUTINE PFISH(X,SG,MG,N,K)
C......................................................................20 MAY 1973
C.... USES OUTPUT FROM PROGRAM FISH TO PRINT CLUSTER DESCRIPTIONS
C.....................................................................
      DIMENSION SG(N,K),MG(N,K),X(N)
      WRITE(6,1) N,K
    1 FORMAT('1 PARTITION OF',I5,'  OBSERVATIONS UP TO ',I5,' CLUSTERS')
      DO 20 J=1,K
      JJ=K-J+1
      WRITE(6,2)JJ,SG(N,JJ)
    2 FORMAT('0THE',I5,'  PARTITION WITH SUM OF SQUARES',F20.6)
      WRITE(6,3)
    3 FORMAT(' CLUSTER   NUMBER OBS   MEAN      S.D.   ')
      IL=N+1
      DO 21 L=1,JJ
      LL=JJ-L+1
      S=0.
      SS=0.
      IU=IL-1
      IL=MG(IU,LL)
      DO 22 II=IL,IU
      S=S+X(II)
   22 SS=SS+X(II)**2
      SN=IU-IL+1
      S=S/SN
      SS=SS/SN-S**2
      SS=(ABS(SS))**(0.5)
      WRITE(6,4) LL,SN,S,SS
    4 FORMAT(I5,5X,3F10.4)
   21 CONTINUE
   20 CONTINUE
      RETURN
      END
```

142

The Ditto Algorithm

7.1 INTRODUCTION

The data in Table 7.1 record the evaluation of various wines from 1961 to 1970. The evaluations are given in terms of five categories—disastrous, poor, average, good, excellent. In many statistical analyses, it is customary to quickly shuffle category variables out of sight by replacing them with interval scale variables—for example, 1, 2, 3, 4, 5 would be reasonable here because the categories are ordered. For clustering, category variables are more natural than interval variables and should be cherished and treated as they lie. A category variable partitions the set of cases. For example, the 1961 evaluations divide the wines into classes—excellent, good, and average. An overall partition of the data is thus a combination of the individual partitions. There is no more reason for a single partition to be adequate to explain the data than for the first principal component to be adequate for interval scale data.

In the ditto algorithm, the center for each cluster is equal to the mode of each variable over cases in the cluster. A measure of partition error is defined similar to that in the K-means algorithm.

In the K-means algorithm, it is necessary to fix the number of clusters in the partition; otherwise the partition of M clusters is best. In this algorithm, the error function does not necessarily decrease as the number of clusters increases, and so the partition size varies during the course of the algorithm.

7.2 DITTO ALGORITHM

Preliminaries. There are M cases, N variables, and a value $A(I, J)$ of the Ith case for the Jth variable. Let $P(M, K)$ denote a partition of the cases. The error of the partition $P(M, K)$ is

$$e[P(M, K)] = M + \sum \{1 \leq J \leq N\} \sum \{1 \leq L \leq K\} \rho[B(0, J), B(L, J)]$$
$$+ \sum \{1 \leq J \leq N\} \sum \{1 \leq I \leq M\} \rho[A(I, J), B(L(I), J)],$$

where $L(I)$ is the cluster containing I, $\rho(A, B) = 1$ if $A \neq B$, and $\rho(A, B) = 0$ if $A = B$. The vector $\{B(L, J), 1 \leq J \leq N\}$ for the Lth cluster is the center of the cluster. It will be chosen to minimize the expression $e[P(M, K)]$; this requirement does not uniquely determine $B(L, J)$, but it is always possible to find $B(L, J)$ minimizing e and equal to a mode of $A(I, J)$ over cases I in the cluster L. The vector $\{B(0, J), 1 \leq J \leq N\}$ is the "grand mode" of the cluster centers.

Table 7.1   Evaluation of Wines 1961–1970

D   disastrous

P   poor

A   average

G   good

E   excellent

		1961	62	63	64	65	66	67	68	69	70
	Red Bordeaux										
MG	Medoc and Graves	E	G	P	G	D	G	G	P	A	G
EP	Saint Emilion and Pomerol	E	A	P	G	P	G	G	P	A	G
	White Bordeaux										
SS	Sauternes	G	G	D	D	D	A	G	P	G	G
GS	Graves	G	G	D	G	D	G	G	P	G	G
RB	Red Burgundy	E	G	A	G	P	G	A	D	G	G
	White Burgundy										
CB	Cote de Beaune	E	G	A	G	A	G	G	A	G	G
CS	Chablis	E	G	A	G	P	G	G	A	G	G
BS	Beaujolais	E	G	P	G	D	G	A	P	G	G
	Red Rhone										
RN	North	E	G	P	G	A	G	G	A	G	G
RS	South	G	A	P	G	A	A	G	A	G	G
WL	White Loire	A	P	P	G	P	A	G	G	G	G
AE	Alsace	G	A	P	E	P	G	G	P	G	G
RE	Rhine	G	A	P	G	P	G	G	P	G	G
ME	Moselle	G	A	P	E	P	G	G	P	G	G
CA	California	-	-	-	P	G	A	G	G	A	G

From *Gourmet Magazine* (August, 1971), pp. 30–33.

There are four stages in the ditto algorithm:

(**i**)   selecting the initial partition;

(**ii**)   moving cases from one cluster to another, or creating new clusters;

(**iii**)   updating the modal values $B(L, J)$ within clusters, and updating the grand mode $B(0, J)$;

(**iv**)   printing out the original data, recording the value $A(I, J)$ as a dot if it agrees with $B[L(I), J]$, and recording $B(L, J)$ as a dot if it agrees with $B(0, J)$. The number of nondot values is $e[P(M, K)]$.

STEP 1.   Initially define a different cluster center $\{B(L, J), J = 1, \ldots, N\}$ for each value taken by a variable. The total number of clusters K will be the number of different values taken by each variable, summed over variables. For the Lth cluster, suppose a variable $J(L)$ takes a value $V(L)$. Then $B(L, J)$ is a mode of values $A(I, J)$ for which $A[I, J(L)] = V(L)$.

Finally, $B(0, J)$ is a mode of $B(L, J)$ over all clusters L $(1 \leq L \leq K)$.

STEP 2.   For each case I $(1 \leq I \leq M)$ allocate I to the first cluster L $(0 \leq L \leq K)$ for which $\sum \{1 \leq J \leq N\} \, \rho[A(I, J), B(L, J)]$ is a minimum.

STEP 3.   Delete clusters containing no objects. Delete clusters containing one object, and allocate the object to the 0 cluster.

STEP 4.   For each cluster L, replace $B(L, J)$ $(1 \leq J \leq N)$ by a mode of values $A(I, J)$ $(I \in L)$ and the value $B(0, J)$. If $B(0, J)$ is a possible value of $B(L, J)$, set $B(L, J) = B(0, J)$. For the cluster 0, replace $B(0, J)$ by the mode of $B(L, J)$ over all clusters and over $A(I, J)$ for cases allocated to 0.

STEP 5.   If any change occurs in Steps 2–4, return to Step 2. Otherwise, replace $A(I, J)$ by a dot, if $I \in L$, $A(I, J) = B(L, J)$ $(1 \leq I \leq M, \ 1 \leq J \leq N)$. Replace $B(L, J)$ by a dot if $B(L, J) = B(0, J)$ $(1 \leq L \leq K)$.

7.3 APPLICATION OF DITTO ALGORITHM TO WINES

STEP 1.   There are 10 variables, taking 28 different values, so initially there are 28 clusters. The first cluster center is computed by using variable $J(1) = 1$ and value $V(1) = E$. The cases I such that $A(I, 1) = E$ are MG EP RB CB CS BS RN. The modal values $B(1, J)$ are

$$B(1, 1) = E, \qquad \text{the mode of E E E E E E E}$$
$$B(1, 2) = G, \qquad \text{the mode of G A G G G G G}$$
$$B(1, 3) = P, \qquad \text{the mode of P P A A A P P}$$

and so on. The next cluster center uses $J(2) = 1$ and $V(2) = G$, the second value taken by the first variable. The matching cases are SS GS RS AE RE ME. The complete set of initial cluster values appears in Table 7.2. Some clusters, such as 20, 26, and 28, are identical. The redundant ones will be eliminated in later steps.

STEP 2.   Each case is now allocated to a cluster—the cluster whose center it best matches. For example, case 1, MG, differs from the center of cluster 4 only in the year 1969, so MG is allocated to cluster 4. The complete allocation is given in Table 7.3. There are many single clusters that will be allocated to the 0 cluster in the next step.

STEP 3.   Delete the many clusters to which no cases are allocated. For example cluster 10, having an identical center to 0, will receive no cases and be eliminated. Delete clusters such as cluster 7 that contain a single object.

STEP 4.   Recompute cluster centers. For example, cluster 4 contains MG and BS. To compute $B(4, 1)$, use the data values E, E and the value $B(0, 1) = E$. The mode of these three values is $B(4, 1) = E$. To consider a less straightforward case, consider $B(9, 4)$. The cluster 9 contains cases SS and GS taking values D and G in variable 4, 1964. The overall mode is $B(0, 4) = G$. Thus the mode of D, G, G is $B(9, 4) = G$.

Finally, the grand mode is recomputed by using cluster centers and cases allocated to the grand mode.

STEP 5.   Steps 2–4 are repeated until there is no change in the clusters, which occurs after two allocations. The array is now prepared for display in a dot diagram. Since MG is allocated to cluster 4, $A(1, 1) = B(4, 1) = E$ and $A(1, 1)$ is replaced by a dot.

Table 7.2 Initial Cluster Centers Applying Ditto Algorithm to Wines

CLUSTER	VARIABLE	VALUE	1961	62	63	64	65	66	67	68	69	70
1	61	E	E	G	P	G	P	G	G	A	G	G
2	61	G	G	A	P	G	P	G	G	P	G	G
3	61	A	A	P	P	G	P	A	G	G	G	G
4	62	G	E	G	P	G	D	G	G	P	G	G
5	62	A	G	A	P	G	P	G	G	P	G	G
6	62	P	G	P	P	G	P	A	G	G	G	G
7	63	P	E	A	P	G	P	G	G	P	G	G
8	63	A	E	G	A	G	P	G	G	A	G	G
9	63	D	G	G	D	G	D	G	G	P	G	G
10	64	G	E	G	P	G	P	G	G	P	G	G
11	64	E	G	A	P	E	P	G	G	P	G	G
12	64	D	G	G	D	D	D	A	G	P	G	G
13	64	P	-	-	-	P	G	A	G	G	A	G
14	65	P	E	A	P	G	P	G	G	P	G	G
15	65	A	G	A	P	G	A	A	G	A	G	G
16	65	D	E	G	P	G	D	G	G	P	G	G
17	65	G	-	-	-	P	G	A	G	G	A	G
18	66	G	E	G	P	G	P	G	G	P	G	G
19	66	A	G	G	P	G	P	A	G	G	G	G
20	67	G	G	G	P	G	P	G	G	P	G	G
21	67	A	E	G	P	G	P	G	G	P	G	G
22	68	P	E	G	P	G	P	G	G	P	G	G
23	68	A	E	G	A	G	P	G	G	A	G	G
24	68	G	A	P	P	G	P	A	G	G	G	G
25	68	D	E	G	A	G	P	G	A	D	G	G
26	69	G	G	G	P	G	P	G	G	P	G	G
27	69	A	E	G	P	G	P	G	G	P	A	G
28	70	G	G	G	P	G	P	G	G	P	G	G
0			E	G	P	G	P	G	G	P	G	G

Similarly, since $B(4, 1) = B(0, 1) = E$, $B(4, 1)$ is replaced by a dot. The final dotted array appears in Table 7.4.

The total number of symbols necessary to represent the data is 41, which should be compared to the 150 data values and to the 28 different values taken by all variables. The story told by the clustering is as follows. The usual grading of wine is "good." Overall there were three poor years, 1963, 1965, and 1968, and one excellent year, 1961. There is a gallimaufry of four wines, St. Emilion and Pomerol, Red Rhone North, White Loire, and California, which vary (differently) from this overall

Table 7.3 Successive Passes of Ditto Algorithm on Wines

CLUSTER MODE

CLUSTER	ALLOCATION 1	1961	62	63	64	65	66	67	68	69	70
4	MG BS	E	G	P	G	D	G	G	P	G	G
7	EP	E	G	P	G	P	G	G	P	G	G
9	SS GS	G	G	D	G	D	G	G	P	G	G
25	RB	E	G	P	G	P	G	G	P	G	G
8	CB CS	E	G	A	G	P	G	G	A	G	G
1	RN	E	G	P	G	P	G	G	P	G	G
15	RS	E	G	P	G	P	G	G	P	G	G
24	WL	E	G	P	G	P	G	G	P	G	G
2	AE RE ME	G	A	P	G	P	G	G	P	G	G
13	CA	E	G	P	G	P	G	G	P	G	G
0		E	G	P	G	P	G	G	P	G	G

CLUSTER	ALLOCATION 2										
4	MG BS	E	G	P	G	D	G	G	P	G	G
9	SS GS	G	G	D	G	D	G	G	P	G	G
8	RB CB CS	E	G	A	G	P	G	G	A	G	G
2	RS AE RE ME	G	A	P	G	P	G	G	P	G	G
0	EP RN WL CA	E	G	P	G	P	G	G	P	G	G

CLUSTER	ALLOCATION 3	No change.
4	MG BS	
9	SS GS	
8	RB CB CS	
2	RS AE RE ME	
0	EP RN WL CA	

pattern. California particularly is quite different. There are four clusters of wines: the Beaujolais cluster (disastrous in 1965), the Sauternes cluster (disastrous in 1963 and 1965), the Chablis cluster (average in 1963 and 1968 when others were poor), the Moselle cluster (not as good as others in 1961 and 1962).

7.4 THINGS TO DO

7.4.1 Running the Ditto Algorithm

This algorithm is especially appropriate for category data. Continuous variables may be converted to this form by division into classes such as low, middle, and high. Variables with large numbers of categories will have little effect on the final classification but will increase the expense of computation, so such variables should be avoided. An unusual feature of the algorithm is the presence of a gallimaufry of objects, each forming a singleton cluster.

Table 7.4 Ditto Diagram of Wines

	1961	62	63	64	65	66	67	68	69	70
CLUSTER 0	E	G	P	G	P	G	G	P	G	G
St. Emilion and Pomerol	.	A	.	.	.	.	.	.	A	.
Red Rhone North	.	.	.	.	A	.	.	A	.	.
White Loire	A	P	.	.	'	A	.	.	.	.
California	–	–	–	P	G	A	.	G	.	.
CLUSTER 4	.	.	.	.	D	.	.	.	.	.
Medoc and Graves	.	.	.	.	.	.	.	.	A	.
Beaujolais	.	.	.	.	.	.	A	.	.	.
CLUSTER 9	G	.	D	.	D	.	.	.	.	.
Sauternes	.	.	.	D	.	A	.	.	.	.
Graves	.	.	.	.	.	.	.	.	.	.
CLUSTER 8	.	.	A	.	.	.	.	A	.	.
Red Burgundy	.	.	.	.	.	.	A	D	.	.
Cote de Beaune	.	.	.	.	A	.	.	.	.	.
Chablis	.	.	.	.	.	.	.	.	.	.
CLUSTER 2	G	A	.	.	.	.	.	.	.	.
Red Rhone South	.	.	.	.	A	A	.	A	.	.
Alsace	.	.	.	.	.	.	.	.	.	.
Rhine	.	.	.	.	.	.	.	.	.	.
Moselle	.	.	.	E	.	.	.	.	.	.

Number of symbols = 41

Number without partitions = 150

Fill in cluster centers from cluster 0. Fill in data values from cluster centers.

A good data set for this algorithm is the sleeping pattern of seventeen monkeys in a vervet troop observed by Struhsaker (Table 7.5). Also see the metamorphosis sequences of British butterflies (Table 7.6).

PROGRAMS

SCALE   converts continuous variables to category variables.

DITTO   computes partition of category data to maximize matching between cases in a cluster and the cluster mode.

DITOUT   using output from ditto, this program prints out dot matrix, where each dot represents an identity between a value in the matrix and the corresponding cluster mode.

Table 7.5 Vervet Sleeping Groups

I = Adult male	IX = Young juvenile female
II = Older adult male	X = Juvenile female
III = Adult male	XI = Subadult female
IV = Adult female	XII = Adult female
V = Juvenile male	XIII = Two young indist. juv. males
VI = Adult female	XIV = Infant male, son of IV
VII = Young juvenile female	IV = Infant female, from XII
VIII = Young juvenile female	XVI = Infant male, from VI

	I	II	III	IV	V	VI	VII	VIII	IX	X	XI	XII	XIII	XIV	XV	XVI	XVII
4Jan64	A	D	A	B	A	E	E	E	C	C	C	A	B	A	B		
30Jan64	B	B	B	B	B	A	A	A	A	A	B	B	B	B	B	B	
5Feb64	B	A	C	A	A	D	D	D	D	D	A	A	A	A	A	A	D
10Feb64	A	B	B	D	B	D	D	D	C	C	A	D	A	A	D	D	D
24Feb64	B	B	B	B	B	B	B	B	B	B	A	A	B	B	B	A	B
25Feb64	A	B	A	B	B	B	B	B	B	B	A	B	B	B	B	B	B
10Mar64	C	C	B	C	C	C	C	C	C	A	B	B	B	C	C	B	C
20Mar64	B	B	A	B	B	B	B	B	B	B	A	A	B	B	B	A	B
1Apr64	D	C	A	D	D	D	D	D	B	A	B	C	A	D	D	C	D
2Apr64	D	A	D	C	D	D	D	D	B	D	B	A	B	C	C	A	D
5Apr64	A	A	C	E	D	B	E	E	E	C	E	E	E	E	E	B	
7Apr64	C	C	B	A	C	B	C	B	A	B	A	B	A	A	A	B	B
18Apr64	C	D	B	A	C	C	C	C	A	A	A	B	A	A	A	B	C
28Apr64	A		A	D	A	D	D	D	A	A	C	B	A	D	D	B	D
1May64	B		D	A	C	C	C	C	A	A	B	B	C	A	A	B	C
5May64	D		C	B	F	A	E	E	E	D	C	E	B	B	C	A	
10May64	B		B	E	B	D	D	D	A	F	A	C	A	B	E	C	D
12May64	D		B	D	D	E	E	E	B	B	C	A	B	D	D	A	E
20May64	A		B	D	C	C	C	C	A	C	A	C	D	C	D	C	C
22May64	E		B	F	F	A	A	A	C	F	C	D	F	F	F	D	A
2Jun64	C		D	C	C	E	E	E	C	C	B	A	C	E	C	A	E
5Jun64	A		D	C	A	D	D	A	D	A	B	A	C	A	C	A	D

From Struhsaker, T. T. (1967). "Behaviour of vervet monkeys and other cercopithecines." *Science* **156**, 1197–1203.

Table 7.6 Times of Appearance of British Butterflies

O (ova); L (larva); P (pupa); I (imago)

	Jan	Feb	Mar	Apr	May	Jun	Jul	Aug	Sep	Oct	Nov	Dec
Adonis Blue	L	L	L	L	PI	LI	P	PI	LI	L	L	L
Bath White	P	P	P	P	LI	L	P	PI	LP	P	P	P
Black Hairstreak	O	O	O	O	L	LPI	I	O	O	O	O	O
Black-veined White	L	L	L	L	LP	PI	LI	L	L	L	L	L
Brinstone	I	I	I	I	LI	LI	PI	PI	I	I	I	I
Brown Argus	L	L	L	L	L	P	PI	I	L	L	L	L
Brown Hairstreak	O	O	O	O	L	LP	PI	I	I	O	O	O
Camberwell Beauty	I	I	I	I	I	L	LP	I	I	I	I	I
Chalk Hill Blue	L	L	L	L	LPI	LP	PI	I	L	L	L	L
Chequered Skipper	L	L	L	P	P	I	O	O	L	L	L	L
Clouded Yellow	-	-	-	-	I	LI	LP	I	LI	LPI	I	-
Comina	I	I	I	I	LI	LP	PI	LI	PI	I	I	I
Common Blue	L	L	L	LP	I	IL	LI	I	I	L	L	L
Dark Green Fritillary	L	L	L	L	L	LP	PI	LI	L	L	L	L
Dingy Skipper	L	L	L	LP	I	LI	LP	I	L	L	L	L
Duke of Burgundy Fritillary	P	P	P	P	PI	IOL	L	L	LP	P	P	P
Essex Skipper	O	O	O	OL	L	P	I	I	O	O	O	O
Gatekeeper	L	L	L	L	L	LP	I	I	L	L	L	L
Glanville Fritillary	L	L	L	LP	LPI	LPI	L	L	L	L	L	L
Grayling	L	L	L	L	L	LPI	PI	LI	LI	L	L	L
Green Hairstreak	P	P	P	P	PI	LI	LP	P	P	P	P	P
Green-veined White	P	P	P	PI	LI	LPI	LPI	LPI	LP	P	P	P
Grizzled Skipper	L	L	L	LP	I	LI	LP	I	L	L	L	L
Heath Fritillary	L	L	L	L	L	LPI	LI	L	L	L	L	L
High Brown Fritillary	L	L	L	L	L	LP	PI	LI	L	L	L	L
Holly Blue	P	P	P	PI	I	LP	PI	LI	LPI	LP	P	P
Large Blue	L	L	L	L	L	LI	I	L	L	L	L	L

[Ford, T. L. E. (1963). *Practical Entomology*, Warne, London, p. 181.] (A subset has been selected from the full list.)

```
      SUBROUTINE SCALE(A,M,N,KL,KK)
C.........................................................20 MAY 1973
C.... SCALES ARRAY TO TAKE INTEGER OR ALPHAMERIC VALUES 1,2,3,... KL
C     MINIMUM AND MAXIMUM VALUES ARE COMPUTED FOR EACH VARIABLE, AND EACH VALUE
C     IS THEN CLASSIFIED INTO ONE OF KL INTERVALS OF EQUAL LENGTH BETWEEN THE
C     MINIMUM AND MAXIMUM.
C.... M = NUMBER OF ROWS
C.... N = NUMBER OF COLUMNS
C.... A = M BY N BORDERED ARRAY
C.... KL = NUMBER OF LEVELS
C.... KK = LEVELLING OPTION
C          KK = 1 UNIFORM OVER ALL DATA VALUES
C          KK = 2 UNIFORM WITHIN VARIABLES
C          KK = 3 SAME AS OPTION 1, CONVERTED TO ALPHAMERIC
C          KK = 4 SAME AS OPTION 2, CONVERTED TO ALPHAMERIC
C.........................................................................
      DIMENSION A(M,N)
      DIMENSION CC(9)
      DATA CC/1H1,1H2,1H3,1H4,1H5,1H6,1H7,1H8,1H9/
      IF(KK.GE.3.AND.KL.GT.9) WRITE(6,1) KL
    1 FORMAT(I5,27H TOO MANY ALPHAMERIC LEVELS
      IF(KK.EQ.1.OR.KK.EQ.3) GO TO 20
C.... COMPUTE MINIMUM AND MAXIMUM
      DO 10 J=2,N
      XMIN=A(2,J)
      XMAX=A(2,J)
      DO 11 I=2,M
      IF(A(I,J).GT.XMAX) XMAX=A(I,J)
      IF(A(I,J).LT.XMIN) XMIN=A(I,J)
   11 CONTINUE
C.... CHANGE A VALUES TO INTEGER OR ALPHAMERIC
      IF(XMIN.EQ.XMAX) XMAX=XMIN+.000001
      ZZ=KL/(XMAX-XMIN)
      DO 12 I=2,M
      K=(A(I,J)-XMIN)*ZZ+1
      IF(KK.EQ.4) A(I,J)=CC(K)
      IF(KK.EQ.2) A(I,J)=K
   12 CONTINUE
   10 CONTINUE
      RETURN
   20 CONTINUE
C.... MINIMUM AND MAXIMUM
      XMIN=A(2,2)
      XMAX=A(2,2)
      DO 21 I=2,M
      DO 21 J=2,N
      IF(A(I,J).LT.XMIN) XMIN=A(I,J)
   21 IF(A(I,J).GT.XMAX) XMAX=A(I,J)
C.... CHANGE A VALUES
      IF(XMIN.EQ.XMAX) XMAX=XMIN+.000001
      ZZ=KL/(XMAX-XMIN)
      DO 22 I=2,M
      DO 22 J=2,N
      K=(A(I,J)-XMIN)*ZZ+1.
      IF(KK.EQ.3) A(I,J)=CC(K)
      IF(KK.EQ.1) A(I,J)=K
   22 CONTINUE
      RETURN
      END
```

```
      SUBROUTINE DITTO(M,N,KL,KC,A,X,LC,LK,Y,Z)
C..............................................................20 MAY 1973
C.... COMPUTES PARTITION OF CATEGORY DATA TO MAXIMIZE MATCHING BETWEEN CASES IN
C          A CLUSTER AND THE CLUSTER MODE.
C.... M = NUMBER OF ROWS
C.... N = NUMBER OF COLUMNS
C.... KL = MAXIMUM NUMBER OF DIFFERENT VALUES TAKEN BY A VARIABLE PLUS ONE.
C.... KC = KL BY N +1
C.... A = M BY N BORDERED DATA ARRAY
C.... X = KC BY N ARRAY OF CLUSTER MODES
C.... LC = 1 BY M ARRAY ASSIGNING CASES TO CLUSTERS
C.... LK = 1 BY KC ARRAY COUNTING CASES IN CLUSTERS
C.... Y = KL BY N ARRAY COUNTING FREQUENCIES IN CLUSTERS
C.... Z = KL BY N ARRAY SPECIFYING DIFFERENT VALUES OF VARIABLES
C..........................................................................
      DIMENSION A(M,N),X(KC,N),LC(M),LK(KC),Y(KL,N),Z(KL,N)
      DIMENSION CC(10)
      DATA CC/1H1,1H2,1H3,1H4,1H5,1H6,1H7,1H8,1H9,2H10/
      DATA RNGE,XMOD/4HRNGE,4HMODE/
C.... PUT LABELS IN VARIOUS ARRAYS
      DO 80 K=2,KL
      KK=(K-1)-((K-2)/10)*10
   80 Z(K,1)=CC(KK)
      DO 81 J=2,N
   81 Z(1,J)=A(1,J)
      Z(1,1)=RNGE
      DO 82 K=2,KC
      KK=(K-1)-((K-2)/10)*10
   82 X(K,1)=CC(KK)
      DO 83 J=2,N
   83 X(1,J)=A(1,J)
      X(1,1)=XMOD
C.... FIND DIFFERENT VALUES TAKEN BY VARIABLES
      DO 9 I=2,M
    9 LC(I)=0
      DO 8 K=1,KC
    8 LK(K)=1
      NC=-1
      DO 10 J=2,N
      DO 10 K=2,KL
   10 Z(K,J)=0
      DO 11 J=2,N
      DO 12 I=2,M
      DO 13 K=2,KL
      IF(Z(K,J).EQ.0) Z(K,J)=A(I,J)
      IF(Z(K,J).EQ.A(I,J)) GO TO 12
   13 CONTINUE
   12 CONTINUE
   11 CONTINUE
C.... COMPUTE MODES
   70 NC=NC+1
      DO 20 J=2,N
      DO 20 K=2,KL
      DO 21 JJ=2,N
      DO 21 KK=2,KL
   21 Y(KK,JJ)=0
      KT=K+1+(J-2)*(KL-1)
      IF(LK(KT).EQ.0) GO TO 20
      DO 23 I=2,M
      IF(NC.EQ.0.AND.A(I,J).NE.Z(K,J)) GO TO 23
      IF(NC.NE.0.AND.LC(I).NE.KT) GO TO 23
      DO 25 JJ=2,N
      DO 25 KK=2,KL
   25 IF(A(I,JJ).EQ.Z(KK,JJ)) Y(KK,JJ)=Y(KK,JJ)+1
   23 CONTINUE
      DO 26 JJ=2,N
      YM=Y(2,JJ)
      KM=2
      DO 27 KK=2,KL
      IF(Y(KK,JJ).LE.YM) GO TO 27
      YM=Y(KK,JJ)
      KM=KK
   27 CONTINUE
   26 X(KT,JJ)=Z(KM,JJ)
   20 CONTINUE
C.... COMPUTE GRAND MODE
      DO 55 J=2,N
      DO 50 K=2,KL
```

152

```
    50 Y(K,J)=0
       DO 51 K=3,KT
       IF(LK(K).EQ.0) GO TO 51
       DO 52 KK=2,KL
    52 IF(X(K,J).EQ.Z(KK,J)) Y(KK,J)=Y(KK,J)+1
    51 CONTINUE
       DO 53 KK=2,KL
       YM=Y(2,J)
       X(2,J)=Z(2,J)
       IF(Y(KK,J).LE.YM) GO TO 53
       YM=Y(KK,J)
       X(2,J)=Z(KK,J)
    53 CONTINUE
    55 CONTINUE
C.... REASSIGN CASES
       DO 30 I=2,M
       DM=10.**10
       KM=1
       DO 32 K=2,KT
       IF(LK(K).EQ.0) GO TO 32
       IF(NC.NE.0.AND.LK(K).LE.1) GO TO 32
       DD=0
       DO 31 J=2,N
    31 IF(A(I,J).NE.X(K,J)) DD=DD+1
       IF(DD.GE.DM) GO TO 32
       KM=K
       DM=DD
    32 CONTINUE
    30 LC(I)=KM
C.... COUNT CASES IN CLUSTERS
       DO 35 K=3,KC
    35 LK(K)=0
       DO 36 I=2,M
       K=LC(I)
    36 LK(K)=LK(K)+1
       DO 37 K=3,KC
       IF(LK(K).NE.0) GO TO 37
       DO 38 J=2,N
    38 X(K,J)=0.
    37 CONTINUE
       IF(NC.LT.5) GO TO 70
       RETURN
       END
```

```
      SUBROUTINE DITOUT(A,M,N,X,KC,LC)
C..............................................................20 MAY 1973
C.... PRINTS OUT DATA MATRIX WITH CLUSTER VALUES DOTTED OUT
C.... DATA MATRIX A AND CLUSTER MATRIX X ARE ALPHABETIC
C.... USE AFTER PROGRAM DITTO
C.... M = NUMBER OF ROWS
C.... N = NUMBER OF COLUMNS
C.... KC = NUMBER OF CLUSTERS
C.... A = M BY N BORDERED ARRAY,DATA VALUES
C.... X = KC BY N BORDERED ARRAY, CLUSTER VALUES
C.... SECOND ROW OF X CONTAINS GRAND CLUSTER
C.... LC = 1 BY M ARRAY ASSIGNING CASES TO CLUSTERS
C.........................................................................
      DIMENSION A(M,N),X(KC,N),LC(M)
      DIMENSION AA(120)
      DATA DOT/1H./
      WRITE(6,1) A(1,1)
    1 FORMAT(14H1DOT MATRIX OF,A5)
      WRITE(6,5)
    5 FORMAT(
     *63H FIRST CLUSTER IS GRAND MODE.  ALL OTHER CLUSTER VALUES ARE
     *63HREPLACED BY  '.' IF THEY AGREE WITH VALUE OF GRAND MODE
     *63H    WITHIN A CLUSTER VALUES ARE REPLACED BY '.' IF THEY AGREE
     *63H WITH CLUSTER VALUE.
     *)
      DO 20 K=2,6
   20 WRITE(6,2)(A(1,J),J=K,N,5)
    2 FORMAT(10X,24A5)
C.... DATA
      DO 30 K=2,KC
      NC=0
      DO 40 I=2,M
      IF(LC(I).NE.K) GO TO 40
      NC=NC+1
      DO 60 J=2,N
   60 IF(K.GE.3.AND.X(K,J).EQ.X(2,J)) X(K,J)=DOT
      IF(NC.EQ.1) WRITE(6,3) K,(X(K,J),J=2,N)
    3 FORMAT(5H0CLUS,I3,3X,120A1)
      DO 50 J=2,N
      AA(J)=A(I,J)
      IF(X(K,J).EQ.DOT) X(K,J)=X(2,J)
   50 IF(A(I,J).EQ.X(K,J)) AA(J)=DOT
      WRITE(6,4) A(I,1),(AA(J),J=2,N)
    4 FORMAT(3X,A5,3X,120A1)
   40 CONTINUE
   30 CONTINUE
      RETURN
      END
```

154

CHAPTER 8

Drawing Trees

8.1 DEFINITION OF A TREE

A tree is a family of clusters such that any two clusters are disjoint or one includes the other. Thus, in the clusters of animals in Table 8.1, cluster 4 and cluster 9 are disjoint and cluster 6 contains cluster 8. It is often convenient to require the set of all objects to be a cluster and to define a single-object cluster for each object.

Table 8.1 Clusters of Animals Forming a Tree

CLUSTER 1.   HUMAN (HN)   MONKEY (MY)   HORSE (HE)   PIG (PG)

                  WHALE (WE)   DOG (DG)   RABBIT (RT)   KANGAROO (KO)

                  CHICKEN (CN)   PENGUIN (PN)   DUCK (DK)   TURTLE (TE)

                  BULLFROG (BG)

CLUSTER 2.   HN MY HE PG WE DG RT KO CN PN DK TE

CLUSTER 3.   CN PN DK TE

CLUSTER 4.   CN PN DK

CLUSTER 5.   CN PN

CLUSTER 6.   HN MY HE PG WE DG RT KO

CLUSTER 7.   HE PG WE DG

CLUSTER 8.   HE PG

CLUSTER 9.   HN MY

8.2   REORDERING TO CONTIGUOUS CLUSTERS

It is always possible to order the objects so that every cluster consists of a set of objects contiguous in the order. Let $O(I)$ ($1 \leq I \leq M$) denote the name of the Ith object in the order.

STEP 1.   Select the first object, $O(1)$, arbitrarily.

STEP 2.   Select the second object, $O(2)$, from the smallest cluster containing $O(1)$ but not included in the set $\{O(1)\}$. If no such cluster exists, select the second object to be any object other than $O(1)$.

STEP 3.   Select the Kth object, $K = 3, 4, \ldots, M$, from the smallest cluster containing $O(K-1)$, but not contained in the set $\{O(1), O(2), \ldots, O(K-1)\}$. If no such cluster exists, select the Kth object to be any object not $O(1), \ldots, O(K-1)$.

8.3   APPLICATION OF REORDERING TO ANIMAL CLUSTERS

The order of objects in cluster 1, Table 8.1, is such that all clusters are contiguous in the order. For example, cluster 7 = {HE, PG, WE, DG} which are the third through sixth objects, respectively, in cluster 1. To illustrate the algorithm, an object other than human will be selected to be the first object.

STEP 1.   Let $O(1)$ = RT, rabbit.

STEP 2.   The smallest cluster containing the rabbit is cluster 6, from which the object $O(2)$ = KO, kangaroo, is selected.

STEP 3.   The smallest cluster containing KO is again cluster 6, from which $O(3)$ = DG. The smallest cluster containing DG is cluster 7, from which $O(4)$ = WE. The smallest containing WE is again 7, from which $O(5)$ = PG. The smallest containing PG is cluster 8, $O(6)$ = HE. The smallest containing HE but not included in $O(1)$, $O(2), O(3), O(4), O(5), O(6)$ is cluster 6, $O(7)$ = MY, and so on. Finally, the order of objects is RT KO DG WE PG HE MY HN PN CN DK TE BG. Every cluster is a contiguous set of objects in this order. For example, cluster 3 is the ninth through twelfth objects.

8.4   NAMING CLUSTERS

Since every cluster consists of a sequence of contiguous objects in some ordering of the objects, a natural way to name the clusters is by the first and last object in the sequence. Thus the human-turtle cluster contains all animals but the bullfrog. This naming procedure is not unique since the ordering of the objects is not unique. The two objects in the name of the cluster have the property that the smallest cluster which includes them both is the cluster named.

Given the ordering of the objects and the two names for each cluster, the complete tree may be recovered. Thus the animal clusters are determined by the order HN MY HE PG WE DG RT KO CN PN DK TE BG and by the named clusters HN-MY, HN-KO, HE-DG, CN-PN, CN-DK, CN-TE, HN-TE, HE-PG.

8.5 *I*-REPRESENTATION OF CLUSTERS WITH DIAMETERS

Preliminaries. Each cluster has a diameter $D(C)$, a positive integer, such that $D(C1) < D(C2)$ if $C1$ is properly included in $C2$. For example, let the smallest clusters have diameter 1, and let every other cluster have a diameter that is one greater than the maximum diameter of the clusters properly contained in it.

This algorithm represents the clusters graphically, using the symbols $-$, I, and $/$, although other symbols are plausible. For example, $'$, $\cdot$, might replace I, $/$. There is one line for each object, with the name of the object left of the 0 position in the line and with various symbols in the positive positions on the line.

STEP 1.   Arrange the objects in order, so that each cluster is a sequence of contiguous objects.

STEP 2.   Write the name of the first object to the left of the 0 position of the first line.

STEP 3.   Let C be the largest cluster such that the first object is the first or last object in C. Write the symbol $-$ in all positions not greater than $D(C)$.

STEP 4.   For every cluster C containing the first object, write the symbol I in position $D(C)$ (if necessary, replace a previously written $-$).

STEP 5.   For all clusters C such that the present object is the last object in C, write the symbol $/$ in position $D(C)$.

STEP 6.   Repeat Steps 2–5 for all objects $(1 \leq I \leq M)$ with the names of the objects listed vertically on the left and with the symbols denoting cluster membership listed on the right.

8.6 *I*-REPRESENTATION OF ANIMAL CLUSTERS

It is necessary to define a diameter for each cluster to determine its horizontal position on the page. The diameter given in Section 8.5 produces a representation occupying a minimum width of page. [A more spread-out representation appropriate for small sets of objects has $D(C)$ equal to the number of objects in C.] For the animals, the diameters are $D(\text{HN-MY}) = 1$, $D(\text{HE-PG}) = 1$, $D(\text{HE-DG}) = 2$, $D(\text{HN-KO}) = 3$, $D(\text{CN-PN}) = 1$, $D(\text{CN-DK}) = 2$, $D(\text{CN-TE}) = 3$, $D(\text{HN-TE}) = 4$, $D(\text{HN-BG}) = 5$. For example, $D(\text{HN-KO}) = \max [D(\text{HE-PG}), D(\text{HE-DG}), D(\text{HN-MY})] + 1 = 3$.

STEP 1.   The order used is the order given in Table 8.1.

STEP 2.   The first object is named human.

STEP 3.   The largest cluster C such that human is first or last is HN-BG. Write the symbol $-$ in positions 1, 2, 3, 4, 5 $= D(\text{HN-BG})$.

STEP 4.   The clusters C that contain human are HN-MY, HN-KO, HN-TE, and HN-BG. Thus the symbol I appears in positions 1, 3, 4, 5 (replacing the symbol $-$).

STEP 5.   Human is the last member of no cluster and so $/$ does not appear.

STEP 6. Steps 2–5 are repeated for monkey. The largest cluster containing monkey as the last object is HN-MY. So the symbol $-$ is written in position 1. The clusters

HN-MY, HN-KO, HN-TE, and HN-BG contain monkey, so the symbol I appears in 1, 3, 4, 5 (replacing $-$ in position 1). The cluster HN-MY has monkey as its last object so the symbol / appears in position 1. In this way, all objects are represented in Table 8.2. Note that during the algorithm, more than one symbol may be assigned to a given position, but the latest symbol is always the one used.

To read the table, look for the symbols /. The vertical tower of I's over every / defines a cluster of objects. Each cluster corresponds to just one / in the table.

Table 8.2 I Representation of Animal Clusters

```
                    5              DIAMETER
        ──────────────┼──────────────────────────────►
   HUMAN · I - I I I
  MONKEY · /   I I I
   HORSE · I I I I I
     PIG · / I I I I
   WHALE ·   I I I I
     DOG · - / I I I
  RABBIT ·     I I I
KANGAROO · - - / I I
 CHICKEN · I I I I I
 PENGUIN · / I I I I
    DUCK · - / I I I
  TURTLE · - - / / I
BULLFROG · - - - - /
```

Each cluster is a vertical tower of I's over a /.

8.7   TREES AND DIRECTED GRAPHS

Suppose that all objects and clusters are nodes of a graph and that each node is linked to the node corresponding to the smallest cluster which includes it, with the link directed from the lesser node to the greater. Then the set of nodes and links form a directed graph, such that each node has at most one link leaving it, and there are no cycles. Any directed graph satisfying these conditions may conversely be represented as a family of clusters. The animal clusters are represented by such a graph in Table 8.3 (see also Figure 8.1).

8.8   LINEAR REPRESENTATION OF TREES

Let $D(C)$ be a diameter, as defined in Section 8.5, such that $D(C)$ is a positive integer and $D(C1) < D(C2)$ whenever $C1 \subset C2$, $C1 \neq C2$. Arrange the objects in order on the line, such that each cluster is a contiguous sequence of objects, and separate two contiguous objects by the diameter of the smallest cluster that contains them both. Clusters are recovered from such a linear representation by associating a cluster with each interval between objects, the cluster consisting of the maximal sequence of objects including the interval and containing no interval of greater size.

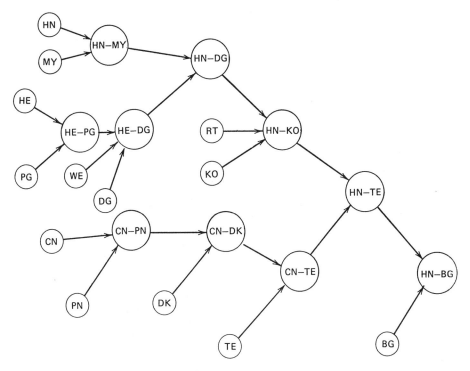

Figure 8.1 Trees and directed graphs.

Table 8.3 Animal Clusters Represented by Directed Graph

NODE	ANCESTOR	NODE	ANCESTOR
HN	HN - MY	HN - MY	HN - KO
MY	HN - MY	HE - PG	HE - DG
HE	HE - PG	HE - DG	HN - KO
PG	HE - PG	HN - KO	HN - TE
WE	HE - DG	CN - PN	CN - DK
DC	HE - DG	CN - DK	CN - TE
RT	HN - KO	CN - TE	HN - TE
KO	HN - KO	HN - TE	HN - BG
CN	CN - PN	HN - BG	
PN	CN - PN		
DK	CN - DK		
TE	CN - TE		
BG	HN - BG		

159

Table 8.4   Linear Representation of Animal Clusters

```
HN -MY ---HE -PG --WE --DG - ---RT --- KO ----CN -PN --DK ---TE -----BG
```

```
HN    MY    HE    PG    WE    DG    RT    KO    CN·   PN    DK    TE    BG
  I     I     I     I     I     I     I     I     I     I     I     I
  I           I     I     I     I     I                 I     I     I
  I                       I     I     I                       I     I
                                I                                   I
                                                                    I
```

The animal clusters are represented in this way in Table 8.4. Also in Table 8.4 is a representation where the spaces between objects are represented by a vertical column of I's. This diagram is intermediate between the linear representation and the representation in Table 8.2. The linear representation is useful when a data vector is associated with each object, and it is desired to incorporate this vector in the representation, as in Table 8.5. Blanks between lines of data indicate the clusters; as a

Table 8.5   Cost and Nutrient Contributions for Selected Foods

FOOD	COST OF SERVING (Cents)	SIZE	% OF DAILY ALLOWANCE				
			PROTEIN	IRON	THIAMINE	RIBOFLAVIN	NIACIN
HAM	28	3 oz	29	21	38	11	29
PORK CHOPS	25	3 oz	29	22	59	12	36
BEEF RIB ROAST	25	3 oz	29	22	4	9	30
BEEF CHUCK ROAST	18	3 oz	32	22	3	10	29
HALIBUT	18	3 oz	33	6	4	4	74
BEEF LIVER	15	3 oz	30	55	18	198	105
EGGS	10	2 eggs	19	22	7	16	1
DRY BEANS	2·	3/4 cup	16	30	8	6	12
BACON	5	2 strips	6	4	7	3	7
PEANUT BUTTER	4	2 tablespoons	12	5	3	2	43

From *Yearbook of Agriculture* (1959).

rule of thumb, the total number of blank lines is about equal to the original number of objects.

8.9   TREES AND DISTANCES

Given a tree with cluster diameters, there are two ways of defining distances between objects. In the first, the distance between two objects is the diameter of the smallest cluster which includes them both:

$$D(I, J) = \min \{C \mid I \in C, J \in C\} \, D(C).$$

Such a distance is an *ultrametric;* that is, it satisfies the inequality $D(I, J) \leq \max [D(I, K), D(J, K)]$. Equivalently, the two largest distances in $D(I, J)$, $D(I, K)$, $D(J, K)$ are equal. Conversely, any ultrametric generates a tree and cluster diameters from which it is recovered using the above definition (see Hartigan, 1967; Johnson, 1967; Jardine, Jardine, and Sibson, 1967). For example, see Table 8.6; the smallest

Table 8.6 Distances Computed from Trees

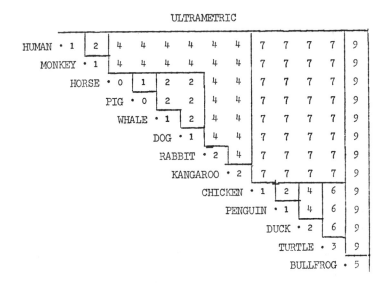

```
                      5         10                    DIAMETER
          ────────────┼─────────┼────────────────────
  HUMAN · / I - I - - I - I
 MONKEY · / /   I     I   I
  HORSE / I I   I     I   I
    PIG / / I   I     I   I
  WHALE · / I   I     I   I
    DOG · / /   I     I   I
 RABBIT · - /   I     I   I
KANGAROO · - / - /     I   I
CHICKEN · / I - I - I I   I
PENGUIN · / /   I   I I   I
   DUCK · - / - /   I I   I
 TURTLE ·.- - / - - / /   I
BULLFROG · - - - - / - - - /
```

ULTRAMETRIC

HUMAN · 1	2	4	4	4	4	4	4	7	7	7	7	9
MONKEY · 1		4	4	4	4	4	4	7	7	7	7	9
HORSE · 0		1	2	2	4	4		7	7	7	7	9
PIG · 0			2	2	4	4		7	7	7	7	9
WHALE · 1				2	4	4		7	7	7	7	9
DOG · 1					4	4		7	7	7	7	9
RABBIT · 2						4		7	7	7	7	9
KANGAROO · 2								7	7	7	7	9
CHICKEN · 1									2	4	6	9
PENGUIN · 1										4	6	9
DUCK · 2											6	9
TURTLE · 3												9
BULLFROG · 5												

Arbitrary diameters have been assigned to the clusters.

cluster including human and monkey is HN-MY, which has a diameter of 2. Thus $D(\text{HN}, \text{MY}) = 2$. The smallest cluster including human and chicken is the HN-TE cluster with a diameter of 7. Also, this cluster is the smallest including monkey and chicken. Thus $D(\text{HN}, \text{CN}) = 7$ and $D(\text{MY}, \text{CN}) = 7$. The two largest distances of $D(\text{HN}, \text{MY})$, $D(\text{HN}, \text{CN})$, $D(\text{MY}, \text{CN})$ are equal, as behooves an ultrametric.

Another plausible definition has the distance between two objects equal to the sum of links connecting the two objects in the directed graph described in Section 8.6. This definition will be stated in terms of cluster diameters, rather than in terms of links. It will be assumed that all simple object clusters are included in the tree and also the full-set cluster. The link distance is defined to be

$$D(I, J) = 2D(C) - D(I) - D(J),$$

where $D(I)$ and $D(J)$ are diameters of objects I and J and where C is the smallest cluster containing I and J. If all single-object clusters have the same diameter, then $D(I, J)$ is an ultrametric.

The link distance is not much different from an ultrametric. A distance $D(I, J)$ is a link distance if and only if, for every K, $D(I, J) - D(I, K) - D(J, K)$ is an ultrametric. It has been suggested by J. S. Farris that a set of empirical distances, thought to fit the link distance model approximately, should be transformed to

$$MD(I, J) - \sum \{1 \le K \le M\} D(I, K) - \sum \{1 \le K \le M\} D(J, K),$$

which will then fit the ultrametric model approximately.

In Table 8.6 is given a method of representing clusters directly on a distance matrix. This is very useful in the routine examination of distance or correlation matrices. The objects are first contiguously ordered relative to the tree. For a square distance matrix, each cluster corresponds to a square contiguous submatrix. This submatrix is outlined. Whenever the distance matrix is symmetrical (the usual case), one half of the matrix and the corresponding outlines are ignored. Each position in the matrix is indexed by a pair of objects. The lines partition the pairs of objects, so that within each member of the partition the pairs all have the same cluster as the minimum cluster including them both. In particular, in the ultrametric model all values within a member of the partition should be equal.

8.10 BLOCK REPRESENTATION OF TREES

There are many ways of representing trees, from the evolutionary trees seen in biology books where the width of the branches has meaning (cluster diameter is the time back to the common ancestor of objects in the cluster) to the stark trees in combinatorics (see Sokal and Sneath, 1963; Bertin, 1967; Lockwood, 1969). A slightly different representation from the I representation is used in biology and is superior in allowing more space in the diagram for text describing the clusters.

STEP 1.   Order the objects so that each cluster is a contiguous sequence of objects.

STEP 2.   On the $2 \times J$th line ($1 \le J \le M$), place the name of object J, then put the symbol I in each position $D(C)$, where C includes J.

STEP 3.   On the ($2 \times J + 1$)th line ($1 \le J \le M$), put the symbol $-$ in every position K ($1 \le K < D(C)$), where C is the smallest cluster containing the objects J and $J + 1$.

STEP 4.   On the first and ($2M + 1$)th lines, put the symbol $-$ in every position K [$1 \le K < D(C)$], where C is the cluster of all objects.

An example of a block representation appears in Table 8.7. Note that each cluster is represented by a maximal tower of I's above a $-$. There is also a block surrounded by I's and $-$'s for each cluster; within this block, descriptive material about the cluster may be written.

Table 8.7 Block Representation of Animal Clusters

```
             ‾ ‾ ‾ ‾ ‾ ‾ ‾
   HUMAN ·  I I   I     I   I
  MONKEY ·  I I _ I     I   I
   HORSE  I I I   I     I   I
     PIG  I I I   I     I   I
   WHALE ·  I I   I     I   I
     DOG ·  I I _ I     I   I
  RABBIT ·  _ I _ I   I   I
KANGAROO ·  _ I _ I _ _ I   I
 CHICKEN ·  I I   I     I   I
 PENGUIN ·  I I _ I     I   I
    DUCK ·  _ I _ I   I   I
  TURTLE ·  _ _ I I _ _ I _ I
BULLFROG ·  _ _ _ _ I _ _ _ I
```

8.11 THINGS TO DO

8.11.1 Graphs

The term tree is used in a slightly different sense in graph theory—to denote an undirected graph with the property that there is a unique path connecting any two nodes. Selecting a particular node to be the root and directing all links toward the root produces a rooted tree, as in Section 8.6. The undirected definition goes naturally with the second distance model in Section 8.8, where the distance between the two nodes is the sum of the distances of the links connecting the two nodes. Fitting this distance model gives no hint about the root, so that a number of different clusterings may conform exactly to the same set of pairwise distances.

Trees have been investigated extensively in graph theory and combinatorics, principally by counting the number of trees of various types but also by considering characteristics of randomly generated trees.

8.11.2 Traversing Trees

A tree is a standard data structure in computing. A tree of K nodes may be represented by an ancestor array $N(1), \ldots, N(K)$, where for each I there is a single link from I to $N(I)$. The root may be identified as the node J for which $N(J) = J$. A very convenient array for traversing the nodes of a tree is the two-dimensional array $\{M(1, I), M(2, I), 1 \leq I \leq K\}$, where $M(1, I)$ is the next node with the same ancestor as I and where $M(2, I)$ is the first node whose ancestor is I. Set $M(1, I) = 0$ and $M(2, I) = 0$ if no node satisfies the definitions.

The array $\{N(I), 1 \leq I \leq K\}$ may be used for certain operations, such as averaging, moving from the end objects toward the root. The array $\{M(1, J), M(2, I), 1 \leq I \leq K\}$ may be used for other operations, such as allocation of objects to the tree or moving from the root toward the end nodes. Show how to traverse all the nodes of the tree by using the arrays $\{N(I), M(1, I), M(2, I), 1 \leq I \leq M\}$. Show how to compute the array $\{N(I), 1 \leq I \leq M\}$ from $\{M(1, I), M(2, I), 1 \leq I \leq M\}$ and vice versa.

8.11.3. Natural trees

Leafy trees are rooted trees, except if their root system is included, when they become unrooted or undirected trees. Veins and arteries are trees directed to the heart. The nervous system is a tree directed to the brain. A river system is a tree directed to a sea, with a few multiple paths in swamps and deltas ignored. Company organization charts are frequently trees, sometimes only slightly related to real channels of power. Genealogies are trees, forward from a person to all his descendants and back to his ancestors. Since there are only two sexes, the backward tree is binary; it is also not a tree once the ancestry is traced far enough back to include relatives by birth.

Trees may be used to describe sentence structure. Trees may be used to organize a book into volumes, chapters, sections, and paragraphs. Schemes such as the Dewey decimal system are trees for naming, storing, and accessing books. It is less important that these be optimal, correct, or rational than that they be unambiguous, widely accepted, and easy to use.

REFERENCES

BERTIN, J. (1967). *Semiology graphique*, Gauthier-Villars, Paris. This book is a monumental work, with many beautiful graphs for all known graphical problems and with some intricate and striking tree diagrams, including a circular tree of the descendants of Genghis Khan (see p. 276).

HARTIGAN, J. A. (1967). "Representation of similarity matrices by trees." *J. Am. Stat. Assoc.* **62**, 1140–1158. Defines a similarity matrix to have "exact tree structure" if it may be defined from a tree with cluster diameters, $S(I, J)$, depending only on the diameter of the smallest cluster containing I and J. The matrix S is then an ultrametric.

JARDINE, C. J., JARDINE, N., and SIBSON, R. (1967). "The structure and construction of taxonomic hierarchies." *Math. Biosci.* **1**, 173–179. A short sharp paper considering ultrametrics as a limiting ($\lambda \rightarrow 0$) case of λ metrics:

$$D(I, J)^{1/\lambda} \leq D(I, K)^{1/\lambda} + D(J, K)^{1/\lambda}.$$

They show that the distance defined from a tree with cluster diameters is an ultrametric, and that conversely an ultrametric generates a tree with cluster diameters. For a given metric distance D, there is a largest distance D^* that is an ultrametric and satisfies $D^*(I, J) \leq D(I, J)$ for all I, J. The corresponding tree is the one obtained from D by the single linkage algorithm.

JOHNSON, S. C. (1967). "Hierarchical clustering schemes." *Psychometrika* **32**, 241–254. Discusses some joining algorithms and considers ultrametrics as the distances for which all algorithms should give the same result.

LOCKWOOD, A. (1969). *Diagrams*, Watson Guptill, New York. Some family trees and evolutionary trees and a tree substitute for bureaucratese, specifying when an old lady is allowed a pension, are given on pp. 125, 126, 127.

SOKAL, R. R., and SNEATH, P. H. A. (1963). *Principles of Numerical Taxonomy*, Freeman, San Francisco. Contains some samples of trees with cluster diameters, called dendrograms by the authors (pp. 197–201).

PROGRAMS

TREE1   prints tree given tree information in boundary form (smallest and largest index of cases in each cluster).

NBTOMT converts tree information in boundary form to tree information in pointer form, specifying for each cluster the ancestor or smallest cluster properly including it.

CNVERT converts tree information in downward pointer form (next cluster at same level and first included cluster are specified for each cluster) to tree information in boundary form.

BLOCK   draws outlines of clusters over distance matrix.

```
      SUBROUTINE TREE1(RL,M,NB,K)
C.................................................................20 MAY 1973
C.... PRINTS TREE ON OBJECTS NAMED IN RL, USING BOUNDARY INFORMATION IN NB.
C.... M = NUMBER OF OBJECTS
C.... K = NUMBER OF CLUSTERS
C.... RL = 1 BY M ARRAY, RL(I)= LABEL OF ITH OBJECT.
C.... NB = 3 BY K ARRAY
C          NB(1,K) = FIRST OBJECT IN CLUSTER, INDEX BETWEEN 1 AND M.
C          NB(2,K) = SECOND OBJECT IN CLUSTER, INDEX BETWEEN 1 AND M
C          NB(3,K) = DIAMETER OF CLUSTER, BETWEEN 0 AND 100
C.................................................................................
      DIMENSION RL(M),NB(3,K),AA(101)
      DATA A,B,C,D,E/1H ,1H-,,1H',1H/,1H /
C.... CHECK NB ARRAY FOR ILLEGAL VALUES
      DO 10 KK=1,K
      I=NB(1,KK)
      J=NB(2,KK)
      L=NB(3,KK)
          IF(I.LT.1.OR.I.GT.J.OR.J.GT.M) WRITE(6,1) KK,I,J,L
      IF(L.LT.0.OR.L.GT.100) WRITE(6,1) KK,I,J,L
   10 CONTINUE
    1 FORMAT(17H ERROR IN CLUSTER,I4,12HFIRST OBJ = ,I3,11HLAST OBJ =
     *,I3,7HDIAM = ,I3)
C.... PRINT TREE, LINE BY LINE.
      DO 20 I=1,M
      DO 21 L=1,101
   21 AA(L)=E
      DO 22 KK=1,K
      IF(NB(1,KK).GT.I) GO TO 22
      IF(NB(2,KK).LT.I) GO TO 22
      J=NB(3,KK)+1
      AA(J)=C
      IF(NB(1,KK).NE.I.AND.NB(2,KK ).NE.I) GO TO 22
      DO 23 L=1,J
   23 IF(AA(L).EQ.E) AA(L)=B
      IF(NB(2,KK).EQ.I) AA(J)=C
      IF(NB(1,KK).EQ.I) AA(J)=A
   22 CONTINUE
      WRITE(6,2) I,RL(I),(AA(J),J=1,101)
    2 FORMAT(I5,4X,A4,1X,101A1)
   20 CONTINUE
      RETURN
      END
```

```
      SUBROUTINE NBTOMT(NB,K,M,KT,MT)
C.......................................................................20 MAY 1973
C.... CONVERTS NB DESCRIPTION OF CLUSTERS INTO MT DESCRIPTION
C.... K = NUMBER OF CLUSTERS, K.GE.1
C.... NB = 3 BY K ARRAY DEFINING K CLUSTERS
C           NB(1,K)=FIRST OBJECT IN CLUSTER,GE 2
C           NB(2,K) = LAST OBJECT IN CLUSTER,LE M.
C           NB(3,K) = CLUSTER DIAMETER, IGNORED IN THIS PROGRAM
C.... M = NUMBER OF OBJECTS
C.... KT = M+K = NUMBER OF NODES IN TREE
C.... MT = KT BY 1 TREE ARRAY, COMPUTED IN PROGRAM.  MT(I).GT.I, EXCEPT AT I=KT
C.......................................................................
      DIMENSION NB(3,K),MT(KT)
C..... CHECK NB ARRAY
      DO 20 I=1,K
      IF(NB(1,I).LT.2) WRITE(6,1) I
      IF(NB(2,I).LT.NB(1,I)) WRITE(6,2)I
   20 IF(NB(2,I).GT.M) WRITE(6,3) I
    1 FORMAT(8H CLUSTER,I5,25H HAS BOUNDARY LESS THAN 2)
    2 FORMAT(8H CLUSTER,I5,45H HAS FIRST BOUNDARY EXCEEDING SECOND BOUNDARY)
     *ARY)
    3 FORMAT(8H CLUSTER,I5,26H HAS BOUNDARY GREATER THAN,I5)
C.... CHECK NB ARRAY FOR OVERLAPS
      DO 21 I=1,K
      DO 21 J=1,I
      I1=NB(1,I)
      I2=NB(2,I)
      J1=NB(1,J)
      J2=NB(2,J)
   21 IF((I1-J1)*(I1-J2)*(I2-J1)*(I2-J2).LT.0) WRITE(6,4) I,J
    4 FORMAT(9H CLUSTERS,I5,4H AND,I5,8H OVERLAP)
C.... CONSTRUCT MT
      DO 30 I=1,KT
   30 MT(I)=0
      DO 22 I=M,KT
      IF(I.EQ.M) GO TO 22
C.... FIND CLUSTER UNASSIGNED, WITH MINIMUM NUMBER OF ELEMENTS
      IM=I
      MIN=M
      DO 23 J=1,K
      IF(NB(1,J).LT.0) GO TO 23
      MM=NB(2,J)-NB(1,J)+1
      IF(MM.GT.MIN) GO TO 23
      IM=J
      MIN=MM
   23 CONTINUE
      IF(MIN.EQ.M) GO TO 22
C.... FIND SMALLEST CLUSTER INCLUDED IN I, NOT YET ASSIGNED
      JL=NB(1,IM)
      JU=NB(2,IM)
      NB(1,IM)=-JL
      DO 24 J=JL,JU
      L=J
   25 IF(MT(L).EQ.0) GO TO 24
      L=MT(L)
      IF(MT(L).EQ.L) GO TO 24
      GO TO 25
   24 MT(L)=I
   22 CONTINUE
      MT(KT)=KT
      DO 40 I=1,K
   40 IF(NB(1,I).LT.0) NB(1,I)=-NB(1,I)
      RETURN
      END
```

```
      SUBROUTINE CNVERT(LL,NC,NB,RL,M)
C.....................................................................20 MAY 1973
C.... USE AFTER LETREE
C.... TRANSFORMS TREE ARRAY LL INTO CLUSTERS NB SUITABLE FOR TREE CONSTRUCTION
C.... LL = 3 BY NC INPUT ARRAY
C          LL IS REAL NOT INTEGER
C          LL(1,I) = NAME OF ITH CLUSTER
C          LL(2,I) = NEXT CLUSTER WITH SAME ANCESTOR
C          LL(3,I) = FIRST CLUSTER WITH ANCESTOR I
C.... NC = NUMBER OF CLUSTERS
C.... NB = 3 BY NC OUTPUT ARRAY DEFINING CLUSTER BOUNDARIES, SUITABLE FOR TREE1
C          NB(1,K)= INDEX OF FIRST OBJECT IN CLUSTER
C          NB(2,K) = INDEX OF LAST OBJECT IN CLUSTER
C          NB(3,K) = LEVEL OF CLUSTER K
C.... M = NUMBER OF OBJECTS, COMPUTED DURING PROGRAM
C.... RL = 1 BY NC ARRAY, RL(I)= NAME OF ITH OBJECT
C......................................................................
      DIMENSION LL(3,NC),NB(3,NC),RL(NC)
      REAL LL
C.... RL(I) INITIALLY IS POSITION OF CLUSTER I IN REORDERING
      M=0
      DO 10 K=1,NC
      RL(K)=0
      IF(LL(3,K).EQ.0) M=M+1
   10 IF(LL(3,K).EQ.0) RL(K)=1
      DO 11 K=1,NC
      KK=NC-K+1
      L2=LL(2,KK)
      IF(L2.NE.0) RL(KK)=RL(KK)+RL(L2)
      L3=LL(3,KK)
      IF(L3.NE.0) RL(KK)=RL(KK)+RL(L3)
   11 CONTINUE
      DO 12 K=1,NC
      KK=LL(3,K)
      IF(KK.EQ.0) GO TO 12
      IK=RL(K)-RL(KK)
   13 RL(KK)=RL(KK)+IK
      KK=LL(2,KK)
      IF(KK.NE.0) GO TO 13
   12 CONTINUE
C.... BOUNDARIES OF CLUSTERS
      DO 50 K=1,NC
      KK=NC-K+1
      NB(2,KK)=10**6
      NB(3,KK)=0
      L3=LL(3,KK)
      IF(L3.EQ.0) NB(2,KK)=RL(KK)
      IF(L3.EQ.0) NB(3,KK)=RL(KK)
      IF(L3.EQ.0) GO TO 50
   51 NB(2,KK)=MIN0(NB(2,KK),NB(2,L3))
      NB(3,KK)=MAX0(NB(3,KK),NB(3,L3))
      L3=LL(2,L3)
      IF(L3.NE.0) GO TO 51
   50 CONTINUE
C.... LABELS BY POSITIONS
      DO 60 K=1,NC
   60 NB(1,K)=RL(K)
      DO 70 K=1,NC
      I=NB(1,K)
   70 RL(I)=LL(1,K)
C.... LEVELS OF CLUSTER
      DO 20 K=1,NC
   20 NB(1,K)=1
      DO 21 K=1,NC
      KK=NC-K+1
      L2=LL(2,KK)
      L3=LL(3,KK)
      IF(L2.NE.0) NB(1,KK)=MAX0(NB(1,L2),NB(1,KK))
      IF(L3.NE.0) NB(1,KK)=MAX0(NB(1,L3)+1,NB(1,KK))
   21 CONTINUE
      DO 22 K=1,NC
      L3=LL(3,K)
      IF(L3.NE.0) NB(1,L3)=NB(1,K)-1
      L2=LL(2,K)
   22 IF(L2.NE.0) NB(1,L2)=NB(1,K)
      DO 80 K=1,NC
      I=NB(1,K)
      NB(1,K)=NB(2,K)
      NB(2,K)=NB(3,K)
   80 NB(3,K)=5*I
      RETURN
      END
```

```
      SUBROUTINE BLOCK(A,M,N,NB,KC)
C.........................................................20 MAY 1973
C.... PRINTS OUT MATRIX A WITH BLOCKS SPECIFIED IN NB
C.... M = NUMBER OF ROWS
C.... N = NUMBER OF COLUMNS
C.... A = M BY N BORDERED ARRAY, A .LT 9999
C.... KC = NUMBER OF CLUSTERS
C.... NB = 4 BY KC
C           NB(1,K) = FIRST ROW IN BLOCK
C           NB(2,K) = LAST ROW IN BLOCK
C           NB(3,K) = FIRST COLUMN IN BLOCK
C           NB(4,K) = LAST COLUMN IN BLOCK
C...................................................................
      DIMENSION A(M,N),NB(4,KC)
      DATA DD/4H----/
      DIMENSION AA(26),AE(26),IA(26)
      DATA DASH,DITTO,COMMA,BLANK,STAR,DOT/1H-,1H',1H,,1H ,1H*,1H./
C.... CHECK BOUNDARY ARRAY NB
      DO 10 K=1,KC
      IF(NB(1,K).LT.2.OR.NB(1,K).GT.NB(2,K).OR.NB(2,K).GT.M) WRITE(6,1)K
      IF(NB(3,K).LT.2.OR.NB(3,K).GT.NB(4,K).OR.NB(4,K).GT.N) WRITE(6,1)K
   10 CONTINUE
    1 FORMAT(25H BAD BOUNDARY IN CLUSTER   ,I3)
      JPP=(N-2)/25+1
      DO 70 JP=1,JPP
      JLP=25*(JP-1)+2
      JUP=25*JP+1
      IF(JUP.GT.N) JUP=N
      JR=JUP-JLP+1
C.... WRITE  TITLES
      WRITE(6,2) A(1,1)
    2 FORMAT(15H1BLOCKED ARRAY A4)
C.... WRITE OUT ARRAY ONE LINE AT A TIME
      WRITE(6,3)(A(1,J),J=JLP,JUP)
    3 FORMAT(6X,25(1X,A4))
      DO 20 I=1,M
      DO 27 L=1,26
      AE(L)=BLANK
   27 AA(L)=BLANK
      IF(I.EQ.1) GO TO 28
C.... FILL IN DATA VALUES
      DO 21 J=JLP,JUP
   21 IA(J-JLP+1)=A(I,J)
C.... FILL IN VERTICAL BOUNDARIES
      DO 23 K=1,KC
      IF(NB(2,K).LT.I.OR.NB(1,K).GT.I) GO TO 23
      JL=NB(3,K)
      JU=NB(4,K)+1
      IF(JL.GE.JLP.AND.JL.LE.JUP) AA(JL-JLP+1)=DITTO
      IF(JU.GE.JLP.AND.JU.LE.JUP) AA(JU-JLP+1)=DITTO
      IF(JU.EQ.JLP+JR) AA(JR+1)=DITTO
   23 CONTINUE
      WRITE(6,6) A(I,1),(AA(J),IA(J),J=1,JR),AA(JR+1)
    6 FORMAT(1X,A4,1X,25(A1,I4),A1)
C.... FILL IN HORIZONTAL BOUNDARIES
   28 CONTINUE
      DO 24 K=1,KC
      IF(NB(1,K).NE.I+1.AND.NB(2,K).NE.I) GO TO 24
      JL=NB(3,K)
      JU=NB(4,K)+1
      J1=JL-JLP+1
      J2=JU-JLP+1
      IF(J1.LE.0) J1=1
      IF(J2.GT.26) J2=26
      IF(J1.GT.26) GO TO 24
      IF(J2.LE.0) GO TO 24
      DO 25 J=J1,J2
      IF(J.EQ.J2) GO TO 25
      AE(J)=DD
   25 IF(AA(J).EQ.BLANK) AA(J)=DASH
      IF(NB(1,K).NE.I+1) GO TO 24
      AA(J1)=COMMA
      AA(J2)=COMMA
   24 CONTINUE
      WRITE(6,7)(AA(J),AE(J),J=1,JR),AA(JR+1)
    7 FORMAT(6X,25(A1,A4),A1)
   20 CONTINUE
   70 CONTINUE
      RETURN
      END
```

168

CHAPTER 9

Quick Tree Calculation

9.1 INTRODUCTION

The data of Table 9.1 consist of the dentition of 67 mammals. Mammals' teeth are divided into four groups with specialized functions, incisors, canines, premolars, and molars. The number of various types on upper and lower jaw provides a simple numerical basis for classifying mammals. Because teeth are likely to appear in fossil remnants, they are also very important in tracing evolutionary changes. (The pattern of cusps on each tooth is used in these evolutionary studies.)

There are a number of tree construction algorithms that are quick in computation and cheap in storage. The first of these is a generalization of the leader algorithm for partitions.

9.2 LEADER ALGORITHM FOR TREES

Preliminaries. A distance $D(I, J)$ is given between any pair of objects $(1 \leq I, J \leq M)$. A decreasing sequence of thresholds $T(1), T(2), \ldots, T(KT)$ is assumed given. There will be KT levels to the tree (with the root at level 0) and a cluster at level J will be within the threshold $T(J)$ of the cluster leader.

The Ith cluster is characterized by

$L(1, I)$,　　the leading object
$L(2, I)$,　　the next cluster with the same ancestor as I, and
$L(3, I)$,　　the first cluster included in I.

Set $L(2, I)$, $L(3, I)$ equal to zero if no cluster satisfies their definition.

The tree is constructed in a single pass through the set of all objects. The clusters are computed along the way. For each new cluster $L(2, I)$ is initially zero, and this value is possibly updated once later. The complete clustering for all objects is contained in the array L, and in the array LC, where LC(I) is the leader of object I at level KT. Each object is treated by looking for the first cluster at level 1, within the threshold of whose leader it is. If there is no such cluster, the object defines a new cluster at level 1; otherwise the object is compared with the leaders of clusters at level 2 contained in the cluster at level 1 and is treated analogously at every level.

STEP 1.　Begin with object number $I = 0$ and total number of clusters NC $= 0$.

STEP 2.　Increase I to $I + 1$. Set the level number $J = 1$. Set the cluster number $K = 1$. If $I = 1$, go to Step 6. If $I = M + 1$, go to Step 8.

Table 9.1   Dentition of Mammals

Mammal's teeth are divided into four groups, incisors, canines, premolars, and molars. In the list below, the dentition of each mammal is described by the number of top incisors, bottom incisors, top canines, bottom canines, top premolars, bottom premolars, top molars, and bottom molars. From Palmer, E. L. [1957] *Fieldbook of Mammals*, Dutton, New York.

opossum	54113344	pocket gopher	11001133	skunk	33113312
hairy tail mole	33114433	kangaroo rat	11001133	river otter	33114312
common mole	32103333	pack rat	11000033	sea otter	32113312
star nose mole	33114433	field mouse	11000033	jaguar	33113211
brown bat	23113333	muskrat	11000033	ocelot	33113211
silver hair bat	23112333	black rat	11000033	cougar	33113211
pigmy bat	23112233	house mouse	11000033	lynx	33113211
house bat	23111233	porcupine	11001133	fur seal	32114411
red bat	13112233	guinea pig	11001133	sea lion	32114411
hoary bat	13112233	coyote	13114433	walrus	10113300
lump nose bat	23112333	wolf	33114423	grey seal	32113322
armadillo	00000088	fox	33114423	elephant seal	21114411
pika	21002233	bear	33114423	peccary	23113333
snowshoe rabbit	21003233	civet cat	33114422	elk	04103333
beaver	11002133	raccoon	33114432	deer	04003333
marmot	11002133	marten	33114412	moose	04003333
groundhog	11002133	fisher	33114412	reindeer	04103333
prairie dog	11002133	weasel	33113312	antelope	04003333
ground squirrel	11002133	mink	33113312	bison	04003333
chipmunk	11002133	ferrer	33113312	mountain goat	04003333
gray squirrel	11001133	wolverine	33114412	muskox	04003333
fox squirrel	11001133	badger	33113312	mountain sheep	04003333

STEP 3.   Compute the distance between I and $L(1, K)$. If this distance does not exceed $T(J)$, go to Step 4. If this distance does exceed $T(J)$, go to Step 5.

STEP 4.   Set KK $= L(3, K)$, $J = J + 1$. If $KK = 0$, set $LC(I) = K$ and return to Step 2. If KK $\neq 0$, set $K =$ KK and return to Step 3.

STEP 5.   Set KK $= L(2, K)$. If KK $\neq 0$, set $K =$ KK and return to Step 3. Set $L(2, K) =$ NC $+ 1$.

STEP 6.   Set NC $=$ NC $+ 1$. Set $L(1, NC) = I$, $L(2, NC) = 0$, and $L(3, NC) =$ NC $+ 1$.

STEP 7.   Set $J = J + 1$. If $J \leq$ KT $+ 1$, go to Step 6. Otherwise, set $L(3, NC) = 0$, $LC(I) = I$, and return to Step 2.

STEP 8.   The tree has been computed, but it is necessary to find an ordering of the objects which occur as leaders so that clusters will be contiguous in the ordering. An ordering vector $O(1), O(2), \ldots, O(NC)$ is defined, where $O(I)$ is the position of the Ith leader in the ordering.

STEP 9. Set $O(I) = 1$ for each cluster I with $L(3, I) = 0$; set $O(I) = 0$, otherwise.

STEP 10. For each cluster K (NC $\geq K \geq 1$) in inverse order, set $O(K) = O(K) + O[L(2, K)] + O[L(3, K)]$, where $O(0) = 0$.

STEP 11. For each cluster K ($1 \leq K \leq$ NC) in usual order, for each K with $L(3, K) \neq 0$, set KK $= O(K) - O[L(3, K)]$. Set $J = L(3, K)$, and increase $O(J)$ by KK. Set $J = L(2, J)$ and increase $O(J)$ by KK, continuing until $J = 0$.

STEP 12. For each K ($1 \leq K \leq$ NC), place the leader of cluster K in position $O(K)$.

9.3 TREE-LEADER ALGORITHM APPLIED TO MAMMALS' TEETH

The measure of distance between two mammals is the sum of absolute deviations between the counts for various teeth types. The seven thresholds are set at 32, 16, 8, 4, 2, 1, 0.

STEP 1. Initialize object number $I = 0$ and cluster number NC $= 0$.

STEP 2. Increase I to 1, and set level number $J = 1$ and cluster number $K = 1$. Since $I = 1$, go to step 6.

STEP 6. Set NC $= 1$, $L(1, 1) =$ object 1, opossum. Set $L(2, 1) = 0$, $L(3, 1) = 2$. Increase J to 2 and set $L(1, 2) =$ opossum, $L(2, 2) = 0$, $L(3, 2) = 3$. Continue until $J = 7$, $L(1, 7) =$ opossum, $L(2, 7) = 0$, and $L(3, 7) = 0$. Return to Step 2 and set LC(I) $= 1$.

STEP 2. Increase I to 2, set level number $J = 1$, and set cluster number $K = 1$.

STEP 3. The distance between object 2, hairy tail mole, and opossum is 7, which does not exceed $T(1) = 32$. Go to Step 4.

STEP 4. Set KK $= L(3, 1) = 2$, $J = 2$. Since KK $\neq 0$ and $K = 2$, return to Step 3.

STEP 3. The distance between object 2, hairy tail mole, and object $L(1, 2) = 1$, opossum, is 7, which does not exceed $T(2) = 16$. Go to Step 4. Continuing, the distance will first exceed threshold at level $J = 4$; go to Step 5.

STEP 5. Set KK $= L(2, 4) = 0$. Set $L(2, 4) =$ NC $+ 1 = 8$. Go to Step 6.

STEP 6. Set NC $= 8$, $L(1, 8) =$ object 2, hairy tail mole, $L(2, 8) = 0$, $L(3, 8) = 9$. This assignment will continue analogously for NC $= 9, 10, 11$. Then set $L(3, 11) = 0$, LC(2) $= 2$, and return to Step 2.

The complete array L is given in Table 9.2, and the corresponding tree is given in Table 9.3.

A defect of the algorithm is apparent in the classification of hairy tail mole, which is in the opossum group but should be classified with house bat. At the time hairy tail mole was classified, house bat had not been classified, and so this choice did not exist. This shows that the tree-leader algorithm, like the partition leader algorithm, is decidedly sensitive to the order of presentation of objects. Some dependence seems inevitable if the objects are to be classified in a single pass.

Table 9.2   Tree-Leader Algorithm Applied to Mammals' Teeth

The cluster, the name of the cluster leader, the next cluster with the same ancestor, and the first descendant cluster are given.

CLUSTER	NAME	NEXT	FIRST	CLUSTER	NAME	NEXT	FIRST
1	OPOSSUM	0	2	47	COYOTE	59	48
2	OPOSSUM	28	3	48	COYOTE	51	49
3	OPOSSUM	21	4	49	COYOTE	0	50
4	OPOSSUM	8	5	50	COYOTE	0	0
5	OPOSSUM	0	6	51	WOLF	54	52
6	OPOSSUM	0	7	52	WOLF	0	53
7	OPOSSUM	0	0	53	WOLF	0	0
8	HAIRY TAIL MOLE	0	9	54	CIVET CAT	0	55
9	HAIRY TAIL MOLE	12	10	55	CIVET CAT	0	56
10	HAIRY TAIL MOLE	0	11	56	CIVET CAT	57	0
11	HAIRY TAIL MOLE	0	0	57	RACCOON	58	0
12	COMMON MOLE	15	13	58	MARTEN	0	0
13	COMMON MOLE	0	14	59	WEASEL	84	60
14	COMMON MOLE	0	0	60	WEASEL	69	61
15	BROWN BAT	0	16	61	WEASEL	63	62
16	BROWN BAT	19	17	62	WEASEL	65	0
17	BROWN BAT	18	0	63	WOLVERINE	67	64
18	SILVER HAIR BAT	0	0	64	WOLVERINE	0	0
19	PIGMY BAT	0	20	65	RIVER OTTER	66	0
20	PIGMY BAT	0	0	66	SEA OTTER	0	0
21	HOUSE BAT	72	22	67	JAGUAR	77	68
22	HOUSE BAT	34	23	68	JAGUAR	0	0
23	HOUSE BAT	81	24	69	FUR SEAL	0	70
24	HOUSE BAT	26	25	70	FUR SEAL	79	71
25	HOUSE BAT	0	0	71	FUR SEAL	0	0
26	RED BAT	0	27	72	WALRUS	0	73
27	RED BAT	0	28	73	WALRUS	0	74
28	ARMADILLO	0	29	74	WALRUS	0	75
29	ARMADILLO	42	30	75	WALRUS	0	76
30	ARMADILLO	0	31	76	WALRUS	0	0
31	ARMADILLO	0	32	77	GREY SEAL	0	78
32	ARMADILLO	0	33	78	GREY SEAL	0	0
33	ARMADILLO	0	0	79	ELEPHANT SEAL	0	80
34	PIKA	47	35	80	ELEPHANT SEAL	0	0
35	PIKA	0	36	81	PECCARY	0	82
36	PIKA	39	37	82	PECCARY	0	83
37	PIKA	38	0	83	PECCARY	0	0
38	SNOWSHOE RABBIT	0	39	84	ELK	0	85
39	BEAVER	0	40	85	ELK	0	86
40	BEAVER	41	0	86	ELK	0	87
41	GRAY SQUIRREL	0	0	87	ELK	0	0
42	PACK RAT	0	43	88	ANTELOPE	0	0
43	PACK RAT	0	44				
44	PACK RAT	0	45				
45	PACK RAT	0	46				
46	PACK RAT	0	0				

9.4   THINGS TO DO

9.4.1   Running the Tree-Leader Algorithm

It will usually be sufficient to use three or four well-chosen thresholds, but in order to choose these a first run should be made with a large number of thresholds from which the final thresholds will be selected. It is plausible to have thresholds decrease geometrically for metric distances—for example, 32, 16, 8, 4, 2, 1. The final tree will

Table 9.3 Tree for Mammals, Based on Dentition

Omitting mammals which have identical dentition to one in tree.

DENTITION

54113344	OPOSSUM	OPOSSUM	OPOSSUM	OPOSSUM	OPOSSUM	OPOSSUM
33114433	HAIRY TAIL MOLE	HAIRY TAIL MOLE	HAIRY TAIL MOLE	HAIRY TAIL MOLE		
32103333	COMMON MOLE	COMMON MOLE	COMMON MOLE			
23113333	BROWN BAT	BROWN BAT	BROWN BAT			
23112333	SILVER HAIR BAT					
23112233	PIGMY BAT	PIGMY BAT				
23111233	HOUSE BAT	HOUSE BAT	HOUSE BAT	HOUSEBAT	HOUSE BAT	
13112233	RED BAT	RED BAT				
23113333	PECCARY	PECCARY	PECCARY			
21002233	PIKA	PIKA	PIKA	PIKA		
21003233	SNOWSHOE RABBIT					
11002133	BEAVER	BEAVER				
11001133	GREY SQUIRREL					
13114433	COYOTE	COYOTE	COYOTE	COYOTE		
33114423	WOLF	WOLF	WOLF			
33114422	CIVET CAT	CIVET CAT	CIVET CAT			
33114432	RACCOON					
33114412	MARTEN					
33113312	WEASEL	WEASEL	WEASEL	WEASEL		
33114312	RIVER OTTER					
33113312	SEA OTTER					
33114412	WOLVERINE	WOLVERINE				
33113211	JAGUAR	JAGUAR				
32113322	GREY SEAL	GREY SEAL				
32114411	FUR SEAL	FUR SEAL	FUR SEAL			
21114411	ELEPHANT SEAL					
04103333	ELK	ELK	ELK	ELK		
04003333	ANTELOPE					
10113300	WALRUS	WALRUS	WALRUS	WALRUS	WALRUS	
00000088	ARMADILLO	ARMADILLO	ARMADILLO	ARMADILLO	ARMADILLO	ARMADILLO
11000033	PACK RAT	PACK RAT	PACK RAT	PACK RAT	PACK RAT	

generate a corresponding "contiguous" ordering of the data, and it is suggested that the algorithm be executed on the objects in this new ordering. This operation should be repeated until there are no further changes in the ordering, in order to reduce the effect of the initial order of presentation. The frequency of car repairs (Table 9.4) is suggested as a trial data set.

9.4.2 Sorting

Suppose that variables are given, each of which takes a small number of different values over the various cases. The first variable partitions the complete set of cases into a number of clusters, the second variable partitions each of these clusters into a number of smaller clusters, and so on, constructing a tree of clusters. Of course, the difficulty here is selecting the variables to be used at various levels of the tree.

Table 9.4 Frequency of Car Repairs

BR = brake system, FU = fuel system, EL = electrical, EX = exhaust, ST = steering, EM = engine, mechanical, RS = rattles and squeaks, RA = rear axle, RU = rust, SA = shock absorbers, TC = transmission, clutch, WA = wheel alignment, OT = other.

	BR	FU	EL	EX	ST	EM	RS	RA	RU	SA	TC	WA	OT
AMC Ambassador 8	+	-	-	-	-	-	-	+	-	-	-	-	-
Buick Special 6	-	-	-	-	-	-	+	-	+	-	-	-	+
Buick Special 8	-	-	-	-	-	-	+	-	-	+	-	+	+
Buick 8 Full	-	-	-	+	-	+	+	-	+	+	-	+	-
Buick Riviera	-	-	+	+	-	-	-	-	-	+	-	-	-
Cadillac													
Chevy II	-	+	-	-	+	-	+	+	+	-	-	+	-
Chevelle 6	-	-	-	-	-	+	+	-	+	-	-	-	-
Chevelle 8	-	+	-	+	+	-	+	-	+	+	-	+	-
Chevrolet Full	-	+	+	+	+	-	+	+	+	+	+	+	-
Corvair 6	-	+	-	-	+	+	-	+	-	+	+	+	+
Corvette	-	-	-	+	-	-	+	+	-	-	+	-	-
Chrysler Newport	+	-	-	-	-	-	-	-	-	-	-	-	-
New Yorker	+	-	-	-	-	-	-	+	-	-	-	-	+
Dodge Full Size	+	-	-	-	-	-	+	-	-	+	-	-	-
Falcon 6	-	-	-	-	-	-	+	-	-	-	+	+	-
Fairlane 6	-	-	-	-	-	-	+	-	-	-	-	-	-
Fairlane 8	-	-	-	+	-	-	+	+	-	-	-	+	-
Ford, Full Size	-	-	-	+	+	-	-	-	-	+	-	+	+
Thunderbird	-	-	+	-	+	+	-	-	-	-	-	+	+
Mercury Full	-	-	-	-	-	-	-	-	-	-	-	+	-
Olds Full	+	+	-	-	-	-	+	-	-	+	-	+	-
Plymouth Full	+	-	-	-	-	-	-	-	-	-	-	-	-
Pontiac Tempest	-	+	-	-	-	-	+	-	+	+	-	+	-
Pontiac Full	+	+	+	-	-	-	+	-	+	+	-	+	-
Rambler Rebel 6	-	-	+	-	-	-	-	+	-	-	+	-	+
Mercedes	-	-	-	-	-	-	-	-	-	+	-	-	-
MG 1100	-	-	+	+	-	-	-	-	-	-	-	-	-
Peugeot	-	-	-	-	-	-	-	-	-	-	+	-	-
Porsche	-	-	-	-	-	-	-	-	-	-	-	-	-
Renault	-	-	-	-	-	-	-	-	-	-	+	-	-
Volvo	-	-	-	+	-	-	-	+	-	-	-	-	-
VW bug	+	-	+	+	+	+	-	-	-	-	+	-	-
VW bus	-	-	+	-	-	+	-	-	+	-	+	-	-

[From *Consumer Reports Buying Guide* (1969).] A + means greater than average frequency of repair in 1962–1967.

174

If the data consist of many category variables (for example, the dentition data where each variable takes values $0, 1, \ldots, 8$), then one solution to the problem is to select the splitting variable at each level from these. The first variable is that one which best predicts the remaining variables. Let $V(1), V(2), \ldots, V(N)$ denote the variables, and let $P[V(I) = K, V(J) = L]$ denote the proportion of times $V(I) = K$ and $V(J) = L$. A measure of predictive power of variable $V(I)$ for variable $V(J)$ is the information

$$\sum \{K, L\} P[V(I) = K, V(J) = L] \log P[V(I) = K, V(J) = L]$$
$$- \sum \{K\} P[V(I) = K] \log P[V(I) = K]$$
$$- \sum \{L\} P[V(J) = L] \log P[V(J) = L].$$

The overall measure of predictive power of $V(I)$ is the sum of this quantity over $J \neq I$, and I is chosen to minimize this sum. At the second level, a variable is chosen that best predicts the remaining variables, given the value of the first variable. And so on.

With this careful selection of variables, more than one pass through the data is required. Actually, one pass will be required for each level of the tree. A version of the above technique for continuous variables selects the first variable to be that variable most correlated with all others, selects the second variable to be most correlated with others given the first, and so on. It is also plausible for continuous variables to select that linear combination of the variables most correlated with all variables to be the first splitting variable. This means that the first splitting variable is the first eigenvector, the second splitting variable is the second eigenvector, and so on. Both the continuous techniques require one pass through the data for variable selection and one pass for classification.

9.4.3 Differential Sorting

There is no particular reason, except perhaps descriptive convenience, to use the same variable for splitting all the clusters at the second level. Thus, for category data, the first variable is selected to best predict all others. This step is then repeated on each of the clusters obtained, giving in general a different splitting variable for each cluster.

In the continuous case, the first eigenvector is used for the first split, and the residual variables, after prediction by the first eigenvector, are retained. Within each cluster, a new first eigenvector is obtained that best predicts these residual variables within the cluster. This procedure is repeated at every level.

9.4.4 Filtering Algorithm

A version of the K-means algorithm appropriate for trees begins with a set of cluster centers, one for each cluster. Suppose that the initial tree of clusters is binary. Each object is successively added to the tree by *filtering*; it is first assigned to the cluster of all objects. This cluster splits into two clusters, and the object is assigned to whichever of these cluster centers it is closest to. This cluster splits in two, and the object is then assigned to whichever of these two cluster centers it is closest to. And so on.

After a complete assignment of all objects, each cluster center is updated to be the mean of all objects in the cluster. The objects are then reassigned. It will be seen that a K-means-type algorithm operates at each division of a cluster into two clusters.

A typical initialization of cluster centers might go as follows. The first cluster center is the mean of all objects. The second cluster center is the mean of all objects that exceed the first cluster center on at least half of the variables, and the third cluster center is the mean of all objects which exceed the first cluster center on at most half of the variables. Each of these two clusters generates two new cluster centers in a similar way, and the process continues until all cluster centers are initialized.

PROGRAMS

LETREE   constructs clusters using the tree-leader algorithm.

```
      SUBROUTINE LETREE(X,N,LL,KC,TH,NT,Y,CN,NC)
C...................................................................20 MAY 1973
C.... CONSTRUCTS LEADER TREE WITH DISTANCE BETWEEN PAIRS OF OBJECTS SPECIFIED IN
C          FUNCTION DIST.      THE VARIOUS LEVELS IN THE TREE ARE DETERMINED BY
C          THE THRESHOLDS IN ARRAY TH.   THRESHOLDS MUST DECREASE.
C.... N = NUMBER OF ELEMENTS IN DATA VECTOR
C.... X = N-VECTOR, FIRST ELEMENT IS CASE NAME
C          THE PROGRAM ACCOMMODATES VECTORS IN SEQUENCE, UPDATING THE NODE
C          ARRAY WHICH DEFINES THE TREE, AND STATING WHICH NODE EACH VECTOR IS
C          ASSIGNED TO.
C.... LL = 3 BY KC NODE ARRAY, COMPUTED BY PROGRAM
C          LL IS REAL NOT INTEGER
C          LL(1,K)= NAME OF NODE
C          LL(2,K)=NEXT NODE WITH SAME ANCESTOR AS K
C          LL(3,K)= FIRST NODE WITH ANCESTOR K
C.... KC = NUMBER OF NODES
C.... TH = THRESHOLD ARRAY
C.... Y = N BY KC LIST OF DATA VALUES OF NODES
C.... NC = ACTUAL NUMBER OF CLUSTERS AFTER PASSING THROUGH PROGRAM
C.... CN = NAME OF CLUSTER WHICH OBJECT IS ASSIGNED.
C...................................................................
      DIMENSION X(N),LL(3,KC),Y(N,KC)
      DIMENSION TH(NT)
      REAL LL
C.... LABELS
      DATA YN,XL/4HLEAD,4HNODE/
      Y(1,1)=YN
      LL(1,1)=XL
      DATA IC/0/
      IF(IC.EQ.0) NC=0
      IC=IC+1
      KK=1
      LEV=1
      IF(NC.EQ.0) GO TO 25
C.... ASSIGN OBJECT TO TREE
   20 CONTINUE
   22 K=KK
      NN=N-1
      D=DIST(X(2),Y(2,K),NN,1,2.)
      IF(D.GT.TH(LEV)) GO TO 21
      LEV=LEV+1
      KK=LL(3,K)
      IF(KK.EQ.0) CN=Y(1,K)
      IF(KK.EQ.0) RETURN
      GO TO 22
   21 KK=LL(2,K)
C.... MAKE A NEW LEADER
      IF(KK.NE.0) GO TO 22
   25 CONTINUE
      IF(NC.GT.0) LL(2,K)=NC+1
      DO 27 KK=LEV,NT
      NC=NC+1
      IF(NC.GT.KC) WRITE(6,1) KC
    1 FORMAT(I5,17H NOT ENOUGH NODES)
      LL(1,NC)=X(1)
      LL(2,NC)=0
      LL(3,NC)=NC+1
      DO 23 J=1,N
   23 Y(J,NC)=X(J)
   27 CONTINUE
      LL(3,NC)=0
      CN=Y(1,NC)
      RETURN
      END
```

Triads

10.1 INTRODUCTION

In clustering the hardware in Figure 10.1, you could seize the nettle and declare the tree to be

$$((NF)T)((PS)B).$$

This tree consists of certain similarity judgments (obviously subjective) about the hardware. Following the data matrix approach, you might list properties of the various objects, define a measure of distance between the vectors of properties, making weighting decisions for each of the properties, and then use a tree construction algorithm to obtain a tree. This procedure is no less subjective than the original one, with the subjective choice of properties to measure and the subjective weighting of properties in the distance function. The properties selected will be based on informal similarity judgements and informal clusterings. For example, the ratio of diameter to length might be used because it is noticed that this measure discriminates well between nails and screws, or whether or not the head is indented might be used because it discriminates well also.

The data matrix approach differs from the direct similarity judgement approach in being more explicit about the factors subjectively selected as important in the clustering. An intermediate form of data, between the data matrix and the tree, are the triads in Table 10.1. For each triple of objects, a judgement is made about which pair are most similar. The algorithm in this section constructs a tree from such triadic data. Direct triadic judgements are not common, partly because $M(M - 1)(M - 2)/6$ judgements are necessary for M objects. However, triads can be constructed from other forms of data, such as distances or data matrices.

10.2 TRIADS ALGORITHM

Preliminaries.   It is necessary to measure the error between a given set of triads and a tree. The given set of triads will be a series of judgements $(IJ)K$; I and J are more similar to each other than either is to K.

A tree may be interpreted as the set of triads: $(IJ)K$ whenever the smallest cluster including I and J is properly included in the smallest cluster containing I and K. This set of triads will include every triple just once only if there are $M - 1$ distinct clusters on M objects—that is, if the tree of clusters is binary.

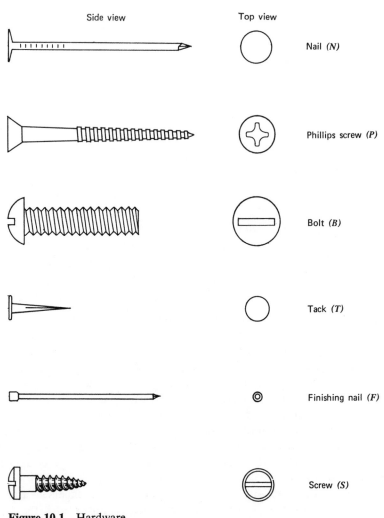

Figure 10.1 Hardware.

Table 10.1 Triads Based on Hardware

(AB)C denotes that A, B are most similar in the triple A, B, C.

NAIL (N)
PHILLIPS SCREW (P)
BOLT (B)
TACK (T)
FINISHING NAIL (F)
SCREW (S)

(PB)N (NT)P (NF)P (PS)N (NT)B (NF)B (BS)N (NF)T (NT)S (NF)S

(PB)T (PB)F (PS)B (TF)P (PS)T (PS)F (TF)B (BS)T (BS)F (TF)S

178

Table 10.2   Trees Obtained During Triads Algorithm (Application to Hardware)

INITIAL TREE	(PN)		
ADD B	(PN)B	(PB)N	
ERROR	1	-1	
ADD T	(((PB)N)T)	((PB)(NT))	
ERROR	0	-4	
ADD F	(((PB)(NT))F)	((PB)((NT)F))	((PB)((NF)T))
ERROR	0	-8	-10
ADD S	(((PB)((NF)T))S)	(((PB)S)((NF)T))	(((PS)B)((NF)T))
ERROR	-6	-18	-20

The tree triads will—some of them—agree with the given triads, and others will disagree. The ones that agree will be called true triads and the ones that disagree false triads. Let T be the number of true triads and F the number of false triads. If T is maximized, a tree with $M - 1$ clusters will be produced. If F is minimized, a tree with just a few clusters will be produced. An error function intermediate between these is $F - T$, and this will be minimized (locally) in the algorithm.

STEP 1.   Begin with an initial tree containing three clusters on the first two objects, the clusters $\{1\}$, $\{2\}$, $\{1, 2\}$. Let I, the object to be assigned, be set initially at $I = 2$.

STEP 2.   Increase I by 1. If $I > M$, stop.
Let C_1 be the cluster of all objects so far in the tree, $1, 2, \ldots, I - 1$. Add the clusters $\{C_1, I\}$ and $\{I\}$ to the tree.

STEP 3.   Compute the number of false minus true triads for each of the following changes to the tree:

(i)   no change;

(ii)   delete C_1;

(iii)   replace C_1 by $\{C^*, I\}$, where C^* is a maximal cluster properly included in C_1. Such a cluster will be called a maximal subcluster of C_1.

If (i) or (ii) have the least error (the least value of $F - T$), return to Step 2.

STEP 4.   Let C^* be the maximal subcluster of C_1 for which the error with C_1 changed to $\{C^*, I\}$ is a minimum. If C^* consists of a single object, return to Step 2. Otherwise, set $C_1 = C^*$, add $\{C_1, I\}$ to the tree, and return to Step 3.

10.3   TRIADS ALGORITHM APPLIED TO HARDWARE

The given triads are in Table 10.1.

STEP 1.   The tree consists of the clusters (N, P), (N), (P).

STEP 2.   First add the object B to the tree. Initially $C_1 = (N, P)$, and $\{N, P, B\}$ and $\{B\}$ are added to the tree.

STEP 3.   (i) The tree with no change is (B), (P), (N), (N, P), (N, P, B). This implies the triad $(NP)B$, which is false. The error is 1.

(ii) Deleting C_1 produces a tree (B), (P), (N), (N, P, B), which generates no triads and has error 0.

(iii) There are two subclusters of $C_1 = (N, P)$—namely, (N) and (P). If C_1 is changed to (N, B), the triad implied is $(NB)P$, which is false. If C_1 is changed to (P, B), the triad implied is $(PB)N$, which is true, and so this tree has error -1.

STEP 4.   Change C_1 to (P, B), move on to the next object, and return to Step 2.
For object T, Step 2 begins with $C_1 = (P, B, N)$.

STEP 3.   **(i)** The tree with no change is $(((PB)N)T)$ which implies the triads $(PB)N$, $(PN)T$, $(BN)T$, and $(PB)T$, two false and two true, so the error is 0.

(ii)   The tree with C_1 deleted is $((PB)NT)$ which implies $(PB)N$, $(PB)T$ for an error of -2.

(iii)   If C_1 is replaced by (PBT), the tree is $(((PB)T)N)$, which has error 0. If C_1 is replaced by (NT), the tree is $((PB)(NT))$, which has error -4.

STEP 4.   Since $C^* = N$ is such that replacing C_1 by (C^*, T) gives minimum error, this replacement is carried out, and the next object is brought in and added to the tree, beginning again at Step 2.

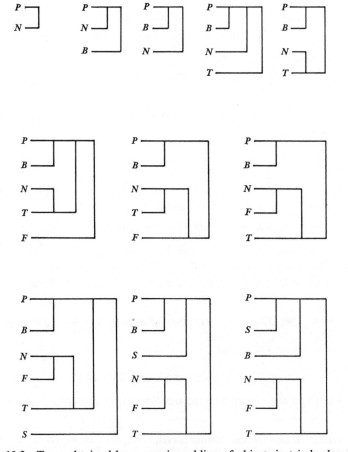

Figure 10.2   Trees obtained by successive adding of objects in triads algorithm.

The complete sequence of trees obtained during the addition process is given in Table 10.2 and also in Figure 10.2. By chance the tree $((PS)B((FN)T))$ fits exactly.

10.4 PROPERTIES OF TRIADS ALGORITHM

The triads algorithm is a local optimization algorithm in defining a measure for error for each tree and in then searching, not exhaustively, through the set of all trees for a satisfactory tree, guided by the error measure. Other measures of error, other search procedures, will be discussed later.

The triads algorithm is very expensive computationally since there are $M(M - 1)(M - 2)/6$ triads, and each triad is examined at least once. The final tree depends on the initial ordering of the data, which is another unsatisfactory feature. If there exists a tree that will generate the observed triads, then this tree will be constructed by the algorithm, so that at least the algorithm works well for "perfect data."

Triads are rather rare as real data. They can be derived from distances in an obvious way. $(IJ)K$ is $D(I, J) \leq D(I, K), D(J, K)$. They can also be derived directly from data matrices, with one triad possible for I, J, K for each variable in the data. The total number of observed triads will then be $M(M - 1)(M - 2)/6$ by N, where N is the number of variables.

The principal difficulty of the triads algorithm is the very large number of triads that must be checked. What is needed is a leader-type algorithm that would allow a single object to represent all the objects in a cluster and so make it unnecessary to look at all objects in the cluster when a new object is assigned to the tree.

10.5 TRIADS-LEADER ALGORITHM

Preliminaries. A binary tree is constructed, with $2M - 1$ clusters on M objects. Each of the objects $2, 3, \ldots, M$ is a leader object for just one of the $M - 1$ clusters containing more than one object. It is convenient to specify the tree by listing for each leader object I the leader objects J, K of the maximal clusters properly included in the cluster for which I is the leader. Write $P(I) = \{J, K\}$; the objects J, K are the *progeny* of I. If a maximal subcluster of I consists of a single cluster—say, J—the progeny of I will be written $\{-J, K\}$, where K is the leader of the other maximal subcluster of I. This notation distinguishes the object J when it appears as a singleton cluster from the object J when it appears as the leader of a multiple cluster. The algorithm assigns each object in turn to the tree, beginning at the largest cluster and moving towards the smaller clusters, branching according to the leading object the object is closest to.

STEP 1. Begin with a tree consisting of the first and second objects, with $P(2) = (-1, -2)$.

STEP 2. If $(12)3$, define $P(3) = (2, -3)$.
If $(13)2$, define $P(3) = (-1, -3)$, $P(2) = (-2, 3)$.
If $(23)1$, define $P(3) = (-2, -3)$, $P(2) = (-1, 3)$.
Let I, the number of objects in the tree, be set to $I = 3$.

STEP 3. Increase I to $I + 1$. If $I > M$, stop. Find that object J $(1 < J < I)$ such that $J \in P(K)$ for no K $(1 \leq K < I)$. (This object is the leading object of the largest cluster.)

STEP 4.   Suppose $P(J) = (K, L)$. If $(KL)I$ and there exists an object JJ such that $P(JJ) = (J, KK)$ for some KK, redefine $P(JJ) = (-I, KK)$. If $(KL)I$, set $P(I) = (J, -I)$ and return to Step 3.

STEP 5.   If $(KI)L$ set $J = K$, and if $(LI)K$ set $J = L$. If $J < 0$, set $P(I) = (-I, J)$ and return to Step 3. Otherwise, return to Step 4.

NOTE.   This algorithm requires only $O(M \log M)$ triads, so if they are being computed from data or distances do not compute them in advance.

10.6   APPLICATION OF TRIADS-LEADER ALGORITHM TO EXPECTATION OF LIFE

In Table 10.3, find the expectations of life for a number of cities, males and females, at ages 0, 25, 50, 75. Triads for three cities might plausibly be computed from the euclidean distances between them with $(IJ)K$ if $D(I, J) < D(I, K), D(K, J)$. It is quicker to compute triads directly from the data. The triad $(IJ)K$ holds if K is most frequently the furthest from I and J, counting over the eight variables. Consider the cities 1, 2, 3, or Montreal, Toronto, and Vancouver. For the first variable, the expectation of males at age 0, the values are respectively 67, 68, 68, so Montreal is furthest for this variable. For the second variable, the values are 44, 45, 46, so Montreal and Vancouver are both "furthest." This appears in the Montreal count as 0.5.

Table 10.3   Expectations of Life in Various Cities (by Age and Sex)

				MALE				FEMALE		
	CITY & YEAR		AGE	0	25	50	75	0	25	50	75
1.	MONTREAL	66		67	44	22	7	73	51	28	9
2.	TORONTO	66		68	45	23	8	75	53	29	10
3.	VANCOUVER	66		68	46	23	8	75	53	30	11
4.	KAOHSIUNG	66		66	43	22	8	71	49	26	9
5.	TAIPEI	66		69	46	24	8	72	50	27	9
6.	DJAKARTA	61		44	36	18	6	46	37	19	6
7.	COPENHAGEN	66		68	46	23	8	75	56	28	9
8.	HELSINKI	66		66	42	21	7	73	50	27	8
9.	EAST BERLIN	61		66	45	22	7	71	49	27	8
A.	WEST BERLIN	61		66	45	22	7	72	50	27	9
B.	WEST BERLIN	67		67	45	22	7	73	50	27	9
C.	DUSSELDORF	66		67	45	22	8	74	51	28	9
D.	HAMBURG	66		68	46	23	7	74	51	27	9
E.	STOCKHOLM	60		69	46	23	7	75	52	28	9
F.	LONDON	67		69	47	23	8	77	54	30	12
G.	SYDNEY	66		67	44	22	7	73	50	27	9

From N. Keyfitz and W. Flieger (1971). *Population*, Freeman, San Francisco.

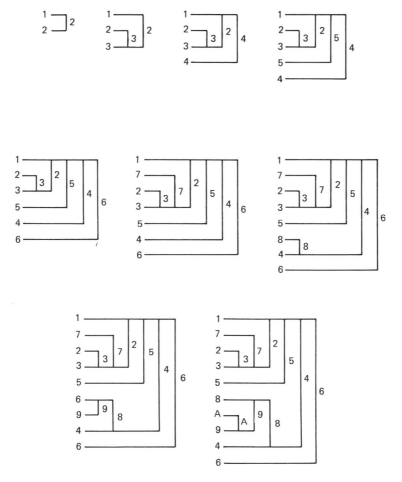

Figure 10.3 Trees obtained during application of triads-leader algorithm.

Overall, Montreal is furthest in $(1 + 0.5 + 1 + 1 + 1 + 1 + 0.5 + 0.5) = 6.5$ cases, so the triad (23)1 holds. (See Figure 10.3 for the sequence of trees obtained.)

STEP 1. Define $P(2) = \{-1, -2\}$.

STEP 2. Since Montreal is most dissimilar among Montreal, Toronto, and Vancouver, (23)1 is the triad. Thus $P(3) = \{-2, -3\}$ and $P(2)$ is changed to $\{-1, 3\}$.

STEP 3. To add object 4, first find the object J such that $J \in P(K)$ for no K. This object is $J = 2$.

STEP 4. $P(J) = \{1, 3\}$. The triad (13)4 holds. Therefore define $P(4) = \{2, -4\}$ and return to Step 3. (The city Kaohsiung is dissimilar to both leading objects Montreal and Vancouver, and so a new branch for Kaohsiung is defined at the highest level.) There is no object JJ such that $P(JJ) = \{2, KK\}$.

Adding object 5 at Step 3, set $J = 4$. At Step 4, $P(4) = \{2, -4\}$. The triad $\{25\}4$ holds; go to Step 5. (Surprisingly, Taipei is more similar to Vancouver than it is to Kaohsiung.) Set $J = 2$.

Return to Step 4. Here $P(J) = P(2) = \{-1, 3\}$. The triad $\{13\}5$ holds. The object $JJ = 4$ has $P(4) = \{2, -4\}$. Therefore $P(4) = \{5, -4\}$ and $P(5) = \{2, -5\}$. Return to Step 3, adding object 6. A full history of the tree construction is given by the record of P as each object is added. At each stage, the most recently defined value of P is the correct one (see Figure 10.3).

10.7 REMARKS ON TRIADS-LEADER ALGORITHM

This algorithm is fast but sloppy. The final clusters are not invariant under the change of order of objects. A binary tree is produced by the algorithm ($M - 1$ nontrivial clusters for M objects), so there is some difficulty in interpreting the large number of final clusters.

The leader of a cluster is selected arbitrarily, which is harmless when the clusters are very distinct but may be important in the common case where clusters are quietly present.

Triads are not frequently used in collecting similarity perception data because of the very large number of triads required for even a few objects [see Levelt, 1967 for an example of triads data (Table 10.4)]. Yet triads are an attractive way to extract

Table 10.4   Relatedness Values in "The Boy Has Lost a Dollar"

Frequencies with which a pair is judged more highly related than other pairs, over many triads and subjects from Levelt (1967).

	the	boy	has	lost	a	dollar
the	•	.99	.43	.29	.19	.16
boy			.63	.65	.16	.31
has				.86	.31	.40
lost					.42	.70
a						.94
dollar						

similarity information because three objects are the minimum number on which to base comparisons between pairwise similarities. How similar is a duck to a pig? Some similarity scale is necessary before a sensible transportable answer can be made. Which does not fit in: a duck, a pig, and a bee? The scale is carried in the question, and the triad (duck, pig) bee can be declared. A child might say (duck, bee) pig since a duck and a bee have wings. This opinion is valid also. Infants' textbooks are full of triads—pick from a hat, a shoe, and a bird the dissimilar object.

The triads-leader algorithm could be used to extract similarity perceptions from subjects without forcing the subjects to respond to all possible triads. The experimenter, in person or through carefully written instructions, must guide the subject into placing objects on the tree. Probably the easiest technique will be to ask the subject to actually construct a tree as in Section 10.3. For just a few objects, all possible trees could be laid out and the subject could be directed to choose among them.

For example, for a four-object questionnaire (which admittedly has only four triads), the instructions might be as follows:

Look at *ABC*.

If $(AB)C$, look at BCD.
    If $(BD)C$, look at BAD and then stop.
    If $(BC)D$ or $(CD)B$, stop.
If $(AC)B$, look at BCD.
    If $(CD)B$, look at CAD and then stop.
    If $(BC)D$ or $(BD)C$, stop.
If $(BC)A$, look at CAD.
    If $(CD)A$, look at BCD and then stop.
    If $(CA)D$ or $(DA)C$, stop.

There is not much of a reduction with four objects, but there would be substantial reductions with, say, seven.

10.8 THINGS TO DO

10.8.1 Running the Triads Algorithm

The triads algorithm should be used on sets containing a few objects—say, $M \leq 10$. On the first 10 typewriters in Table 10.5, construct triads as outlined in Section 10.4. A good experiment requires a trusting friend who will be requested to make triadic similarity judgements directly on six candy bars or six politicians. There will be thirty-five triads from which a tree may be constructed.

Now ask him to place the objects successively in a tree using the technique in Section 10.7. How different are the two trees?

10.8.2 Complete Searches

Since the triads are so numerous and complete searches are also very time consuming, except for small numbers of objects, it is practicable to think of evaluating the true and false triads for every tree and choosing the best ones. This technique will only be used for small numbers of objects.

Let the array $\{A(I), 1 \leq I \leq K\}$ define a tree with K nodes, where $A(I) > I$ for $I < K$ and $A(I)$ is the ancestor of the node I. For a tree on M objects, the first M nodes will be the objects, and so $A(I) > M$ for $I \leq M$. The number of arrays such that $A(I) > I$ for $I < K$, $A(I) > M$ for $I \leq M$, and $A(K) = K$ is $(K - M)^M (K - M - 1)!$. Each tree corresponds to at least one array (although the correspondence is not one to one). Thus all trees on M objects may be generated simply by generating the arrays A.

The array A is also useful in searching through trees. For example, the best ancestor for the node I may be found by looking at all trees with $M, I < A(I) \leq K$. The best ancestor for each node is found in succession, until no improvement occurs by change of ancestor of any single node. (This is analogous to the K-means algorithm, with more clusters available for objects or clusters to be attached to.)

10.8.3* Axioms

For any tree, for any four objects I, J, K, L,

$$(IJ)K \Rightarrow (JI)K$$

and

$$(IJ)K, (JL)I \Rightarrow (JL)K.$$

Table 10.5 Portable Typewriters

HT: height in inches
WH: width in inches
DH: depth in inches-
WT: weight in pounds
PL: platen length
KS: number of keys
EP:. erasure plate
TC: two carriage releases
MR: margin release
CL: carriage lock

PE: pica or elite type
VH: vertical half spacing
TA: tabulator
TP: touch pressure control
PR: platen release
HH: horizontal half spacing
PI: page end indicator
PG: paper guide
PB: paper bail
PS: paper support

	HT	WH	DH	WT	PL	KS	PE	TA	TP	PR	HH	VH	PI	PG	PB	PS	EP	TC	MR	CL
OLYMPIA(10)	15.75	15	7.75	21	19.38	44	1	1	1	1	1	1	1	1	1	1	1	0	0	1
OLYMPIA(13)	16.0	19	8	25.5	13.25	44	1	1	1	1	1	1	1	1	1	1	1	1	0	1
SMITH C. RONA(12)	14.5	17.75	6	19.5	11.88	44	1	1	1	1	1	0	1	1	1	1	1	1	0	0
SEARS(12)	14.5	17.75	6	19.5	11.88	44	1	1	1	1	1	1	1	1	0	1	1	1	0	0
HERMES 3000	14.25	13.25	6	17.0	9.75	44	1	1	1	1	1	1	0	1	1	0	1	1	0	1
WARDS 510	14.75	16.25	6.5	19.5	9.88	44	1	1	1	1	1	1	0	1	1	0	1	1	1	1
WARDS 513	15.0	19.25	6.5	21	12.88	44	1	1	1	1	1	1	1	1	1	1	1	1	1	1
SMITH CORONA ST.	14.5	15.0	6	17.5	9.13	44	1	1	1	1	0	0	0	1	0	0	1	1	0	0
ROYAL SAFARI	15.25	16.75	6.75	17.5	10	44	1	1	1	1	1	1	1	1	1	1	1	1	0	0
FACIT	14.75	17.0	5.5	18	9.88	44	1	1	1	1	1	1	0	0	0	0	1	1	1	1
BROTHER 900	14.75	16.25	6.5	19.5	12.88	44	1	1	1	1	1	1	0	1	1	0	1	1	1	1
BROTHER 905	15.0	19.25	6.5	21	9.38	44	1	1	1	1	1	1	0	1	1	1	1	1	1	1
OLYMPIA	12.75	12.5	3.5	12	9.13	44	1	0	1	1	1	1	0	0	0	1	1	0	1	1
WARDS 440T	14.0	17.25	5.5	15	9.75	44	1	1	1	0	1	1	0	1	1	1	1	0	1	1
OLIVETTI 44	17.25	14.75	7.5	21	9.75	43	1	1	1	1	1	1	0	0	1	1	1	1	0	1
OLIVETTI 21	17.5	14.75	6.5	21.5	9.38	43	1	1	0	1	1	1	1	1	1	1	1	1	0	1
REMINGTON II	15.5	15.0	7.75	20.5	9.63	44	1	1	1	1	1	1	0	0	1	1	0	1	0	1
OLIVETTI 32	14.25	12.0	4.25	11.5	9.0	43	1	1	1	1	1	1	1	1	1	1	1	1	0	1
REMINGTON 1040	14.75	13.75	6.25	14.5	9.25	43	1	1	1	1	1	1	0	0	1	0	1	1	0	1
BROTHER 885	13.0	12.75	4.0	11	9.13	44	1	0	1	0	1	1	0	0	1	1	1	0	1	1

From *Consumers' Reports Buying Guide*, (1967), Consumers' Union, Mount Vernon, N.Y.

If a set of triads satisfies these conditions, is there necessarily a tree that implies all the triads?

10.8.4* Ranks

Show that a tree generates for each I an ordering $\leq$ of all objects by their similarity to I. (Different objects J, K may have $J \leq K, K \leq J$). Suppose that you are given such an ordering for each object I; when is the ordering such that a tree might have generated it?

These orderings, by similarities to various objects, are more efficient than triads and so might be a better way to extract similarity judgements in psychological experiments. A good algorithm is needed for constructing a tree from a set of orderings.

10.8.5 Probability Models

Suppose a diameter $D(C)$ is associated with each cluster C in a tree. The probability of observing the triad $(IJ)K$ is $1 - D(I, J)/D(I, K)$, where $D(I, J)$ denotes the diameter of the smallest cluster containing $D(I, J)$ and $D(I, J) \leq D(I, K) = D(J, K)$. For any given tree, the diameters may be estimated by maximum likelihood. Over many different trees, the maximum likelihood for each tree may be used as a criterion.

REFERENCES

LEVELT, W. J. M. (1967). "Psychological representations of syntactic structure," in *The Structure and Psychology of Language*, T. G. Bever and W. Weksel, eds., Holt, Rinehart and Winston, New York. The sentence "The boy has lost a dollar" was presented to 24 students. They were instructed to judge the degrees of relationship between the words in the sentence by selecting the pair out of each triplet of words that were most strongly related and also by selecting the least strongly related pair. For each word pair, the frequency is computed with which the pair is judged more highly related than some other pair over all triads and students. These frequencies appear in Table 10.4. The frequencies conform to the intuitive grammatical tree structure

((the boy)((has lost)(a dollar))).

PROGRAMS

SEARCH   searches through trees, changing ancestor of each node in succession, to most reduce $\alpha T - (1 - \alpha)F$, where T is the number of given triads predicted by the tree and F is the number of given triads incorrectly predicted.

TRIAD   used by SEARCH to compute the number of true and false triads predicted by a tree.

MTREE   quick tree printing routine.

TRDIST   produces triads from a distance matrix.

ULTRA   computes ultrametric from tree.

```
      SUBROUTINE SEARCH(MT,M,K,T,A,TR)
C...................................................................20 MAY 1973
C.... SEARCHES THROUGH TREES , MOVING ONE NODE AT A TIME, TO MAXIMIZE
C     A*TRUE-(1-A)*FALSE, WHERE TRUE IS THE NUMBER OF TRIADS IN T CORRECTLY
C     PREDICTED BY THE TREE.  USES MTREE AND TRIAD.
C.... MT = 1 BY K ARRAY, MT(I).GT.I IS ANCESTOR OF I EXCEPT MT(K)=K.
C.... M = NUMBER OF OBJECTS
C.... K = NUMBER OF NODES, GREATER THAN M
C.... T = M BY M BY M BORDERED TRIAD ARRAY
C.... A = VALUE BETWEEN 0 AND 1 , SAY A =0.5
C.... TR = M BY M BORDERED ARRAY, ULTRAMETRIC CONFORMING TO MT.
C...................................................................
      DIMENSION MT(K),T(M,M,M),TR(M,M)
C.... INITIALIZE TREE
      DO 10 I=1,K
   10 MT(I)=K
      DO 11 I=2,M
       TR(I,I)=I
      DO 11 J=2,M
   11 IF(I.NE.J) TR(I,J)=K
C.... FIND BEST ANCESTOR FOR EACH OBJECT
      DO 20 I=2,K
      MM=M+1
      EE=-10.**10
      JJ=MT(I)
      DO 21 J=I,K
      IF(J.LE.M) GO TO 21
      IF(J.LE.I) GO TO 21
      MT(I)=J
      CALL TRIAD(I,MT,M,K,TR,T,TT,FT)
      EJ=A*TT-(1.-A)*FT
      IF(EJ.LE.EE) GO TO 21
      EE=EJ
      JJ=J
   21 CONTINUE
      MT(I)=JJ
      CALL TRIAD(K,MT,M,K,TR,T,TT,FT)
      TT=TT/3
      FT=FT/3
      WRITE(6,1) I,JJ,FT,TT
    1 FORMAT(I5     , 13H HAS ANCESTOR,I5,7H FALSE=,F5.0,6H TRUE=,F5.0)
      CALL MTREE(MT,M,K)
   20 CONTINUE
      RETURN
      END
```

```
      SUBROUTINE TRIAD(II,MT,M,K,TR,T,TT,FT)
C.................................................................20 MAY 1973
C.... COMPUTES FALSE TRIADS AND CORRECT TRIADS IN ARRAY T, ACCORDING TO ARRAY TR
C     COMPUTES ERRORS INVOLVING II TH OBJECT.
C     COMPUTES ERRORS FOR ALL OBJECTS IF II.GT.M.
C.... II = OBJECT
C.... M = NUMBER OF OBJECTS
C.... TR = M BY M BORDERED ARRAY, ULTRAMETRIC FROM MT
C.... T = M BY M BY M BORDERED ARRAY
C        T(I,J,K)=NUMBER OF TIMES IJ ARE MOST SIMILAR OF IJ,IK,KJ.
C.... FT = NUMBER OF FALSE TRIADS
C.... TT = NUMBER OF TRUE TRIADS
C.................................................................................
      DIMENSION MT(K),TR(M,M),T(M,M,M)
      IL=II
      IU=II
      IF(II.GE.2.AND.II.LE.M) GO TO 60
      IL=2
      IU=M
   60 CONTINUE
      FT=0
      TT=0
C.... UPDATES ULTRAMETRIC TR
      DO 20 I=IL,IU
      CALL ULTRA(I,TR,MT,M,K)
   20 CONTINUE
      DO 10 I=IL,IU
      DO 10 J=2,M
      DO 10 L=2,M
      IF(I.EQ.J.OR.J.EQ.L.OR.L.EQ.I) GO TO 10
      IF(TR(I,J).LT.TR(I,L)) TT=TT+T(I,J,L)
      IF(TR(I,J).LT.TR(I,L)) FT=FT+T(I,L,J)+T(L,J,I)
      IF(TR(J,L).LT.TR(I,J)) TT=TT+T(L,J,I)*0.5
      IF(TR(L,J).LT.TR(I,J)) FT=FT+(T(I,L,J)+T(I,J,L))*0.5
   10 CONTINUE
      RETURN
      END




      SUBROUTINE MTREE(MT,M,K)
C.................................................................20 MAY 1973
C.... QUICK TREE PRINT-OUT
C.... M = NUMBER OF OBJECTS
C.... K = NUMBER OF NODES OF TREE, COUNTING OBJECTS
C.... MT = K BY 1 ARRAY DEFINING TREE, MT(I).GT.I EXCEPT FOR I=5.
C.................................................................................
      DIMENSION MT(K)
      DIMENSION MS(50)
      NC=1
      DO 20 I=2,M
   20 MS(I)=I
   30 WRITE(6,1)(MS(I),I=1,M)
    1 FORMAT(50I2)
      IF(NC.EQ.0) RETURN
      NC=0
      DO 21 I=2,M
      J=MS(I)
      MS(I)=MT(J)
      IF(MS(I).LE.J.AND.J.NE.K) RETURN
   21 IF(MS(I).NE.K) NC=1
      GO TO 30
      END
```

```
      SUBROUTINE TRDIST(D,M,T)
C..............................................................20 MAY 1973
C.... PRODUCES TRIADS FROM A DISTANCE MATRIX
C.... M = NUMBER OF ROWS
C.... D = M BY M BORDERED DISTANCE MATRIX
C.... T = M BY M BY M BORDERED TRIADS
C..........................................................................
      DIMENSION D(M,M),T(M,M,M)
      DO 20 I=2,M
      DO 20 J=2,M
      DO 20 K=2,M
      T(I,J,K)=0
      IF(D(I,J).LT.D(J,K).AND.D(I,J).LT.D(K,I)) T(I,J,K)=1
   20 CONTINUE
      RETURN
      END




      SUBROUTINE ULTRA(I,D,MT,M,K)
C..............................................................20 MAY 1973
C.... COMPUTES ULTRAMETRIC FROM MT, ITH ROW AND COLUMN
C.... D = M BY M BORDERED ARRAY, ULTRAMETRIC
C.... MT =  K BY 1 TREE ARRAY
C..........................................................................
      DIMENSION D(M,M),MT(K)
      IT=I
      DO 20 J=2,M
      II=IT
      JJ=J
   21 IF(II.LT.JJ) II=MT(II)
      IF(II.GT.JJ) JJ=MT(JJ)
      IF(II.NE.JJ) GO TO 21
      D(J,IT)=II
   20 D(IT,J)=II
      RETURN
      END
```

CHAPTER 11

Single-Linkage Trees

11.1 INTRODUCTION

It causes eye strain to look at a distance matrix like Table 11.1, the airline distances between principal cities of the world. A correct approach to representing these data would associate each city with a position on the globe, thus reducing the present $30 \times 29/2$ distances to the 30×2 coordinate values. The usual data for clustering form a data matrix with many objects and some variables measured on them. The first step in clustering defines distances between each pair of objects. Thus the first step in understanding the data is backward, in that many more numbers are required for the distances than for the original data matrix!

Correlation matrices may be converted to distances by setting $1 - \text{correlation} = \text{distance}$. This distance is the square of the euclidean distance between standardized variables. More generally, if $\text{COV}(I, J)$ denotes the covariance between variables I and J, $\text{COV}(I, I) + \text{COV}(J, J) - 2\text{COV}(I, J)$ is the square of euclidean distance between variables with means removed. Correlation and covariance matrices are the most frequent sources of distance matrices for clustering variables.

A standard approach to clustering separates the question of defining distances from the question of constructing clusters based on the distances (see, for example, Sokal and Sneath, 1963; Johnson, 1967; Hartigan, 1967, Jardine et al., 1967; Gower and Ross, 1969; Estabrooke, 1966; Sneath, 1957). The algorithms assume a distance matrix to be given. Of these algorithms, the oldest, the most famous, and the most fruitful mathematically is the single-linkage algorithm. The algorithm joins together the two closest objects to form a cluster, then the next two closest objects to form a cluster, then the next two closest objects, and so on. If the two objects to be joined lie in different clusters obtained in previous steps, the two clusters are joined instead. The term single linkage is used because two clusters are joined if any of the distances between the objects in the different clusters is sufficiently small—that is, if there is a single link between the clusters. The algorithm is explicitly described in Sneath (1957).

11.2 SINGLE-LINKAGE ALGORITHM

Preliminaries. The M objects will be arranged in order so that each cluster is a contiguous sequence of objects. The Ith object in this new order will be denoted by $O(I)$. A *gap* $G(I)$ is associated with the Ith object in the order. These gaps determine the boundaries of the clusters.

Table 11.1 Airline Distances Between Principal Cities of the World

```
        AZORES  •│AZ
       BAGHDAD  •│39 BD                          PY 54 33 59 33│ 31 37 93 88 84│
        BERLIN  •│22 20 BN                       PS 57  7 56│ 72 50 57105 61│
        BOMBAY  •│59 20 39 BY                    RO 57 66│ 18 69113 84115│
  BUENOS AIRES  •│54 81 74 93 BS                 RE 63│ 74 57 57101 61│
         CAIRO  •│33  8 18 27 73│CO              SF│59  7 61 74 52│
      CAPETOWN  •│57 49 60 51 43│ 45 CN          SO 64117 71107│
       CHICAGO  •│32 64 44 81 56│ 61 85 CH       SE 57 77 48│
          GUAM  •│89 63 71 48104│ 71 88 74 GM    SI 49 11│
      HONOLULU  •│73 84 73 80 76│ 88115 43 38 HU SY 48│
      ISTANBUL  •│29 10 11 30 76│  8 52 55 69 81│ IL           TO│
        JUNEAU  •│46 61 46 69 77│ 63103 23 51 28│ 55 JU
        LONDON  •│16 25  6 45 69│ 22 60 40 75 72│ 16 44 LN
        MANILA  •│83 49 61 32111│ 57 75 81 16 53│ 57 59 67 MA
     MELBOURNE  •│120 81 99 61 72│ 87 64 97 35 55│ 91 81105 39 ME
   MEXICO CITY  •│45 81 61 97 46│ 77 85 17 75 38│ 71 32 56 88 84│ MY
      MONTREAL  •│24 58 37 75 56│ 54 79  8 77 49│ 48 26 33 82104│ 23 ML
        MOSCOW  •│32 16 10 31 84│ 18 63 50 61 70│ 11 46 16 51 90│ 67 44 MW
   NEW ORLEANS  •│36 72 51 89 49│ 68 83  8 77 42│ 62 29 46 87 93│  9 14 58 NS
      NEW YORK  •│25 60 40 78 53│ 56 78  7 80 50│ 50 29 35 85104│ 21  3 47 12 NY
   PANAMA CITY  •│38 78 59 97 33│ 71 70 23 90 53│ 68 45 53103 90│ 15 25 67 16 22
         PARIS  •│16 24  5 44 69│ 20 58 41 76 75│ 14 47  2 67104│ 57 34 16 48 36
RIO DE JANEIRO  •│43 69 62 83 12│ 61 38 53116 83│ 64 76 57113 82│ 48 51 72 48 48
          ROME  •│21 18  7 38 69│ 13 52 48 76 80│  9 53  9 65 99│ 64 41 15 55 43
 SAN FRANCISCO  •│50 75 57 84 64│ 75103 19 58 24│ 67 15 54 70 79│ 19 25 59 19 26
      SANTIAGO  •│57 88 78100  7│ 80 49 53 98 69│ 81 73 72109 70│ 41 54 88 45 51
       SEATTLE  •│46 68 51 77 69│ 68102 17 57 27│ 61  9 48 67 82│ 23 23 52 21 24
      SHANGHAI  •│72 44 51 31122│ 52 81 70 19 49│ 49 49 57 12 50│ 80 70 42 77 73
        SYDNEY  •│121 83100 63 73│ 90 69 92 33 51│ 93 77106 39  4│ 81100 90 89100
         TOKYO  •│73 52 56 42114│ 60 92 63 16 39│ 56 40 60 19 51│ 70 65 47 69 68
```

From *The World Almanac* (1966), p. 510 (in hundreds of miles).

STEP 1. Let $O(1)$ be any object. Let $G(1) = \infty$.

STEP 2. Let $O(2)$ be the object closest to $O(1)$. Let $G(2)$ be the distance between $O(2)$ and $O(1)$.

STEP 3. For each I $(3 \leq I \leq M)$ let $O(I)$ be the object, not among $O(1)$, $O(2), \ldots, O(I-1)$, that is closest to one of $O(1), O(2), \ldots, O(I-1)$. That is, for some K $(1 \leq K \leq I-1)$

$$D[O(I), O(K)] \leq D(J, L),$$

where J ranges over $O(1), \ldots, O(I - 1)$ and L ranges over the remaining objects. The gap $G(I)$ is set equal to this minimum distance $D[O(I), O(K)]$.

STEP 4. The cluster $O(L1) - O(L2)$, containing objects $O(L1)$, $O(L1 + 1), \ldots$, $O(L2 - 1)$, $O(L2)$ is associated with gap $G(I)$, where $(L1, L2)$ is the maximal interval including I, such that $G(J) \leq G(I)$ for all J with $L1 < J \leq L2$. If all gaps are different, there will be $M - 1$ such clusters; if some gaps are equal, there may be less.

11.3 APPLICATION OF SINGLE-LINKAGE ALGORITHM TO AIRLINE DISTANCES

Initially, only the last five cities from Table 11.1 will be considered. The searching of the distance matrix is shown in Table 11.2. The ordering of the objects and construction of the tree is shown in Table 11.3. The reordered distance matrix appears in Table 11.4.

Table 11.2 Application of Single-Linkage Algorithm, to Airline Distances—Searching the Matrix

```
SANTIAGO   SO                                    SO
SEATTLE    64   SE                               64   S̶E̶
SHANGHAI   117  57   SI                          117  57 │ SI
SYDNEY     71   77   49   SY                     71   77 │ 49   SY
TOKYO      107  48   11   48   TO                107  48 │ 11   48   TO

           BEGIN WITH SE                         FIND OBJECT CLOSEST TO SE, WHICH
                                                 IS TO, WITH GAP 48
```

```
SO                                    SO
64   S̶E̶                               64   S̶E̶
117  57 │ SI                          117  5̶7̶│ S̶I̶
71   77 │ 49   SY                     71   77 │ 49   SY
107  4̶8̶│ 11   48   T̶O̶                107  4̶8̶│ 1̶1̶│ 48   S̶O̶

ELIMINATE DISTANCES BETWEEN SELECTED   ELIMINATE DISTANCES BETWEEN.
OBJECTS.  FIND OBJECT CLOSEST TO SE, TO,   FIND CLOSEST OBJECT, SY. GAP = 48.
WHICH IS SI.  GAP = 11.
```

```
SO
64   S̶E̶
117  5̶7̶  S̶I̶
71   7̶7̶  4̶9̶  S̶Y̶
107  4̶8̶  1̶1̶  4̶8̶  T̶O̶

ONLY SO IS LEFT.  GAP = 64.
```

Table 11.3 Application of Single-Linkage Algorithms to Airline Distances—Constructing Tree

```
∞
O(1)      SE - - - I - - I
G(2)      48        I     I
O(2)      TO - I    I     I
G(3)      11   I    I     I
O(3)      SI - /    I     I
G(4)      48        I     I
O(4)      SY - - - /      I
G(5)      64              I
O(5)      SO - - - - - - /
```

STEP 1. The first object is selected arbitrarily to be SE. Set $O(1) = \text{SE}$, $G(1) = \infty$.

STEP 2. The distances of SE to the other objects are 64, 57, 77, 48. The smallest of these is 48, $D(\text{SE, TO})$. Thus, $O(2) = \text{TO}$, $G(2) = 48$.

STEP 3. First find that object which is closest to SE or TO. Eliminate the distance $D(\text{SE, TO})$, but check all other distances to SE and TO (underlined in Table 11.2), 64, 57, 77, 107, 11, 48. The smallest of these is 11, $D(\text{TO, SI})$. Thus $O(3) = \text{SI}$, $G(3) = 11$. Now eliminate the distances $D(\text{SE, SI})$ and $D(\text{TO, SI})$ and check all other distances to SE, TO, SI (underlined in Table 11.2), 64, 117, 107, 77, 49, 48. The smallest of these is 48, $D(\text{TO, SY})$. Thus $O(4) = \text{SY}$, $G(4) = 48$. Finally, $O(5) = \text{SO}$, $G(5) = 64$.

STEP 4. The ordered objects and gaps are listed in Table 11.3. There is a cluster corresponding to each gap. For the gap $G(3) = 11$, the cluster is the largest sequence $O(L1) - O(L2)$ such that $L1 \leq 3 \leq L2$ and $G(J) \leq G(I)$ for all J ($L1 < J \leq L2$). For $L1 = 2$ and $L3 = 3$, these conditions are satisfied, but for no larger interval. The cluster corresponding to $G(3) = 11$ is TO–SI. The cluster corresponding to $G(2) = 48$ is $\text{SE} - \text{SY}$, and this is also the cluster corresponding to $G(4) = 48$. This is a case when clusters coincide for two equal gaps, although gaps may be equal without the coincidence of corresponding clusters. Finally, the cluster $\text{SE} - \text{SO}$ corresponds to the gap $G(5) = 64$.

The single-linkage algorithm is applied to the airline data for 30 cities in Table 11.5. The overall clustering conforms to known geography. There are five principal clusters of cities in Europe, North America, Asia, South America, South Africa, and

Table 11.4 Reordered Distance Matrix—Single Linkage Applied to Airline Distances

SEATTLE	SE				
TOKYO	48	TO			
SHANGHAI	57	11	SI		
SYDNEY	77	48	49	SY	
SANTIAGO	64	107	117	71	SO

Table 11.5   Single-Linkage Algorithm Applied to Airline Distances between 30 Cities

```
                               DIAMETER
         _ _ _ _ 5 _ _ _ _ 10 _ _ _ _ 15 _ _ _ _ 20 _ _ _ _ 25 _ _ _ _ 30 _ _ _ _ 35 _ _ _ _ 40
      AZORES •  - - - - - - - - - - - - - - I - - - I - - - I - - - - - - I - - - - - - I I
       PARIS •  - I - - I - I - - I I         I         I         I                 I         I I
      LONDON •  - /      I    I      I I         I         I         I                 I         I I
      BERLIN •  - - - - /    I      I I         I         I         I                 I         I I
        ROME •  - - - - - - /        I I         I         I         I                 I         I I
      MOSCOW •  - - - - - - - - - / I         I         I         I                 I         I I
    ISTANBUL •  - - - - - - - I      I         I         I         I                 I         I I
       CAIRO •               I         I         I         I         I                 I         I I
     BHAGDAD •  - - - - - - - / - - / - - - - / I         I                 I         I I
      BOMBAY •  - - - - - - - - - - - - - - - - - /         I                 I         I I
    MONTREAL •  - - I - - - I I I - - - - - I - I         I                 I         I I
    NEW YORK •  - - /        I I I         I    I         I                 I         I I
     CHICAGO •  - - - - - - / I I         I    I         I                 I         I I
 NEW ORLEANS •  - - - - - - - / I         I    I         I                 I         I I
 MEXICO CITY •  - - - - - - - - / I    I         I                 I         I I
 PANAMA CITY •  - - - - - - - - - - - - - / I         I                 I         I I
     SEATTLE •  - - - - - - I - I         I         I                 I         I I
SAN FRANCISCO •  - - - - - - / I         I         I                 I         I I
      JUNEAU •  - - - - - - - - / - - - - - - - - / I                 I         I I
    HONOLULU •  - - - - - - - - - - - - - - - - - - - - - - - - /                 I         I I
    SHANGHAI •  - - - - - - - - - - - I I - - - I                 I         I I
       TOKYO •  - - - - - - - - - - / I         I                 I         I I
      MANILA •  - - - - - - - - - - - / I                 I         I I
        GUAM •  - - - - - - - - - - - - - /                 I         I I
    SANTIAGO •  - - - - - - I - - - - I                 I         I I
BUENOS AIRES •  - - - - - - / I                 I         I I
RIO DE JANEIRO •  - - - - - - - - - - / - - - - - - - - - - - - - - - - - - /         I I
   CAPE TOWN •  - - - - - - - - - - - - - - - - - - - - - - - - - - - - - - - - - - - - / I
      SYDNEY •  - - - I                                                                 I
   MELBOURNE •  - - - / - - - - - - - - - - - - - - - - - - - - - - - - - - - - - - - - - - - /
```

Australia. These clusters must be picked out from a confusing array of 27 clusters. (There would have been 29 except for two coincidences of gap values.) The usual output of the single-linkage algorithm is a binary tree of $M - 1$ clusters (of more than one object) for M objects. The binary tree is convenient to construct but difficult to interpret, since many of the clusters differ only slightly for each other. For example, in the European cluster there is the sequence of clusters

((((Paris, London) Berlin) Rome) Moscow).

Should all of these be accepted as different clusters, or should only the European cluster (Paris, London, Berlin, Rome, Moscow) be offered? This question will be discussed later in Section 11.12.

11.4 COMPUTATIONAL PROPERTIES OF SINGLE LINKAGE

The single-linkage algorithm produces clusters invariant under reordering of the objects, although the initial object may be selected arbitrarily. For example, if the first object selected in Table 11.5 were San Francisco, the next would be Seattle, then Juneau, then one of Montreal–Panama City, and so on. The computations require $O(N^2)$ comparisons; roughly, each element in the distance matrix is looked at once. It is not necessary to store and compute the whole distance matrix before beginning the algorithm—for 3000 objects and three variables, you had better compute the distances as you need them.

The clusters obtained are invariant under monotonic transformation of the distance matrix.

11.5 SPIRAL SEARCH ALGORITHM

Preliminaries. In many clustering algorithms, including the single-linkage algorithm, the principal computing expense is in finding the closest object of some set to a given object. Normally this requires $O(MN)$ computations, where M is the number of objects searched over and N is the number of variables used in the distance calculations. The spiral search technique avoids computation of the larger distances entirely by searching over objects that are close to the given object on all variables. It is most useful for M very large and N small.

The object which is Ith largest on the Jth variable is denoted by $O(I, J)$. At each stage, there will be two candidate objects $O1(J)$, $O2(J)$ corresponding to the Jth variable. All objects closer than these to the given object on the Jth variable will have been considered. If the minimum distance of $O1(J)$, $O2(J)$ to the given object is $T(J)$ on the Jth variable, then all objects not considered have a distance overall which exceeds $[T(1)^2 + \cdots + T(N)^2]^{1/2}$. The search therefore stops when an object already considered has a distance to the given object less than this lower bound.

STEP 1. For $1 \leq J \leq N$, let the rank of the given object on the Jth variable be $K(J)$. Initial candidate objects for the Jth variable are

$$O1(J) = O[K(J) - 1, J]$$

and

$$O2(J) = O[K(J) + 1, J].$$

The closest of these to the given object in the Jth variable is distant $T(J)$. Initially the closest object overall to the given object is not defined, so set DMIN $= \infty$. Also, if $K(J) - 1 \leq 0$, assume that $O1(J)$ is infinitely distant and, if $K(J) + 1 > M$, assume that $O2(J)$ is infinitely distant.

STEP 2. For each J in succession ($1 \leq J \leq N$), select $O1(J)$ or $O2(J)$, whichever is closer to the given object on the Jth variable. If the distance overall, of this selected object from the given object is less than DMIN, set OMIN equal to this object, and set DMIN equal to this new minimum distance. If $O1(J)$ is used, redefine $O1(J)$ to be the next lowest object on the Jth variable. If $O2(J)$ is used, redefine $O2(J)$ to be the next highest object on the Jth variable. Redefine $T(J)$ to be the minimum distance of $O1(J)$ and $O2(J)$ to the given object on the Jth variable. If DMIN$^2 \leq T(1)^2 + \cdots + T(N)^2$, stop. Otherwise, continue repeating Step 2.

NOTE. Since ordering the objects on all variables takes $O(M \log M)N$ comparisons, the above calculation saves time only when many searches for closest objects must be made, all using the same initial orderings of the objects.

11.6 APPLICATION OF SPIRAL SEARCH ALGORITHM
TO BIRTHS AND DEATHS

The birth and death rates of 70 countries are given in Table 11.6. The orders of countries by birth and death rates are given in Table 11.7. In the single-linkage algorithm, the first step would be to find the closest object to some initial country—say, Algeria, object 1. Apply the spiral search algorithm as follows:

Table 11.6 Birth and Death Rates per 1000 Persons

		BIRTH	DEATH				BIRTH	DEATH
1.	ALGERIA	36.4	14.6		36.	ARGENTINA	21.8	8.1
2.	CONGO	37.3	8.0		37.	BOLIVIA	17.4	5.8
3.	EGYPT	42.1	15.3		38.	BRAZIL	45.0	13.5
4.	GHANA	55.8	25.6		39.	CHILE	33.6	11.8
5.	IVORY COAST	56.1	33.1		40.	COLOMBIA	44.0	11.7
6.	MALAGASY	41.8	15.8		41.	ECUADOR	44.2	13.5
7.	MOROCCO	46.1	18.7		42.	PERU	27.7	8.2
8.	TUNISIA	41.7	10.1		43.	URUGUAY	22.5	7.8
9.	CAMBODIA	41.4	19.7		44.	VENEZUELA	42.8	6.7
10.	CEYLON	35.8	8.5		45.	AUSTRIA	18.8	12.8
11.	CHINA	34.0	11.0		46.	BELGIUM	17.1	12.7
12.	TAIWAN	36.3	6.1		47.	BRITAIN	18.2	12.2
13.	HONG KONG	32.1	5.5		48.	BULGARIA	16.4	8.2
14.	INDIA	20.9	8.8		49.	CZECHOSLOVAKIA	16.9	9.5
15.	INDONESIA	27.7	10.2		50.	DENMARK	17.6	19.8
16.	IRAQ	20.5	3.9		51.	FINLAND	18.1	9.2
17.	ISRAEL	25.0	6.2		52.	FRANCE	18.2	11.7
18.	JAPAN	17.3	7.0		53.	E. GERMANY	17.5	13.7
19.	JORDAN	46.3	6.4		54.	W. GERMANY	18.5	11.4
20.	KOREA	14.8	5.7		55.	GREECE	17.4	7.8
21.	MALAYSIA	33.5	6.4		56.	HUNGARY	13.1	9.9
22.	MONGOLIA	39.2	11.2		57.	IRELAND	22.3	11.9
23.	PHILLIPINES	28.4	7.1		58.	ITALY	19.0	10.2
24.	SYRIA	26.2	4.3		59.	NETHERLANDS	20.9	8.0
25.	THAILAND	34.8	7.9		60.	NORWAY	17.5	10.0
26.	VIETNAM	23.4	5.1		61.	POLAND	19.0	7.5
27.	CANADA	24.8	7.8		62.	PORTUGAL	23.5	10.8
28.	COSTA RICA	49.9	8.5		63.	ROMANIA	15.7	8.3
29.	DOMINICAN R	33.0	8.4		64.	SPAIN	21.5	9.1
30.	GUATEMALA	47.7	17.3		65.	SWEDEN	14.8	10.1
31.	HONDURAS	46.6	9.7		66.	SWITZERLAND	18.9	9.6
32.	MEXICO	45.1	10.5		67.	U.S.S.R.	21.2	7.2
33.	NICARAGUA	42.9	7.1		68.	YUGOSLAVIA	21.4	8.9
34.	PANAMA	40.1	8.0		69.	AUSTRALIA	21.6	8.7
35.	UNITED STATES	21.7	9.6		70.	NEW ZEALAND	25.5	8.8

From *Reader's Digest Almanac* (1966).

Table 11.7   Countries Ordered by Birth Rates and Death Rates

BIRTH	DEATH	RANK	BIRTH	DEATH	RANK	BIRTH	DEATH	POSITION
56	16	70	67	48	46	12	11	22
20	24	69	68	63	45	1	22	21
65	26	68	64	29	44	2	54	20
63	13	67	69	10	43	22	40	19
48	20	66	35	28	42	34	52	18
49	37	65	36	49	41	9	39	17
46	12	64	57	69	40	8	57	16
18	17	63	43	14	39	6	47	15
37	19	62	26	55	38	3	46	14
55	21	61	62	70	37	44	45	13
53	44	60	27	68	36	33	38	12
60	18	59	17	64	35	40	41	11
50	23	58	70	51	34	41	53	10
51	33	57	24	35	33	38	1	9
47	67	56	15	66	32	32	3	8
52	61	55	42	31	31	7	6	7
54	27	54	23	56	30	19	30	6
45	43	53	13	60	29	31	7	5
66	25	52	29	8	28	30	9	4
58	2	51	21	65	27	28	50	3
61	34	50	39	15	26	4	4	2
16	59	49	11	58	25	5	5	1
14	36	48	25	32	24			
59	42	47	10	62	23			

Country 56, Hungary, is 70th in birth rate. (See Table 10.6)

STEP 1.   Algeria is 21st in birth rates and 9th in death rates, so $K(1) = 21$, $K(2) = 9$. Initial candidate objects are

$$O1(1) = 12, \qquad O2(1) = 2, \qquad T(1) = 0.1$$

and

$$O2(1) = 53, \qquad O2(2) = 3, \qquad T(2) = 0.7.$$

These are the objects immediately adjoining Algeria in birth and death rates.

STEP 2.   For $J = 1$, the closest object is $O1(1) = 12$. Then $D(1, 12) = 8.5$. The new object $O1(1) = 10$, $T(1) = 0.6$, DMIN = 8.5 and $[T(1)^2 + T(2)^2]^{1/2} = 0.9$, so Step 2 is repeated. For $J = 2$, the closest object is $O2(2) = 3$. Then $D(1, 3) = 5.7$, $O2(2) = 6$, and $T(2) = 0.9$. Set DMIN = 5.7. For $J = 1$, the closest object is $O1(1) = 10$, which is further from object 1 than 5.7. Set $O1(1) = 25$, $T(1) = 0.9$. Continuing in this way, no change in DMIN occurs until object 11 is reached, DMIN = 4.3. This is still greater than the stopping threshold, so the spiral continues. Then again at object 39, DMIN = 3.9. After several more objects are checked, arrive at $O1(1) = 29$,

$O1(2) = 34$, $O2(1) = 47$, and $O2(2) = 30$. The thresholds are $T(1) = 3.4$, $T(2) = 2.4$, and now

$$\text{DMIN}^2 \leq T(1)^2 + T(2)^2,$$

so the search stops. The closest object to Algeria is object 39, Chile. The total number of objects considered was 17, which is a substantial reduction on the 70 required without spiraling.

11.7 SINGLE-LINKAGE CLUSTERS FROM PARTITIONS

The gaps obtained in the single-linkage algorithm of Section 11.2 may be interpreted as diameters of the clusters corresponding to them. If a cluster C corresponds to a gap G, any pair of objects I, J in the cluster C are connected by a chain of objects—say, $I = I(1), I(2), \ldots, I(K) = J$—such that $D[I(L), I(L + 1)] \leq G$ for every L ($1 \leq L < K$). Another way of generating single-linkage clusters is to link any pair of objects whose distance is no greater than G. The objects and links form a graph whose maximal connected components are single-linkage clusters. At each gap level G, these components form a partition of the data. For example, at $G = 26$ in Table 11.5 the partition of cities would be Europe and North America, Asia, South America, South Africa, and Australia. The clusters at gap level G^* ($<G$) will each be included in a cluster at gap level G. (For if two objects are connected by links of length no greater than G^*, they are also connected by links of length no greater than G.)

Single-linkage clusters are the outcome of exact optimization of a partition $P(M, K)$ using the error function (contrived for the occasion)

$$e[P(M, K)] = \max_{1 \leq L \leq K} \text{DIAM}\,[C(L)],$$

where $\text{DIAM}(C)$ is the minimum value of G such that all pairs of objects in C are connected by a chain of links no greater than G. The optimal partition will be the maximal connected components of the graph, with objects linked when their distance is no greater than G for some G.

Single-linkage clusters comprise optimal partitions with the error function $e[P(M, K)] = \sum_L \text{DIAM}\,[C(L)]$ but not with the error function $e[P(M, K)] = \sum_L \text{DIAM}\,[C(L)]\text{NUMB}[C(L)]$, where $\text{NUMB}(C)$ is the number of objects in the cluster C. This last error function is the most natural analog to the sum-of-squares criterion used in the K-means algorithm.

11.8 JOINING AND SPLITTING

Begin with M clusters, each consisting of a single object. Define the distance between any two clusters as the minimum distance among pairs of objects, one in each cluster. Find the closest pair of clusters, and construct a new cluster by joining them. Thus at each step the total number of clusters is reduced by one, and after $M - 1$ steps a single cluster consisting of all objects remains. During the steps, $2M - 1$ clusters will have been computed (ignoring the annoying case when two (or more) clusters are closest to another cluster and are both joined to it in one step), and these are just the single-linkage clusters. This way of approaching single-linkage clusters is conceptually simple but not convenient for computation, because it does not provide an order for the objects in which clusters are contiguous sequences of objects. Other "joining" algorithms suggest themselves, with the distance between clusters defined as the maximum distance among pairs of objects in the two clusters, or the average distance, or other combinations of distances (see, for example, Sokal and Sneath, 1963; Johnson, 1967; Lance and Williams, 1966).

In the other direction, consider splitting the set of all objects into two clusters such that the distance between the clusters (defined as the minimum over pairs) is maximized. Then split each of these clusters in the same way, and so on until only clusters consisting of single objects remain. The clusters obtained in this way also are single-linkage clusters. Note that joining and splitting give the same clusters with this minimum distance between clusters, but the clusters are not necessarily the same with other definitions of distance between clusters.

11.9   ULTRAMETRICS

Define the *chain distance* between any two objects as the minimum value G such that the objects are connected in the graph obtained by linking objects no more than G apart. Denote this by $DCH(I, J)$. Single-linkage clusters are then characterized by the property that, for every I, J in C and every K in C and L not in C,

$$DCH(I, J) < DCH(K, L).$$

The diameter of a cluster C is the maximum value of $DCH(I, J)$ over I and J in C.

The chain distance is an ultrametric (see Chapter 8); that is, for any triple I, J, K,

$$DCH(I, J) \leq \max [DCH(I, K), DCH(J, K)].$$

The tree corresponding to the ultrametric is just the single-linkage tree with the definition of cluster diameter given above.

Which ultrametric D^* best approximates the given distance D? The measure of error considered in Hartigan (1967) is $e(D, D^*) = \sum_{I, J} [D(I, J) - D^*(I, J)]^2$, which corresponds to the distance between clusters being equal to the average distance between pairs of objects in the two clusters. Suppose that $D^*(I, J) \leq D(I, J)$. For any error $e(D, D^*)$ that is increasing with $D(I, J) - D^*(I, J)$, the ultrametric minimizing $e(D, D^*)$ is $D^* = DCH$. The single-linkage tree is thus optimal, but under the rather unusual constraint that $D^*(I, J) \leq D(I, J)$. This result is hinted at in Hartigan (1967), but it is stated and demonstrated more explicitly in Jardine, Jardine, and Sibson (1967).

11.10   STRUNG-OUT CLUSTERS

Single-linkage clusters are famously strung out in long sausage shapes, in which objects far apart are linked together by a chain of close objects. Of course, the clusters are not necessarily convex. A cluster of objects may, on the average, be close to some other cluster, but because of a few "bridge" objects it may be connected to some far-off third cluster (see Figure 11.1). This problem can be approached by using the joining algorithm with both minimum and average definitions of distance.

Sometimes clusters *are* far-flung sausage shapes with high densities of objects within each cluster. The single-linkage algorithm will be better at discovering such shapes than the average or maximum method.

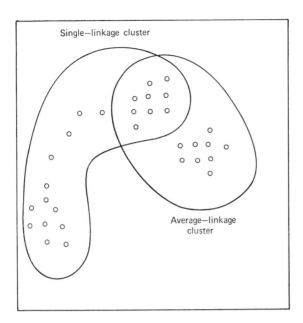

Figure 11.1 Strung-out single-linkage clusters.

11.11 MINIMUM SPANNING TREES

A number of cities are to be connected by a road network, consisting of links between selected pairs of cities, which has a fixed cost per mile. What is the network of minimum cost, such that a path exists between every pair of cities? First, it is clear that there will be only one path between each pair of cities, because otherwise a link between cities could be removed, which would reduce costs. The network is thus an *undirected* tree. This should not be confused with *directed* trees of clusters.

The tree contains $M - 1$ links for M cities, and the minimum spanning tree is the tree for which the sum of the distances over the $M - 1$ links is a minimum (see Kruskal, 1956).

For objects and distances between them, a minimum spanning tree may be constructed by selecting an arbitrary object, linking to it the closest object to it (giving a tree on two objects), linking to this tree whichever object among the remainder is closest to an object in the tree (giving a tree on three objects), and so on, building up the tree one object at a time. The minimum spanning tree is not necessarily unique, because the closest object to the present tree at each stage in the algorithm may not be unique. Note that the order in which objects are added to the tree is the order in which the clusters are contiguous in the single-linkage algorithm of Section 11.2.

The single-linkage clusters may be obtained from the minimum spanning tree by a number of methods (Gower and Ross, 1969). For example, by deleting all links from the tree of length greater than G, the maximal single-linkage clusters of diameter no greater than G are obtained. To obtain all the single-linkage clusters, remove the largest link from the tree to obtain two connected clusters, then the second largest to obtain two more clusters, and so on. (There will be a division into three or more clusters if the largest link is not unique.)

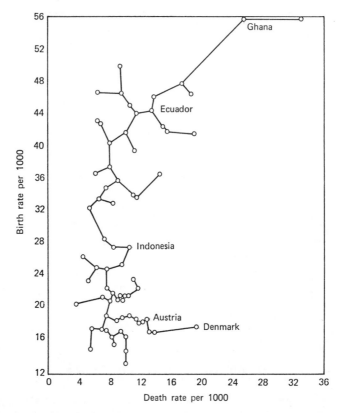

Figure 11.2   Minimum spanning tree for births and deaths (Table 11.6).

In just two dimensions, the judgment of closest distances by eye makes it feasible to construct minimum spanning trees and thence single-linkage clusters, for up to 100 points by hand. The birth and death rates data are so treated in Figure 11.2. All 69 single-linkage clusters may be constructed from this tree by removing links in order of their sizes. There are perhaps two or three real clusters in the data: (Ghana and Ivory Coast) with extremely high birth and death rates, the developing countries with high birth rates, the developed countries with low birth rates.

11.12.*   REALITY OF CLUSTERS

An important, difficult, and unsolved problem is the selection of just a few significant clusters from the many produced by the single-linkage and other joining algorithms. In particular, are there any "real" clusters other than the set of all objects? Is the tree different from those obtained from objects randomly selected from a multivariate normal distribution?

Estabrooke (1966) proposes a measure of isolation for a single cluster C, which he calls the "moat" of C, equal to the diameter of the smallest cluster including C minus the diameter of C. A similar measure, with distances converted to ranks, is proposed by Ling in an unpublished Ph.D. thesis (Yale University). Ling, using results on connectedness of random graphs due to Erdos and Renyi (1961), obtains exact and approximate distribution theory for his isolation index under the null hypothesis that any of the $[M(M-1)/2]!$ permutations of the ranked distances are equally probable. He acknowledges that this null hypothesis is not the most plausible one, since $D(I, J)$ and $D(J, K)$ small imply $D(I, K)$ small for a metric distance and this property is violated by the uniform distribution over permutations.

To test the reality of the tree overall, it is necessary to use all the cluster diameters in some way. A sequence of cluster diameters arises naturally from the algorithm in Section 11.2 in the gaps $G(2), \ldots, G(M)$. This sequence is not unique because the initial object may be arbitrarily chosen. Suppose that the objects were sampled from a unimodal distribution and that the initial object were chosen at the mode. Except for sampling errors, the gaps will be steadily increasing as objects are successively added from regions where the density of the distribution is decreasing. On the other hand, if the objects were sampled from a bimodal distribution, the gaps will increase as objects at the outside of the first cluster are brought in, then decrease as objects in the second cluster are brought in, and then increase again. The reality of the clusters may thus be tested by checking departures from monotonicity of the gap sequence. (See Figure 11.3, for the relation between the modality of distribution and the modality of the gap sequence.)

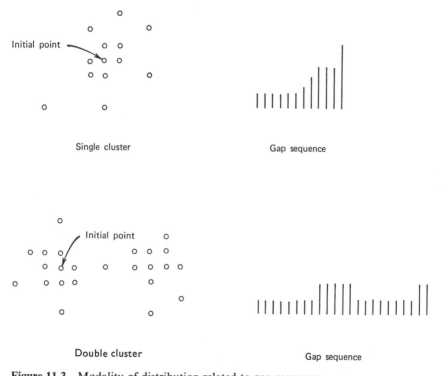

Figure 11.3 Modality of distribution related to gap sequence.

For M cases and N variables, assume that the gap powers $G^N(2), \ldots, G^N(M)$ are independent exponentials with $G^N(I)$ having expectation $A(I)$, where $A(2) \leq \cdots \leq A(M)$. (This assumption is justified for large samples from a unimodal distribution; DMIN, the smallest distance from a given point to one of the sample points, is the minimum of a large number of independent random variables, from which it follows that $DMIN^N$ times the density is approximately exponentially distributed with constant expectation. Nearly all the gaps are such minimum distances.) The log likelihood of the observed gaps is $\sum -\log A(I) - G^N(I)/A(I)$. This log likelihood is maximized by averaging neighboring gaps (following Barlow et al., 1972).

STEP 1.   Initially set $A(I) = G^N(I)$ $(2 \leq I \leq M)$.

STEP 2.   For each I $(1 < I < M)$, if $A(I) \leq A(I + 1)$ increase I by one and go back to Step 2; if $A(I) > A(I + 1)$, find the minimal J such that $A(J) \geq A(J + 1) \geq \cdots \geq A(I)$ and the maximal K such that $A(I + 1) \geq A(I + 2) \geq \cdots \geq A(K)$, and replace each of $A(J), A(J + 1), \ldots, A(K)$ by the average value of $A(J), \ldots, A(K)$. Set $I = J - 1$, and begin Step 2 again.

The log likelihood under this monotonic constraint, subtracted from the log likelihood without constraint, is $\sum \log [A(I)/G^N(I)]$. The larger this value, the less plausible is the monotonic hypothesis; the value is always nonnegative, and it is zero only if the sequence is nondecreasing. The criterion for departure from monotonicity is then $\sum \log [A(I)/G^N(I)]$, where the $A(I)$ are selected as above. The actual distribution of this quantity will depend on the real distribution of gaps rather than the exponentials hypothesized. Under the exponential model, its asymptotic distribution [given that the $A(I)$'s are nondecreasing but that only a finite number of different $A(I)$'s occur] is normal with mean $M\alpha_1$ and variance $M\alpha_2$, where $\alpha_1 = -\int_0^\infty (\log x)e^{-x} dx$, and $\alpha_2 = \int_0^\infty (\log x)^2 e^{-x} dx - \alpha_1^2$. These are well-known integrals derivable from the gamma function; $\alpha_1 = \gamma = 0.577$ (Euler's constant) and $\alpha_2 = \pi^2/6 = 1.645$.

Thus,

$$\sum \log [A(I)/G^N(I)] \approx N(\gamma M, \pi^2 M/6).$$

To test this procedure on the cities data (Table 11.1), begin with the modal city Paris. The best monotonic fit to the squares gaps is computed by averaging over sequences of neighboring values that are monotonically decreasing. The tree can be redrawn with these new diameters; there are the small clusters {[(Paris) London] Berlin}; then a large cluster, Europe; a still larger cluster, Europe and North America;

Table 11.8   Gaps in Single-Linkage Tree on Cities (From Table 10.5)

OBSERVED GAPS

2	5	7	10	11	8	8	16	20	24
24	9	7	16	15	9	8	7	4	31
16	12	11	31	7	38	39	4		

OBSERVED GAPS SQUARED

4	25	49	100	121	64	64	256	400	576
576	81	49	256	225	81	64	49	16	961
256	144	121	961	49	1444	1521	16		

MONOTONIZED GAPS

4	25	49	87	87	87	87	219	219	219
219	219	219	219	219	219	219	219	219	300
300	300	300	505	505	994	994	994		

then South America is added on; and finally Australia and South Africa are added. The monotonicity of the fitted gaps forces the clusters also to be monotonic (see Table 11.8).

11.13 DENSITY-CONTOUR TREE

If objects are points distributed in an N-dimensional space (one dimension for each variable), clusters may be thought of as regions of high density separated from other such regions by regions of low density. It is easiest to first formalize this definition for a distribution of points (consisting of an infinite number of points), characterized by a density $f(\mathbf{x})$ at each point $\mathbf{x}$. The number $f(\mathbf{x})$ is proportional to the number of objects per unit volume at the point $\mathbf{x}$. A *density-contour* cluster at level f_0 is a subset C of the N-dimensional space, such that C is maximal among connected sets satisfying $f(\mathbf{x}) \geq f_0$ for $x \in C$. It is easy to show that such clusters form a tree. For a cluster C at level f_0, the density inside C is no less than f_0, but for every path connecting $\mathbf{x}$ in C to $\mathbf{y}$ outside C the density somewhere on the path is less than f_0. The cluster C thus conforms to the informal requirement that C is a high-density region surrounded by a low-density region. An example of density-contour clusters is given in Figure 11.4.

As the level f_0 is reduced, the cluster C at level f_0 gradually expands until suddenly, at some splitting level f_0^*, it coalesces with other clusters to form a much larger cluster. These "rigid" clusters, which cannot be expanded smoothly, are important in determining the tree structure. For example, if there is only one such cluster, the density is unimodal. A cluster C is *rigid* if every cluster properly including it contains a cluster disjoint from C. The rigid density-contour clusters form a tree that is a subset of the tree of all density-contour clusters.

The essential components of the above definitions are the density f, defined at each point, and the paths between points in the space. The definitions may be generalized to spaces in which these two components are present, and such a generalization is

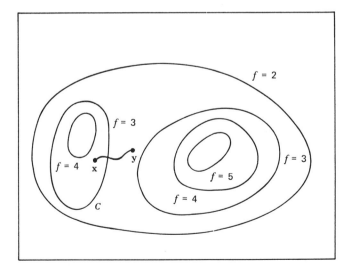

Figure 11.4. Density-contour clusters. C is a density-contour cluster at level 3. All densities inside C are no less than 3 and, for $\mathbf{x}$ in C and $\mathbf{y}$ outside C, some density on every path between $\mathbf{x}$ and $\mathbf{y}$ is less than 3.

advisable because the infinite number of points do not appear in practical clustering problems. Consider then a space S of a finite number of objects with some pairs of objects linked together. (This is just an undirected graph.) A *path* between object I and object J is a sequence $I = I(1), I(2), I(3), \ldots, I(K) = J$ such that $I(L)$ is linked to $I(L + 1)$ for each L $(1 \leq L < K)$. A *connected set* C is a set of objects such that each pair of objects in the set are linked by a path whose members all lie in the set C. With these definitions of path and connectedness, the definitions of density-contour cluster and rigid cluster apply to finite sets of objects, provided that a density is given for each object.

Table 11.9 contains per capita income in the 50 states. The density is supposed to be proportional to the number of points per unit volume. Here each dollar earned in a state corresponds to a point in that state, and the volume of a state is the number of people in it. The volume measurement is always chosen as some comparable count to the point count, so the density is a ratio of two counts. There is no need for the density to refer to the objects to be clustered in any particular way, in order for the

Table 11.9   U.S. Per Capita Income, 1964 (in Dollars)

DE	DELAWARE	3426	NB	NEBRASKA	2302
CT	CONNECTICUT. . .	3250	FL	FLORIDA	2280
NV	NEVADA	3248	VA	VIRGINIA	2224
NY	NEW YORK	3139	AZ	ARIZONA	2218
AK	ALASKA	3128	MT	MONTANA.	2183
CA	CALIFORNIA . . .	3092	TX	TEXAS.	2175
IL	ILLINOIS	3003	UH	UTAH	2174
NJ	NEW JERSEY . . .	2962	VT	VERMONT	2144
MA	MASSACHUSETTS. .	2922	ME	MAINE.	2130
MD	MARYLAND	2888	OK	OKLAHOMA	2095
MI	MICHIGAN	2733	ND	NORTH DAKOTA . .	2012
WA	WASHINGTON . . .	2634	ID	IDAHO.	2012
OH	OHIO	2623	NM	NEW MEXICO . . .	2010
OR	OREGON	2602	WV	WEST VIRGINIA .	1962
MO	MISSOURI	2595	GA	GEORGIA	1933
CO	COLORADO	2591	NC	NORTH CAROLINA .	1900
HI	HAWAII	2579	LA	LOUISIANA . . .	1864
PA	PENNSYLVANIA .	2575	TN	TENNESSEE . . .	1852
IN	INDIANA	2529	SD	SOUTH DAKOTA . .	1832
WI	WISCONSIN . . .	2492	KY	KENTUCKY	1811
RI	RHODE ISLAND . .	2479	AL	ALABAMA	1737
WY	WYOMING	2475	SC	SOUTH CAROLINA .	1647
MN	MINNESOTA . . .	2373	AR	ARKANSAS	1633
IA	IOWA	2370	MS	MISSISSIPPI . .	1444
NH	NEW HAMPSHIRE. .	2340			
KS	KANSAS	2311			

algorithm to be applicable. In this case, the density is the ratio of the number of dollars to the number of people, but the objects to be clustered are states and the connectivity properties are defined on states. The contour-density clusters will be regions of neighboring states where the densities are high surrounded by regions of states where the densities are low.

In general, there will be M objects, a density $F(I)$ ($1 \leq I \leq M$) for each object, and links specified between some pairs of objects. This information is converted to a distance

$$D(I, J) = -\min [F(I), F(J)] \qquad \text{if } I \text{ and } J \text{ are linked,}$$
$$= 0 \qquad \text{if } I \text{ and } J \text{ are not linked.}$$

The single-linkage clusters will be the maximal connected sets of objects of density greater than C, say. The single-linkage algorithm thus generates the density-contour clusters with this measure of distance.

From per capita income in Table 11.9 and from the links between neighboring states in Table 11.10, the single-linkage algorithm produces the clusters in Table 11.11. Only

Table 11.10   Links between States

STATE	BORDERS ON	STATE	BORDERS ON
DE - - -	NJ PA MD	KS - - -	MO OK CO NB
CT - - -	NY MA RI	NB - - -	CO WY SD IA MO KS
NV - - -	CA OR ID UT AZ	FL - - -	GA AL
NY - - -	PA VT MA CT NJ	VA - - -	KY WV MD TN NC
AK - - -		AZ - - -	NM UT NV CA
CA - - -	OR AZ NV	MT - - -	ND SD ID WY
IL - - -	IA WI IN KY MO	TX - - -	NM OK AR LA
NJ - - -	DE PA NY	UT - - -	AZ NV ID WY CO
MA - - -	RI CT NY VT NH	VT - - -	NY MA NH
MD - - -	VA WV PA DE	ME - - -	NH
MI - - -	WI IN OH	OK - - -	TX NM CO KS MO AR
WA - - -	OR ID	ND - - -	MT SD MN
OH - - -	IN MI PA WV KY	ID - - -	UT NV OR WA MT WY
OR - - -	WA ID NV CA	NM - - -	TX OK CO AZ
MO - - -	IL TN AR OK KS NB IA KY	WV - - -	VA KY OH PA MD
CO - - -	NM UT WY KS OK NB	GA - - -	FL AL TX NC SC
HI - - -		NC - - -	SC GA TN VA
PA - - -	NJ DE MD OH NY WV	IA - - -	AR MS TX
IN - - -	IL MI OH KY	TN - - -	MS AR KY VA NC GA AL MO
WI - - -	IA MN MI IL	SD - - -	NB WY MT ND MN IA
RI - - -	CT MA	KY - - -	TN MO IL IN OH WV VA
WY - - -	CO UT ID MT SD NB	AL - - -	MS TN GA FL
MN - - -	ND SD IA WI	SC - - -	GA NC
IA - - -	NB MO IL WI MN SD	AR - - -	LA MS TX OK MO TN
NH - - -	VT ME MA	MS - - -	LA AR TN AL

See Table 10.9 for state code.

Table 11.11 Rigid Clusters of States from Density-Contour Algorithm Applied to Per Capita Income

-GAP	STATE							-GAP	STATE			
0	DE	I	I	I	I	I	I	2174	AZ I		I	I
2962	NJ	I	I	I	I	I	I	2218	NV I		I	I
2962	NY	I	I	I	I	I	I	3092	CA I		I	I
3139	CT	I	I	I	I	I	I	2602	OR I		I	I
2922	MA	I	I	I	I	I	I	2602	WA /		I	I
2888	MD	/	I	I	I	I	I	2144	VT		I	I
2575	PA		I	I	I	I	I	2130	ME		I	I
2575	OH	I	I	I	I	I	I	2095	OK		I	I
2733	MI	/	/	I	I	I	I	2095	TX		I	I
2529	IN			I	I	I	I	2012	ND		I	I
2529	IL	I		I	I	I	I	2012	ID		I	I
2595	MO	/		I	I	I	I	2010	NM		I	I
2492	WI			I	I	I	I	1962	WV		/	I
2479	RI			I	I	I	I	1900	NC			I
2373	MN			I	I	I	I	1900	GA I			I
2370	IA			I	I	I	I	1933	FL /			I
2340	NH		/	I	I	I		1864	IA			I
2311	KS				I	I	I	1852	TN			I
2311	CO	I			I	I	I	1832	SD			I
2475	WY	/			I	I	I	1811	KY			I
2302	NB				I	I	I	1737	AL			I
2224	VA				I	I	I	1647	SC			I
2183	MT		/		I	I		1633	AR			I
2174	UT				I	I		1444	MS			/
								0	AK			
								0	HI			

the rigid clusters are shown; each rigid cluster is such that the smallest cluster properly including it contains a cluster disjoint with the rigid cluster. The principal features of the tree are a northeastern cluster of Delaware through Michigan, a northern cluster of Delaware through Montana, a western cluster of Arizona through Washington, and the isolated clusters of Alaska and Hawaii. (Sorry, Alaska and Hawaii!)

11.14 DENSITIES AND CONNECTEDNESS, DISTANCES GIVEN

Given a set of objects and links between some pairs of them, a natural measure of density at object I is the number of objects J to which it is linked. Thus density-contour clusters may be constructed from any set of objects on which links are defined. In particular, if the distance is given for all pairs of objects, each pair of objects will be linked if their distance is less than some threshold DO. As a rule of thumb, DO might be the minimum distance which connects the data (that is, the maximum gap in the single-linkage algorithm.)

An example of density-contour clusters in two dimensions is given in Figure 11.5. Note that there are only two rigid clusters, which corresponds to the intuitive visual clustering.

11.15 THINGS TO DO

11.15.1 Running the Single-Linkage Algorithm

This algorithm requires a distance between every pair of points. It costs about M^2 comparisons between distances, so it is practicable only for small numbers of points—say, M < 1000. It is possible to avoid some of the $M(M - 1)/2$ distance calculations by trickery like spiral searches. The clusters obtained have several global characterizations that make them easier to understand than many clusterings. A definite defect is that the single-linkage clustering depends only on the small distances in the distance matrix, and so it will be relatively unstable if these distances are not accurately determined. The average joining algorithm gives more compact and more stable clusters.

Mutation distances (Table 11.12) and stock yields (Table 11.13) are proposed as trial data sets.

Table 11.12   Mutation Distances

```
MAN              0.
MONKEY           1. 0.
DOG             13.12. 0.
HORSE           17.16.10. 0.
DONKEY          16.15. 8. 1. 0.
PIG             13.12. 4. 5. 4. 0.
RABBIT          12.11. 6.11.10. 6. 0.
KANGAROO        12.13. 7.11.12. 7. 7. 0.
PEKIN DUCK      17.16.12.16.15.13.10.14. 0.
PIGEON          16.15.12.16.15.13. 8.14. 3. 0.
CHICKEN         18.17.14.16.15.13.11.15. 3. 4. 0.
KING PENGUIN    18.17.14.17.16.14.11.13. 3. 4. 2. 0.
SNAPPING TURTLE 19.18.13.16.15.13.11.14. 7. 8. 8. 8. 0.
RATTLESNAKE     20.21.30.32.31.30.25.30.24.24.28.28.30. 0.
TUNA            31.32.29.27.26.25.26.27.27.27.26.27.27.38. 0.
SCREWWORM FLY   33.32.24.24.25.26.23.26.26.26.26.28.30.40.34. 0.
MOTH            36.35.28.33.32.31.29.31.30.30.31.30.33.41.41.16. 0.
BAKER'S MOULD   63.62.64.64.64.64.62.66.59.59.61.62.65.61.72.58.59. 0.
BREAD YEAST     56.57.61.60.59.59.59.58.62.62.62.61.64.61.66.63.60.57. 0.
SKIN FUNGUS     66.65.66.68.67.67.67.68.66.66.66.65.67.69.69.65.61.61.41. 0.
```

From Fitch and Margoliash, *Science* (1967). A distance between two species is the number of positions in the protein molecule cytochrome-*c*, where the two species have different amino acids.

Table 11.13   Yield of Stocks

Yield equals the cash dividend by the average price of stock (from Moody's *Handbook of Common Stocks*).

	59	60	61	62	63	64	65	66	67	68	69
Aetna Life & Casualty	1.4	1.6	1.2	1.4	1.2	1.0	1.2	1.5	2.1	2.1	2.6
Allied Chemical	2.8	3.4	3.1	3.9	3.6	3.3	3.6	4.5	4.8	4.5	3.9
American Airlines	3.5	4.7	4.3	5.1	3.7	2.6	2.2	2.1	2.0	2.7	2.6
American Broadcasting Company	3.7	2.7	1.9	2.8	3.0	2.9	2.4	2.1	1.9	2.7	2.6
American Cyanamid	2.9	3.2	3.6	4.0	3.3	3.2	2.8	3.2	3.9	4.4	4.3
American Express	2.6	2.6	2.0	2.5	2.8	3.2	2.3	1.7	1.0	1.0	1.4
American Motors	3.9	4.5	5.8	6.6	5.1	6.3	5.6	0	0	0	0
American Telephone & Telegraph	3.8	3.5	2.8	3.1	2.8	2.8	3.1	3.9	3.9	4.5	4.5
Anaconda Company	3.7	4.5	4.6	5.7	5.3	5.1	5.4	5.7	5.1	4.3	4.7
Atlantic Richfield Co.	5.3	4.3	4.8	4.5	4.0	3.5	3.4	3.1	1.9	0	1.8
Bank America	4.1	4.3	3.3	3.7	3.2	2.9	3.3	3.8	4.0	2.9	3.2
Bethlehem Steel	4.4	5.1	5.4	6.1	4.7	4.0	3.9	5.3	4.3	5.1	5.6
Burroughs Corporation	2.7	3.0	2.8	2.6	3.6	3.9	2.6	1.4	0.7	0.5	0.4
Chase Manhattan	3.8	3.9	3.2	3.4	2.9	2.6	2.9	3.5	3.4	3.0	3.5
Chrysler	1.6	2.7	2.1	1.8	1.3	1.9	2.4	4.4	4.5	3.3	4.5
Coca Cola	4.5	3.7	2.6	2.8	2.7	2.4	2.1	2.3	1.9	1.6	1.7
Columbia Broadcasting	3.0	3.5	3.8	3.6	2.4	2.5	3.0	2.3	2.2	2.7	2.8
Consolidated Edison	4.4	4.7	3.8	4.0	3.9	3.7	4.0	4.8	5.4	5.4	6.0
Dow Chemical	1.4	1.6	1.9	2.8	2.6	2.4	2.4	3.0	2.8	3.0	3.4
Du Pont	4.6	4.8	5.3	5.1	4.2	3.3	2.5	3.0	3.1	3.4	3.9
Firestone	1.9	2.5	2.3	2.7	2.8	2.7	2.6	3.0	2.8	2.6	2.8
Ford	3.9	3.9	3.3	3.8	3.6	3.6	3.7	5.0	5.1	4.4	5.1
General Electric	2.3	2.4	2.8	3.0	2.5	2.6	2.2	2.6	2.8	2.9	3.0
General Foods	2.8	2.2	1.8	2.3	2.3	2.3	2.5	3.0	3.3	3.1	3.3
General Mills	3.0	4.2	3.4	4.3	3.2	2.8	2.6	2.5	2.3	2.1	2.4
General Motors	3.3	3.7	3.8	5.2	5.9	4.8	5.6	6.6	6.5	4.2	3.8
A & P	2.4	2.8	2.5	2.7	3.8	3.8	3.8	4.7	5.2	4.9	4.3
Greyhound	4.9	4.6	4.6	4.0	3.1	2.6	3.4	4.6	4.7	4.3	4.8
Gulf Oil	2.5	3.1	2.9	3.9	3.6	3.1	3.3	3.8	3.6	3.4	3.9
Imperial Chemical	2.0	2.5	2.3	3.0	2.2	2.6	3.1	3.3	4.9	3.2	3.2
I.B.M.	0.5	0.6	0.5	0.7	1.0	1.1	1.3	1.3	0.9	0.8	1.1
International Harvester	4.1	5.4	4.9	4.8	4.3	3.7	3.7	4.1	4.9	5.2	5.8
International Tel. & Tel.	2.5	1.9	2.2	2.1	1.8	2.0	2.0	1.5	0	1.6	1.8
Johnson & Johnson	1.3	1.5	1.1	1.2	1.1	1.1	1.0	1.0	0.8	0.7	0.6

11.15.2   Density Contour and Single Linkage

If all densities are unity and two objects are linked if their distance is less than some threshold DO, then the clusters obtained are all single-linkage clusters. If the density at object *I* is the number of objects within some threshold distance of object *I*, the clusters will not be single-linkage clusters but will differ from them in inhibiting the formation of long thin clusters. Two objects will lie in the same density-contour

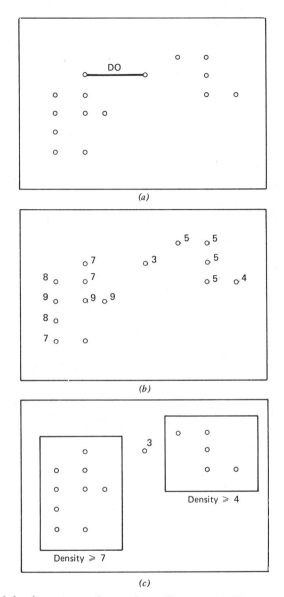

Figure 11.5 Rigid density-contour clusters from distances. (*a*) Choose threshold DO to be minimum connecting distance. (*b*) Density equals number of objects within threshold. (*c*) Rigid density-contour clusters.

cluster if they may be connected by a chain of objects in which each object is close to its predecessor and also has sufficiently high density. See Figure 11.5.

These facts should be demonstrated by constructing an example in two dimensions.

11.15.3 Modes

For a set of objects I $(1 \leq I \leq M)$, densities $F(I)$ are given and each object is connected to some other objects. An object is a mode if its density is a maximum of the densities of the objects to which it is connected. Show that the maximal density-contour cluster containing a specific mode, and no other, is a rigid cluster, if such a cluster exists.

11.15.4 Density Estimates

A natural density estimate is the number of objects within a threshold distance of the given object. Another natural density estimate is the inverse of the threshold distance necessary to find K objects within threshold. This second estimate is more satisfactory in sparse regions of the distribution where the first estimate will produce undiscriminating unit densities for all objects.

11.15.5 Reducing the Number of Clusters

A possible way to reduce the number of single-linkage clusters is to consider only rigid clusters. Another way is to truncate the original distance matrix to just a few levels—say, $D(1), \ldots, D(L)$. Thus, if $D(K1) < D(I, J) \leq D(K2)$, the value of $D(I, J)$ is changed to $D(K2)$. Show that all the clusters obtained from the new distance matrix are single-linkage clusters.

11.15.6* Convexity

Single-linkage clusters are serpentine, and average distance clusters tend to be globular. An intermediate restraint on clusters is that they be convex. A joining algorithm that constructs convex clusters goes as follows. Begin with a number of convex sets $C(1), \ldots, C(K)$ containing $N(1), \ldots, N(K)$ points, respectively. The diameter of a cluster $C(I)$ is $N(I) \log [N(I)/V(I)]$, where $V(I)$ is the volume of the Ith cluster. Joining $C(I)$ and $C(J)$ produces a new cluster C^*, the smallest convex set containing $C(I)$ and $C(J)$. Those clusters are joined which least increase the total cluster diameter, and the joins are continued until a single cluster remains. Initializing $C(1), \ldots, C(K)$ is tricky. Try this technique in one dimension.

11.15.7 Avoiding distance calculations

If a distance D is a metric, much distance calculation may be avoided by first using a quick partition algorithm of the leader type. If the threshold is T, and I has leader $L(I)$, then $D(I, J) \geq D(L(I), L(J)) - 2T$. In looking for small distances, it will usually be sufficient to know only the distances between leaders.

REFERENCES

BARLOW, R. E., BARTHOLOMEW, D. J., BREMNER, J. M., and BRUNK, H. D. (1972). *Statistical Inference under Order Restrictions*, Wiley, New York.

ERDÖS, P., and RENYI, A. (1961). "On the evolution of random graphs." *Bull. Inst. Int. Stat., Tokyo*, 343–347. Graphs are built up by randomly adding links. They consider many probabilistic problems, such as the number of links required for the graph to have K connected components.

ESTABROOKE, G. J. (1966). "A mathematical model in graph theory for biological classification." *J. Theor. Biol.* **12**, 297–310. A cluster is c-connected if connected in the graph obtained by linking points of similarity greater than c. A c-cluster is c-connected, but not c'-connected for $c' > c$. The moat of C is the difference between c and the smallest value of c' such that the cluster is c'-connected.

GOWER, J. C., and ROSS, G. J. S. (1969). "Minimum spanning trees and single linkage cluster analysis." *Appl. Stat.* **18**, 54–64. The single-linkage clusters may be obtained from the minimum spanning tree by successively eliminating links, largest first. The clusters are the maximal connected components at each step.

HARTIGAN, J. A. (1967). Representation of similarity matrices by trees. *J. Am. Stat. Ass.* **62**, 1140–1158.

JARDINE, C. J., JARDINE, N., and SIBSON, R. (1967). "The structure and construction of taxonomic hierarchies." *Math. Biosci.* **1**, 173–179. The relationship between a tree of clusters and an ultrametric is established. The single-linkage ultrametric d^*, using chain distance, minimizes $\sum F(d^* - d)$, where F is decreasing, under the constraint $d^* \leq d$.

JOHNSON, S. C. (1967). "Hierarchical clustering schemes." *Psychometrika* **32**, 241–254. A one-to-one relationship is established between a tree of clusters with diameters and the ultrametric distance, defined, between two objects, as the diameter of the smallest cluster containing them. Joining algorithms are considered—one in which the distance between clusters is the minimum, the other with maximum distance. These produce clusters whose diameters are, in the first case, the chain diameter and, in the second, the maximum distance between pairs of objects in the cluster. These two algorithms are invariant under monotonic transformation of the distances.

KRUSKAL, J. B. (1956). "On the shortest spanning subtree of a graph, and a traveling salesman problem." *Proc. Amer. Math. Soc.* **7**, 48–50. The minimum-length spanning tree is constructed by adding the shortest link, then the second shortest, and so on, omitting a link whenever it introduces a cycle.

LANCE, G. N., and WILLIAMS, W. T. (1966). "A general theory of classificatory sorting strategies. 1. Hierarchical systems." *Comp. J.* **9**, 337–380. Many hierarchical procedures are "joining" algorithms in which two objects I and J are amalgamated to form a new object IJ whose distance to any other object K is given by a rule of form

$$D(IJ, K) = \alpha D(I, J) + \beta D(I, K) + \gamma D(K, J),$$

where α, β, and γ may depend on the number of objects already amalgamated into I, J, and K.

LING, R. (1971). Ph.D. thesis, Yale University. A set C of objects is a (K, r)-cluster if each object in C is a distance less than r from K other objects in C, if each pair of objects are connected by a chain of objects in C with successive objects less than r apart, and if no cluster including C has these two properties. Single-linkage clusters are $(1, r)$-clusters.

Considerable probability theory is given for the evolution of $(1, r)$-clusters from randomly ranked distances, and some empirical results are presented in the K-linked case.

McQUITTY, L. L. (1960). "Hierarchical syndrome analysis." *Educ. Psychol. Meas.* **20**, 293–303. He describes an algorithm in which two objects are joined and the similarity of the new coalesced objects to any other object is the minimum similarity of the two combined objects. In distance terms, this corresponds to the distance between clusters as the maximum distance over pairs of objects in the two clusters.

SNEATH, P. H. A. (1957). "The application of computers to taxonomy." *J. Gen. Microbiol.* **17**, 201–206. The measure of similarity used between two strains of bacteria is the number of characters held in common, divided by the total number of characters held by either. The two most similar strains are joined, then the next most similar, and so on. If the next most similar lie in groups already formed, the groups are coalesced.

SOKAL, R. R., and SNEATH, P. H. A. (1963). *Principles of Numerical Taxonomy*, Freeman, San Francisco. On p. 180, a description of the single-linkage algorithm: The most similar objects are coalesced, then the next most similar, and so on, with two groups coalesced whenever any pair of objects in them are the next most similar. On p. 182, it is suggested that the distance between clusters should be the average distance between the pairs of objects, rather than the minimum. On p. 192, there is a discussion of the serpentine character of single-linkage clusters.

WISHART, D. (1969). "A generalization of nearest neighbor which reduces chaining effects," in *Numerical Taxonomy*, A. J. Cole, ed. Academic, London. Ball-like clusters, such as arise in the K-means algorithm or in the joining algorithms with a maximum distance between clusters, sometimes result in cluster boundaries cutting across regions of high density. An example is given for the Hertzprung–Russell diagram, plotting color against brightness, which clusters stars into four types, main sequence, super giants, giants, and dwarfs. These are long sausage clusters which are not detected at all by a ball-oriented algorithm.

On the other hand, the single-linkage algorithm can easily join quite different clusters if a few "bridge" objects are fortuitously placed between them. To reduce the effect of these noisy objects, use the following algorithm:

(i)   Select a frequency threshold K.

(ii)   Associate with each object I the distance $PMIN(I)$, the smallest distance such that K objects are within distance $PMIN(I)$ of object I.

(iii)   Begin a new cluster with the object for which $PMIN(I)$ is a minimum.

(iv)   Incorporate objects in the tree in the order of the $PMIN(I)$. When object I is incorporated, there will be a number of clusters partitioning objects already incorporated. Begin a new cluster with object I and amalgamate any pair of clusters which are linked by a distance less than $PMIN(I)$.

(v)   All clusters formed along the way form the tree.

PROGRAMS

SLINK   computes single-linkage tree, reorders distance matrix according to order of objects consistent with tree.

```
      SUBROUTINE SLINK(D,M,NB)
C.................................................................20 MAY 1973
C.... COMPUTES SINGLE LINKAGE TREE FROM DISTANCE MATRIX D.  REORDERS DISTANCE .
C.... M = NUMBER OF OBJECTS
C.... D = M BY M BORDERED SYMMETRIC ARRAY
C.... NB = 3 BY M CLUSTER ARRAY
C            NB(1,K) = FIRST OBJECT IN CLUSTER
C            NB(2,K) = LAST OBJECT IN CLUSTER
C            NB(3,K) = CLUSTER DIAMETER
C.... ONLY M-2 CLUSTERS ARE PRODUCED
C.................................................................
      DIMENSION D(M,M),NB(3,M)
      DO 30 NEXT = 1,M
      J=NEXT+1
      IF(NEXT.EQ.1) GO TO 50
      IF(NEXT.EQ.M) GO TO 30
      DMIN=10.**10
      IMIN=NEXT
      DO 31 I=2,NEXT
      IF(D(I,I).GE.DMIN) GO TO 31
      DMIN=D(I,I)
      IMIN=I
   31 CONTINUE
      NB(3,J)=100.*DMIN
      I=D(1,IMIN)
C.... INTERCHANGE OBJECT I AND J
      DO 32 K=1,M
      A=D(I,K)
      D(I,K)=D(J,K)
   32 D(J,K)=A
      DO 33 K=1,M
      A=D(K,I)
      D(K,I)=D(K,J)
   33 D(K,J)=A
      DO 36 K=2,NEXT
      IF(D(1,K).EQ.I) D(1,K)=2
   36 IF(D(1,K).EQ.J) D(1,K)=I
C.... UPDATE CLOSEST OBJECT TO GRABBED OBJECTS
   50 CONTINUE
      DO 34 I=2,J
      D(1,J)=J
      IF(D(1,I).GT.J) GO TO 34
      D(1,I)=I
      D(I,I)=10.**10
      DO 35 K=J,M
      IF(K.EQ.J) GO TO 35
      IF(D(I,K).GE.D(I,I)) GO TO 35
      D(I,I)=D(I,K)
      D(1,I)=K
   35 CONTINUE
   34 CONTINUE
   30 CONTINUE
C.... FIND BOUNDARIES OF CLUSTERS
      NB(3,2)=10**10
      DO 40 K=2,M
      NB(1,K)=K
      NB(2,K)=K
      DO 41 L=K,M
      IF(L.EQ.K) GO TO 41
      IF(NB(3,L).GT.NB(3,K)) GO TO 42
   41 NB(2,K)=L
   42 CONTINUE
      DO 43 L=2,K
      LL=K-L+2
      IF(L.EQ.2) GO TO 43
      IF(NB(3,LL).GT.NB(3,K)) GO TO 44
   43 CONTINUE
   44 NB(1,K)=LL
   40 CONTINUE
      MM=M-2
      DO 45 K=1,MM
      DO 45 L=1,3
   45 NB(L,K)=NB(L,K+2)
C.... SET MAXIMUM DIAMETER EQUAL TO 100
      NMAX = 0
      DO 46 K=1,MM
   46 IF(NMAX.LT.NB(3,K)) NMAX=NB(3,K)
      DO 47 K=1,MM
   47 NB(3,K)=(NB(3,K)*100)/NMAX
C.... REPAIR DISTANCE MATRIX
      DO 60 I=1,M
      D(1,I)=D(I,1)
   60 IF(I.NE.1) D(I,I)=0
      RETURN
      END
```

215

CHAPTER 12

Distance and Amalgamation Algorithms

12.1 INTRODUCTION

In Table 12.1, the outcome is given of the first five matches of each of the Ivy League colleges in the 1965 football season. In order to predict future matches, it is desirable to identify similar colleges; for example, if Brown is similar to Yale, it might be predicted that Dartmouth will beat Yale about 35–9, since it has already beaten Brown by this amount.

The single-linkage technique proceeds by computing distances, somehow, between every pair of colleges, joining the closest pair, and then treating this cluster in later steps as a single object by defining its distance to all other objects in some way.

A generalization of this technique assumes the following to be given:

(i) a rule for computing a distance between any pair of objects;

(ii) a rule for amalgamating any two objects to form a third object. Algorithms using these two rules will be called distance and amalgamation algorithms. There still remains to be specified the exact method of tree construction. The first method considered is a joining technique, which finds the two closest objects using the distance rule, then amalgamates these two objects to form a new object using the amalgamation rule, and repeats the step on the reduced set of objects.

12.2 JOINING ALGORITHM

Preliminaries. It is assumed that a rule is given for computing distances between any two objects, and another rule is given for amalgamating any two objects to form a third object. The algorithm proceeds by finding the closest pair of objects, removing them from the set of objects, and replacing them by a single amalgamated object. There are two difficulties—the first is the problem of searching over all pairs of objects at each step, the second is keeping track of the sequence of amalgamations so the tree can be conveniently reconstructed. A number of clusters will be constructed by amalgamation in the course of the algorithm. Initially, each object is regarded as a cluster. At each step, the value IMIN(I) denotes the closest cluster to the Ith one (among clusters not yet amalgamated), and the value DMIN(I) denotes the distance of I to IMIN(I). Once the Ith cluster is removed by amalgamation, DMIN(I) = ∞ and IMIN(I) equals the cluster absorbing I.

216

Table 12.1 Ivy League Football, First Half of 1965 Season

BROWN - - - - - BN		HOLY CROSS - - - HC
BUCKNELL - - - BL		LAFAYETTE - - - LE
COLGATE - - - - CE		LEHIGH - - - - - LH
CONNECTICUT - - CT		PENNSYLVANIA - - PA
COLUMBIA - - - CA		PRINCETON - - - PN
CORNELL - - - - CL		RHODE ISLAND - - RI
DARTMOUTH - - - DH		RUTGERS - - - - RS
HARVARD - - - - HD		TUFTS - - - - - TS
NEW HAMPSHIRE - NH		YALE - - - - - - YE

BROWN--------	6-14(RI)	0- 7(PA)	0- 3(YE)	9-35(DH)	6- 0(CE)
COLUMBIA------	0-14(LE)	0-31(PN)	6-21(HD)	21- 7(YE)	12- 7(RS)
CORNELL-------	0- 0(CE)	49-13(LH)	27-36(PN)	3- 3(HD)	14-24(YE)
DARTMOUTH-----	56- 6( )	27- 6(HC)	24-19(PA)	35- 9(BN)	14- 0(HD)
HARVARD-------	17- 7(HC)	33- 0(TS)	21- 6(CA)	3- 3(CL)	0-14(DH)
PENNSYLVANIA--	20-14(LH)	7- 0(BN)	19-24(DH)	16-13(BL)	0-51(PN)
PRINCETON-----	32- 6(RS)	31- 0(CA)	36-27(CL)	27- 0(CE)	51- 0(PA)
YALE----------	6-13(CT)	0- 7(CE)	3- 0(BN)	7-21(CA)	24-14(CL)

STEP 1.   For each I $(1 \leq I \leq M)$ let IMIN(I) be that object which is closest to I, and let DMIN(I) be its distance to I. Set $K = M$.

STEP 2.   Let DM be the minimum value of DMIN(I) $(1 \leq I \leq K)$. Let J, L be clusters $(J < L)$ such that IMIN(J) = L, IMIN(L) = J, and DMIN(J) = DMIN(L) = DM.

STEP 3.   Define a new cluster, the $(K + 1)$th, to contain all objects in cluster J or cluster L. Set DMIN(J) = DMIN(L) = ∞, IMIN(J) = IMIN(L) = $K + 1$.

STEP 4.   Let $D(K + 1, I)$ be the distance between the $(K + 1)$th cluster and the Ith cluster, computed whenever DMIN(I) $< \infty$. If $D(K + 1, I) <$ DMIN(I), set DMIN(I) = $D(K + 1, I)$, IMIN(I) = $K + 1$. Let DMIN($K + 1$) be the minimum value among the $D(K + 1, I)$ and let IMIN($K + 1$) be the corresponding I value.

For each I $(1 \leq I \leq K)$ with DMIN(I) $< \infty$, if IMIN(I) = J or IMIN(I) = L, recompute the closest cluster IMIN(I) and its distance DMIN(I).

STEP 5.   Increase K by 1 and return to Step 2 unless $K = 2M$.

STEP 6.   Initially, POS(I) = 1 for $1 \leq I \leq M$, and POS(I) = 0 for $M < I < 2M$. Eventually POS(I) is the position of the Ith object in the ordering, such that each cluster is contiguous.

STEP 7.   For each I $(1 \leq I < 2M - 1)$ replace POS[IMIN(I)] by POS[IMIN(I)] + POS(I). After this step, POS(I) is the number of objects in the Ith cluster.

STEP 8.   For each I $(2M - 1 > I \geq 1)$, in reverse order, replace POS(I) by POS[IMIN(I)], and change POS[IMIN(I)] to POS[IMIN(I)] $-$ POS(I).

12.3 JOINING ALGORITHM APPLIED TO IVY LEAGUE FOOTBALL

The distance is computed between each pair of schools as the average squared difference between the scores over those schools which they have both played. For example, the distance between Brown and Cornell, since they played only Yale in common, is $[\frac{1}{2}(0 - 14)^2 + \frac{1}{2}(3 - 24)^2]$. Notice that the score by the opposing school and the score against the opposing school are both included, although it might be plausible to use only the difference of the scores.

To amalgamate a pair of schools, associate with the amalgamated schools all scores of all schools played by either amalgamated school. If a school was played by both, list both scores. If the amalgamated schools played each other, list the score in each order.

For example, to amalgamate Cornell and Harvard, the list is

0–0(CE)	49–13(LH)	27–36(PN)	3–3(HD)	14–24(YE)
17–7(HC)	33–0 (TS)	21–6 (CA)	3–3(CL)	0–14(DH).

With this amalgamation rule, the distance may be simply computed between clusters. Suppose cluster I is amalgamated with cluster J, cluster I has R scores in common with cluster K, and cluster J has S scores in common with cluster K. Then the amalgamated cluster—L, say—is distance $D(L, K)$ from K, where $(R + S)D(L, K) = RD(I, K) + SD(J, K)$. The distance matrix may thus be updated directly without reference to the lengthening list of scores for each cluster. It is actually more convenient to work with the sums of squares $RD(I, K)$ and the number of common opponents R, since the updating is then by simple addition.

STEP 1.   For the first object, BN, the closest object is object 8, YE, which is distance 51.5 from BN. (All distances are given in the first matrix in Table 12.2.) Therefore IMIN(1) = 8, DMIN(1) = 51.5. The closest object to object 2, CA, is object 8, YE, and its distance is 392. Therefore IMIN(2) = 8, DMIN(2) = 392. The list of closest objects and corresponding distances is given in Table 12.3. Set $K = 8$.

STEP 2.   The minimum value of DMIN(I) $(1 \leq I \leq 8)$ is 0. Clusters 3 and 5 both have this minimum distance to each other. So $J = 3$, $L = 5$.

STEP 3.   The ninth cluster is now defined, consisting of clusters 3 and 5, the objects CL and HD. Set DMIN(3) = DMIN(5) = ∞, IMIN(3) = IMIN(5) = 9. Clusters 3 and 5 will not be considered again in the amalgamation process, but the final value of IMIN is used later in reconstructing the tree.

STEP 4.   It is necessary to compute $D(9, I)$ $(1 \leq I \leq 8)$, the distance of the new cluster to all other clusters except the ones removed by amalgamation. For the definitions of amalgamation and distance used in this problem, these new distances may be computed by weighted averages of the old. For example, cluster 3, CL, played two

Table 12.2   Sequence of Distance Updates in Applying Joining Algorithm to Football Data (Table 12.1)

SUM OF SQUARED DIFFERENCES BETWEEN SCORES/NUMBER OF COMMON OPPONENTS

	BN	CA	CL	DH	HD	PA	PN	YE
BN	0							
CA	457/1	0						
CL	673/2	1425/3	0					
DH	2072/2	505/1	130/1	0				
HD	522/1	450/1	0/1	493/2	0			
PA	319/2	400/1	1796/2	915/2	461/1	0		
PN	3091/2	2323/2	891/2	1090/1	1801/2	5202/1	0	
YE	103/2	392/1	249/2	1105/1	983/2	16/1	2108/3	0

AMALGAMATE CL and HD

	BN	CA	CL-HD	DH	PA	PN	YE
BN	0						
CA	457/1	0					
CL-HD	1195/3	1875/4	0				
DH	2072/2	505/1	623/3	0			
PA	319/2	400/1	2257/3	915/2	0		
PN	3091/2	2323/2	2692/4	1090/1	5202/1	0	
YE	103/2	392/1	1232/4	1105/1	16/1	2108/3	0

AMALGAMATE YE and PA

	BN	CA	CL-HD	DH	PA-YE	PN
BN	0					
CA	457/1	0				
CL-HD	1195/3	1875/4	0			
DH	2072/2	505/1	623/3	0		
PA-YE	422/4	792/2	3489/7	2020/3	0	
PN	3091/2	2323/2	2692/4	1090/1	7310/4	0

AMALGAMATE BN and PA-YE

	BN-YE	CA	CL-HD	DH	PN
CA	1249/3				
CL-HD	4684/10	1875/4	0		
DH	4092/5	505/1	623/3	0	
PN	10401/6	2323/2	2692/4	1090/1	0

AMALGAMATE DH and CL-HD

	BN-YE	CA	CL-DH	PN
CA	1249/3	0		
CL-DH	8776/15	2380/5	0	
PN	10401/6	2323/2	3782/5	0

AMALGAMATE CA and BN-YE

	BN-CA	CL-DH	PN
CL-DH	11156/20		
PN	12724/8	3782/5	

schools in common with cluster 1, BN, and cluster 5, HD, played just one school in common with cluster 1, BN. Thus,

$$4D(9, 1) = 2D(3, 1) + D(5, 1)$$

and

$$D(9, 1) = 1195/3.$$

The updated distance matrices, after each amalgamation, are given in Table 12.2.

With the particular definitions of distance used here, it is not possible for $D(9, I)$ to be less than DMIN(I), since $D(9, I)$ is a weighted average of distances not less than DMIN(I).

The value DMIN(9) is the minimum of the $D(9, I)$, which from Table 12.2 shows DMIN(9) = 208, IMIN(9) = 4.

Some clusters have one of the amalgamated clusters 3 or 5 as their closest cluster, and a new closest cluster must be discovered. Clusters 4, DH, and 7, PN, are both closest to 3. They are both closest to the new cluster 9, so IMIN(4) = IMIN(7) = 9, and DMIN(4) = 208, DMIN(7) = 673.

The full record of these changes in IMIN and DMIN due to amalgamation is carried in Table 12.3.

STEP 5.   Increase K to 9 and return to Step 2.

REPEAT STEPS 2–5.   The minimum value of DMIN(I) is 16 for $J = 6$, PA, and $L = 8$, YE. The 10th cluster consists of PA and YE. The distance matrix is updated by weighted averaging. The closest cluster to 10 is cluster 1, BN, DMIN(10) = 105. Clusters 1 and 2 originally had cluster 8 as their closest cluster, and it is replaced by cluster 10. Increase K to 10, and return to Step 2.

Eventually, a single cluster will remain; the amalgamations in order are $9 = (3, 5)$, $10 = (6, 8)$, $11 = (1, 10)$, $12 = (4, 9)$, $13 = (11, 2)$, $14 = (12, 13)$, and $15 = (7, 14)$. These amalgamations, in their order, may be discovered from the final values of IMIN. For example, IMIN(3) = IMIN(5) = 9, IMIN(6) = IMIN(8) = 10, and so on.

STEP 6.   Initially, POS(I) = 1 for $1 \le I \le 8$, and POS(I) = 0 for $8 < I < 16$.

STEP 7.   For $I = 1$, IMIN(1) = 11, so POS(11) is replaced by POS(11) + POS(1) = 1. For $I = 2$, IMIN(I) = 13 and POS(13) becomes 1. At $I = 3$, POS(9) becomes 1. At $I = 4$, POS(12) becomes 1. At $I = 5$, POS(9) becomes 2. Eventually POS(I) = 1 for $1 \le I \le 8$, and POS(9) = 2, POS(10) = 2, POS(11) = 3, POS(12) = 3, POS(13) = 4, POS(14) = 7, and POS(15) = 8.

STEP 8.   At $I = 14$, POS(14) becomes 8, POS(15) becomes 1. At $I = 13$, POS(13) becomes 8, POS(14) becomes 4. At $I = 12$, POS(12) becomes 4, POS(14) becomes 1. Eventually POS(1) = 6, POS(2) = 5, POS(3) = 3, POS(4) = 2, POS(5) = 4, POS(6) = 7, POS(7) = 1, and POS(8) = 8. These are the positions of the objects such that the clusters are contiguous; the objects are ordered

$$7 \quad 4 \quad 3 \quad 5 \quad 2 \quad 1 \quad 6 \quad 8.$$

The tree can be quickly reconstructed from the IMIN array. Here

$$[7([4(35)]\{2[1(68)]\})].$$

In conclusion, with these data, there are about three clusters, {Princeton}, {Cornell, Dartmouth, Harvard}, and {Brown, Columbia, Pennsylvania, Yale}, that might vividly be described as the powerhouse, the pretenders, and the pushovers. The use of the clusters is in giving an extended list of scores in making predictions. For example, how well will Dartmouth do against Princeton? The only school similar to Dartmouth that Princeton has played is Cornell (the score, 36–27). So, the prediction is Princeton 36, Dartmouth 27. As it turned out, the score was Princeton 14, Dartmouth 28. Note, how accurately the Dartmouth score is predicted!

Table 12.3 Sequence of Closest Distances in Applying Joining Algorithm to Football Data (Table 12.1)

	INITIALLY CLUSTER	IMIN	DMIN
BN	1	8	52
CA	2	8	392
CL	3	5	0
DH	4	3	130
HD	5	3	0
PA	6	8	16
PN	7	3	445
YE	8	6	16

JOIN CL and HD CLUSTER	IMIN	DMIN
1	8	52
2	8	392
3	9	∞
4	9	208
5	9	∞
6	8	16
7	9	673
8	6	16
9	4	208

JOIN PA and YE CLUSTER	IMIN	DMIN
1	10	105
2	10	396
3	9	∞
4	9	208
5	9	∞
6	10	∞
7	9	673
8	10	∞
9	4	208
10	1	105

JOIN BN and PA-YE CLUSTER	IMIN	DMIN
1	11	∞
2	11	412
3	9	∞
4	9	208
5	9	∞
6	10	∞
7	9	673
8	10	∞
9	4	208
10	11	∞
11	2	412

JOIN CL-HD and DH CLUSTER	IMIN	DMIN
1	11	∞
2	11	412
3	9	∞
4	12	∞
5	9	∞
6	10	∞
7	12	756
8	10	∞
9	12	∞
10	11	∞
11	2	412
12	2	476

JOIN BN-YE and CA CLUSTER	IMIN	DMIN
1	11	∞
2	13	∞
3	9	∞
4	12	∞
5	9	∞
6	10	∞
7	12	756
8	10	∞
9	12	∞
10	11	∞
11	13	∞
12	13	656
13	12	656

JOIN BN-CA and CL-DH	JOIN BN-DH and PN
IMIN(12) = IMIN(13) = 14	IMIN(7) = IMIN(14) = 15

What about Harvard-Yale? Against Harvard cluster schools, Yale has 24–14 (Cornell). Against Yale-cluster schools, Harvard has 21–6 (Columbia). The suggestion is Harvard-Yale, 14–24 or 21–6; to make a single estimate, 17.5–15. The actual score was 13–0.

With the distance and amalgamation rules used, it was unnecessary to refer to the actual scores once the initial distance matrix was computed. The amalgamation rule implies the updating of the distance matrix (by using also the number of schools played by each team). For very large numbers of schools—say, 500—the 500×500 distance matrix would be impractical, and it would be cheaper to store the data and compute distances when they are needed directly from the data.

12.4   REMARKS ON JOINING ALGORITHM

The joining algorithm, with various definitions of distance and amalgamation rules, includes many common types of algorithms. In particular, the single-linkage algorithm is a joining algorithm. The data for each object are a vector of distances to all other objects. When two objects are combined, the new vector is the minimum distance to each object of the two amalgamated objects. Other amalgamation rules take weighted or unweighted averages of the two distances or take the maximum. A general class of amalgamation rules on distances has been described by Lance and Williams (1966).

The algorithm is not really suitable for large data sets, since M^2 distances must be computed and examined (although only the smallest distances need be looked at, so there may be some savings in using the spiral search technique). The final tree is invariant under permutation of the original objects, provided all distances computed during the algorithm are different. If all objects were originally identical, the binary tree would depend on their order.

The algorithm produces $2M - 1$ clusters, which must be reduced somehow. There is the automatic pruning which eliminates all clusters except the "rigid" clusters; the size of the minimum cluster properly including a rigid cluster exceeds the size of the rigid cluster by at least 2. It would be better to be stingy in constructing the tree in the first place. The basic joining operation, instead of joining two objects, would join many. The distance between two objects is an acceptable basis for joining two, but something more is necessary in joining many—a definition of cluster diameter. The algorithms would proceed—find the cluster of smallest diameter, amalgamate the objects in the cluster to form a new object, and repeat. It is only rarely possible to search over all 2^n clusters for that of smallest diameter, which makes this method computationally less clear cut than the distance methods. Sometimes the definition of diameter reduces the number of possible clusters.

In one dimension, let $X(1), \ldots, X(M)$ denote the ordered objects. Let C have diameter max $\{I, J \in C\} |X(I) - X(J)|/N(C)$, where $N(C)$ is the number of elements in the cluster C. The only clusters which need be considered are convex, consisting of objects $\{X(I), X(I + 1), \ldots, X(J)\}$ in an interval, and there are only $M(M - 1)/2$ of these.

12.5   ADDING ALGORITHM

Preliminaries.   It is assumed that a rule is given for computing the distance between any pair of objects and that a rule is given for amalgamating two objects to become a new object. The adding algorithm builds up a tree by successively adding objects to it. The objects are added in the order initially given, and, as a result, the tree may depend on this initial order.

After the construction, there will be $2M - 1$ clusters; the original objects will be numbered $1, 2, \ldots, M$ and the clusters $M + 1, \ldots, 2M - 1$. There will be representative objects associated with each cluster. The tree structure is described by the vector $C1(I)$, $C2(I)$ which specifies for the Ith cluster the two maximal proper sub-clusters ($M < I < 2M$). The algorithm adds the Ith object by branching down the tree from the root, at each step moving toward that cluster to which it is closest. Its operation is similar to that of the triads- and tree-leader algorithms.

STEP 1.   Define $C1(M + 1) = 1$, $C2(M + 1) = 2$. Amalgamate 1 and 2 to form the object $M + 1$.

STEP 2.   Add the Ith object ($3 \leq I \leq M$). Set $KK = K = M + I - 2$.

STEP 3.   Set $JJ = C1(KK)$, $LL = C2(KK)$.

(i)   If $KK \leq M$ or $\min\{D(I, JJ), D(I, LL)\} \geq D(JJ, LL)$, define $C1(M + I - 1) = KK$, $C2(M + I - 1) = I$; if $C1(K) = KK$, set $C1(K) = M + I - 1$; if $C2(K) = KK$, set $C2(K) = M + I - 1$. Amalgamate KK and I to form the new object $M + I - 1$. Return to Step 2.

(ii)   Otherwise, set $K = KK$.

STEP 4.   Amalgamate I with K to form a new object K. Define $J = C1(K)$, $L = C2(K)$.
If $D(I, J) \leq D(I, L)$, set $KK = J$. If $D(I, J) > D(I, L)$, set $KK = L$. Return to Step 3.

12.6   ADDING ALGORITHM APPLIED TO QUESTIONNAIRE (TABLE 12.4)

Since some questions were asked positively ("The course content was about right") and some negatively ("I would prefer not to have my work criticized in public"), it is reasonable to prescale the questions so that they are all answered positively. A negative variable V is therefore transformed to $6 - V$. Squared euclidian distance is then used on the transformed variables (or questions). Two variables are amalgamated by weighted averaging. The variable corresponding to a cluster will be the average of all variables in the cluster.

STEP 1. $M = 31$, $C1(32) = 1$, $C2(32) = 2$. The object 32 is the average of variables 1 and 2 for each of the eight students.

STEP 2.   Add the third object, $I = 3$. Set $K = KK = 32$.

STEP 3.   Set $JJ = C1(32) = 1$, $LL = C2(32) = 2$.
$D(1, 2) = 7$, $D(1, 3) = 15$, $D(2, 3) = 17$, which is equivalent to choice (i).
Thus $D(JJ, LL) \leq D(I, JJ)$, $D(I, LL)$. Define $C1(33) = 32$, $C2(33) = 3$. Amalgamate 32 and 3 to form object 33, (11 15 11 15 9 14 13 11)/3. Return to Step 2. The tree is now [(12)3].

STEP 2 REPEATED.   Add the fourth object, $I = 4$. Set $K = KK = 33$.

STEP 3.   Set $JJ = C1(33) = 32$, $LL = C2(33) = 3$.
$D(4, 3) = 20$, $D(4, 32) = 13.75$, $D(3, 32) = 14.75$.
The otherwise choice (ii) occurs, $K = 33$.

STEP 4.   Amalgamate 33 and 4 to form a new object 33, (16 18 13 19 13 19 17 15)/4. Set $J = C1(33) = 32$. Set $L = C2(33) = 3$. Since $D(4, 32) < D(4, 3)$, set $KK = 32$. Return to Step 3.

Table 12.4 Questionnaire About Data Analysis Course

This questionnaire was given to eight students after a data analysis course in 1969. The responses are coded as follows: 1, strongly disagree; 2, disagree; 3, neutral; 4, agree; 5, agree strongly.

QUESTIONS	RESPONSES							
1. I learned a lot.	4	5	4	5	4	5	4	5
2. Statisticians can get along without most of this stuff.	4	1	2	1	3	2	2	2
3. If I had known what it was like I wouldn't have taken it.	1	1	3	1	4	1	1	4
4. The course should be given before consulting work.	5	3	2	4	4	5	4	5
5. I think I will be able to use most of the material.	4	4	4	4	4	5	4	4
6. The course was interesting.	5	5	4	4	4	5	5	4
7. I was thrown into the computer without preparation.	4	1	2	3	3	3	1	2
8. A working knowledge of Fortran should be a prerequisite.	4	5	4	2	3	4	5	3
9. More emphasis should be placed on complete packages.	3	2	3	1	3	3	2	2
10. My mathematical background was sufficient.	5	5	4	5	3	5	5	4
11. I didn't know enough statistical theory to understand.	2	1	2	1	4	2	1	2
12. Too much was assumed about elementary data analysis.	3	2	2	2	4	2	2	2
13. I was adequately prepared.	3	5	4	3	2	4	5	3
14. I had to do too much work myself	3	3	5	2	1	2	1	3
15. I spent too much time at the Computer Center fiddling.	4	5	5	4	1	4	1	3
16. The computing was more fun than the statistics.	3	4	3	3	5	5	1	3
17. There weren't enough interesting data sets.	3	4	5	4	3	4	2	4
18. There should be more packaged programs	3	2	3	1	4	4	2	2
19. I want to write my own subroutines	4	4	1	2	2	2	3	4
20. We shouldn't waste a session on student projects.	1.1	4	2	2	2	1	3	
21. I would prefer not to have my work criticized in public.	1	1	3	2	1	2	1	2
22. The level of student participation is about right.	4	4	1	4	3	4	3	3
23. We went through too much, too fast, to grasp anything.	4	5	4	4	5	4	1	2
24. I want more correlations, and factor analysis.	4	2	3	4	4	4	2	3
25. We spent too much time on clustering.	3	2	4	2	4	4	2	2
26. We should consider only real variables, let 0-1, category go.	3	4	4	2	1	2	1	3
27. We need to discuss time series.	4	3	3	4	1	4	4	4
28. Too much emphasis on trivial data analysis like plots.	2	2	4	4	1	2	2	2
29. More time on regression and analysis of variance.	4	4	2	3	1	5	3	4
30. A more mathematical treatment please.	3	4	4	3	4	4	3	4
31. The course content was about right.	4	4	4	3	4	4	4	4

STEP 3. Set $JJ = C1(32) = 1$, $LL = C2(32) = 2$. Then $D(1, 2) = 7$, $D(1, 4) = 10$, $D(2, 4) = 24$. Therefore (i) holds, with $D(1, 2) \leq D(1, 4)$, $D(2, 4)$. Set $C1(34) = 32$, $C2(34) = 4$, amalgamate 32 and 4 to form 34, (11 13 10 14 11 14 12 14)/3. Since $C1(33) = 32$, set $C1(33) = 34$. Return to Step 2.

The algorithm proceeds by adding each object in turn to the tree, beginning at the root and branching at each choice toward the closest cluster. An object is computed for each cluster, which at each stage is the average of all original objects in the cluster.

These constructed objects are necessary in the allocation of new objects, but they are also useful in interpreting the final clusters.

The result of the algorithm for the first eight questions are given in Table 12.5. There are about two clusters: first, the questions 1, 5, 6, 8, the first three of which all ask for an overall evaluation of the course and naturally go together. The second cluster contains questions 2, 7, 9. Questions 7, 9 both ask about computer preparation. The algorithm has been successful in identifying similar groups of questions.

Table 12.5 Output Tree of Adding Algorithm on Questionnaire (Table 12.4)
Responses marked with a dash reflected about 3.

```
        CLUSTER NUMBER 35 36 38 37 39 32 34 33

   1.  LEARNED----/-I--I--I--------I--I--I
   5.  USE--------/-/  I  I        I  I  I
   6.  INTEREST---/----/  I        I  I  I
   8.  FORTRAN----/-------/        I  I  I
 - 2.  WITHOUT----/----------I--I  I  I  I
 - 7.  THROWN-----/----------/  I  I  I  I
 - 9.  PACKAGES---/-------------/--/  I  I
   4.  CONSULTING-/-------------------/  I
 - 3.  TAKEN------/----------------------/
```

```
OBJECTS CORRESPONDING TO CLUSTERS (ORIGINAL OBJECTS UNDERLINED)
                     RESPONSES
```

01 ------	4	5	4	5	4	5	4	5
05 ------	4	4	4	4	4	5	4	4
-35 -----	4	4.5	4	4.5	4	5	4	4.5
06 ------	5	5	4	4	4	5	5	4
--36 ----	4.3	4.7	4	4.3	4	5	4.3	4.3
08 ------	4	5	4	2	3	4	5	3
---38 ---	4.3	4.5	4	3.8	3.8	4.8	4.5	4
02 ------	2	5	4	5	3	4	4	4
07 ------	2	5	4	3	3	3	5	4
-37 -----	2	5	4	4	3	3.5	4.5	4
09 ------	3	4	3	5	3	3	4	4
--39 ----	2.3	4.7	3.7	4.3	3	3.7	4.3	4
----32 --	3.4	4.7	3.9	4.0	3.4	4.1	4.4	4
04 ------	5	3	2	4	4	5	4	5
-----34 -	3.6	4.5	3.6	4.0	3.5	4.3	4.4	4.1
03 ------	5	5	3	5	2	5	5	2
------35	3.8	4.6	3.6	4.1	3.3	4.3	4.4	3.9

Table 12.6 Sequence of Trees in Applying Adding Algorithm to Questionnaires (first Eight Questions)

```
1 --32            1 --32--33          1 --32--34--33
2 --32            2 --32--33          2 --32--34--33
                  3 ------33          4 ------34--33
                                      3 ----------33
```

```
1 --35--32--34--33          1 --35--36--32--34--33
5 --35--32--34--33          5 --35--36--32--34--33
2 ------32--34--33          6 ------36--32--34--33
4 ----------34--33          2 ----------32--34--33
3 ------------33            4 --------------34--33
                            3 ----------------33
```

```
1 --35--36--32--34--33          1 --35--36--38--32--34--33
5 --35--36--32--34--33          5 --35--36--38--32--34--33
6 ------36--32--34--33          6 ------36--38--32--34--33
2 --37------32--34--33          8 ----------38--32--34--33
7 --37------32--34--33          2 --37----------32--34--33
4 --------------34--33          7 --37----------32--34--33
3 ----------------33           4 --------------------34--33
                               3 ----------------------33
```

```
1 --35--36--38--32--34--33
5 --35--36--38--32--34--33
6 ------36--38--32--34--33
8 ----------38--32--34--33
2 --37--39------32--34--33
7 --37--39------32--34--33
9 ------39------32--34--33
4 ------------------34--33
3 ----------------------33
```

The clusters may be interpreted by looking at the object corresponding to each cluster—here the mean over all original objects in the cluster. For example, cluster 36 containing 1, 5, 6 corresponds to an object, or question, with responses between 4 and 5 indicating that the course was acceptable, overall, to all students. The responses for the "computer preparation" cluster are all about 4, except for the first student, who was dissatisfied, and the fifth student, who was neutral. For the cluster of all questions, the responses are simply averages over all questions. (Note that this average would be senseless if the negative questions are not reflected about 3.) The fifth student has a neutral response on average, but it is much more revealing to find out what aspects of the course displease him and which please him, and for this it is necessary to identify clusters of questions and examine averages over the clusters.

It is natural to think of clustering students, as well as questions, and to try to relate the clusters of students to clusters of questions. One group of students wants more mathematical treatment, another group wants more time on linear methods, and so on. Two-way clustering techniques to perform this analysis will be discussed in later chapters.

12.7 THINGS TO DO

12.7.1 Running Distance and Amalgamation Algorithms

These algorithms are applicable to a very wide variety of data types, with the user specifying distance functions and amalgamation rules appropriate to the data type. The amalgamation rule results in a representative object for each cluster that is very useful in interpreting the results.

The joining algorithm is the recommended procedure for less than 100 objects. Its principal defect is the binary tree necessarily obtained; 99 clusters will rarely be justified by the data and will confuse interpretation. The adding algorithm is recommended for more than 1000 objects, since it completes the classification in one pass. The results will be dependent on input order of the objects; this dependence might be reduced by an initial ordering by distance from a mean object. Another trick for reducing order dependence is to add the objects, say, three times and then to eliminate all but the latest set of objects. (See nails and screws, Table 12.7, and cakes, Table 12.8, for trial data sets.)

12.7.2 Weighted and Unweighted Averages

The first-thought distance between points is euclidean distance, and the corresponding amalgamation rule is the average of the two points. The K-means algorithm and analysis-of-variance techniques would suggest weighting the average by the number of original objects in the clusters corresponding to the points. The unweighted average is more attractive because clustering philosophy suggests discounting points discovered very similar to other points. This it might be argued that, if 10 original data points were identical, only one should be used in the clustering. The weighted-averages technique is more likely to split a single dense cluster into two clusters. The difference between these distances should be explored by considering examples in two dimensions.

12.7.3 Contingency Tables

A variable $V1$ takes values $1, \ldots, M$, and a variable $V2$ takes values $1, \ldots, K$. The number of times $V1 = I$ and $V2 = J$ is $N(I, J)$. The measure for dependence is the log likelihood ratio

$$\sum \{1 \leq I \leq M, 1 \leq J \leq K\} \log [N(I, J)N(\cdot, \cdot)/N(I, \cdot)N(\cdot, J)],$$

where

$$N(I, \cdot) = \sum \{1 \leq I \leq K\} N(I, J),$$

$$N(\cdot, J) = \sum \{1 \leq I \leq M\} N(I, J),$$

and

$$N(\cdot, \cdot) = \sum \{1 \leq I \leq M\} N(I, \cdot).$$

Table 12.7 Nails and Screws

Codes.	Thread:		YES=Y NO=N			
	Head:		FLAT=F CUP=U CONE=O ROUND=R CYLINDER=Y			
	Head Indentation:		NONE=N STAR=T SLIT=L			
	Bottom:		SHARP=S FLAT=F			
	Length:		(HALF INCHES)			
	Brass:		YES=Y NO=N			

	THREAD	HEAD	HEAD INDENTATION	BOTTOM	LENGTH	BRASS
TACK	N	F	N	S	1	N
NAIL1	N	F	N	S	4	N
NAIL2	N	F	N	S	2	N
NAIL3	N	F	N	S	2	N
NAIL4	N	F	N	S	2	N
NAIL5	N	F	N	S	2	N
NAIL6	N	U	N	S	5	N
NAIL7	N	U	N	S	3	N
NAIL8	N	U	N	S	3	N
SCREW1	Y	O	T	S	5	N
SCREW2	Y	R	L	S	4	N
SCREW3	Y	Y	L	S	4	N
SCREW4	Y	R	L	S	2	N
SCREW5	Y	Y	L	S	2	N
BOLT1	Y	R	L	F	4	N
BOLT2	Y	O	L	F	1	N
BOLT3	Y	Y	L	F	1	N
BOLT4	Y	Y	L	F	1	N
BOLT5	Y	Y	L	F	1	N
BOLT6	Y	Y	L	F	1	N
TACK1	N	F	N	S	1	Y
TACK2	N	F	N	S	1	Y
NAILB	N	F	N	S	1	Y
SCREWB	Y	O	L	S	1	Y

The distance between two rows is the reduction in log likelihood ratio due to combining two rows, which is just the log likelihood ratio for the part of the table consisting of the two rows. The amalgamation rule adds the two rows.

The joining algorithm may be applied to the rows, or to the columns, or to both at once.

12.7.4 Decreasing Diameters

The diameter of a cluster is the distance between objects joined to form the cluster. For some measures of distance, the diameter may not increase with the cluster. Consider the unweighted-averages euclidean distance rule.

Table 12.8   Ingredients in Cakes

From *The New York Times Cookbook* (1961), Craig Claiborne, Ed., Harper and Row, New York.

LEGEND:	UNIT INGREDIENT	LEGEND:	UNIT INGREDIENT
AE:	Teaspoon, almond essence	LR:	teaspoon, lemon rind
BM:	cup, buttermilk	MK:	cup, milk
BP:	teaspoon, baking powder	NG:	teaspoon, nutmeg
BR:	cup, butter	NS:	cup, nuts
BS:	one, bananas	RM:	ounce, rum
CA:	tablespoon, cocoa	SA:	teaspoon, soda
CC:	pounds, cottage cheese	SC:	cup, sour cream
CE:	ounce, chocolate	SG:	tablespoon, shortening
CI:	cup, crushed ice	SR:	cup, gra ulated sugar
DC:	tablespoon, dried currants	SS:	quart, strawberries
EG:	one, eggs	ST:	teaspoon, salt
EY:	one, egg yolk	VE:	teaspoon, vanilla extract
EW:	one, egg white	WR:	cup, water
FR:	cup, sifted flour	YT:	ounce, yeast
GN:	tablespoon, gelatin	ZH:	ounce, zwiebach
HC:	cup, heavy cream	CS:	cup, crumbs
LJ:	tablespoon, lemon juice	CT:	teaspoon, cream of tartar

```
CAKE                   :    INGREDIENTS
ANGEL                  :  1FR 1.5SR 10EW 1.25CT .25ST 1VE .25AE
BABAS AU RHUM          :  .25MK .25BR .6YT .25WR 2EY .25SR 1EG .5LR 2DC 1.75FR 2RM
SWEET CHOCOLATE        :  4CE .5WR 1BR 2SR 4EG 1VE 2.5FR 1SA 0.5ST 1BK
BUCHE DE NOEL          :  1FR 0.5ST 1.3SR 4EG 1VE
CHEESECAKE             :  6ZH 1SR .25BR 1.5CC .25FR .25ST 6EG 1SC 1LR 1LJ
RUM CHEESECAKE         :  2GN 1SR .25ST 2EG 1MK 1LJ 1RM 1.5CC 1HC
BLENDER CHEESECAKE     :  1GN 1LJ 1LR 0.5WR .3GS 2EY 0.5CC 1CI 1SC
ONE BOWL CHOCOLATE     :  2FR 2BP 0.5SA 0.25ST 10CA 1.5SR 10SG .5WR .7MK 2EG 1VE
RED DEVIL'S FOOD       :  1.75FR 1.5SR 0.3CA 1.25SA 1ST 0.5SG 1MK 2EG 1VE
SOUR CREAM FUDGE       :  2FR 1.5SR 1SA 1ST 0.3SG 1SC 3CE 2EG 1VE .25WR
HUNGARIAN CREAM        :  3FR 3BP 1ST 3EG 1.5AE 2FR 1.5SR 2HC
CRUMB AND NUT          :  1SG 1SR 4EG 2VE 3CS 1NS 3BP 1MK
SPICED POUND           :  1BR 1.5NG 0.5ST 1.7SR 5EG 2FR
STRAWBERRY ROLL        :  4EG .75BP 1ST .75SR 1VE .75FR 0.5CC 1SS
SAVARIN                :  1YT 0.25WR 1.5FR .06SR 0.5ST 3EG .25MK .7BR 1LR
BANANA SHORTCAKE       :  3FR 0.5ST 4BP .13SR 0.6BR 1MK 3BS 1HC
STRAWBERRY SHORTCAKE:  2FR 3BP .75ST .6SR .2CC .12BR 1EG 0.5MK 1SS 1SC
SPONGE                 :  .75FR 1BP .25ST 4EG .75SR 0.5AE 2WR
```

12.7.5   Probability Models

An initial log likelihood is given for each observation—say, $L(I, \theta)$ for the Ith observation. The distance between I and J is

$$\max_{\theta} L(I, \theta) + \max_{\theta} L(J, \theta) - \max_{\theta} [L(I, \theta) + L(J, \theta)].$$

The amalgamation rule specifies $L(I, \theta) + L(J, \theta)$ as the log likelihood for the new cluster.

REFERENCES

LANCE, G. N., and WILLIAMS, W. T. (1966). "A general theory of classification sorting strategies 1. Hierarchical systems." *Comp. J.* **9**, 373–380. Clusters are constructed by joining, beginning with M clusters each consisting of a single object and ending with a single cluster containing M objects. The pair of clusters joined at each stage is the closest pair available at that stage. The distance between clusters, given the distance between objects, may be defined in a number of ways. A clustering strategy is "combinatorial" if, for the cluster k obtained by joining clusters i and j, the distance d_{hk} between k and any other cluster h is a function only of d_{hi}, d_{hj}, d_{ij}, and of n_i, the number of objects in i, and n_j, the number of objects in j.

A particular form of this function, which includes many standard algorithms, is

$$d_{hk} = \alpha_i d_{hi} + \alpha_j d_{hj} + \beta d_{ij} + \gamma \, |d_{hi} - d_{hj}| \, .$$

For example, $\alpha_i = \alpha_j = -\gamma = 0.5$, gives

$$d_{hk} = \min \, (d_{hi}, d_{hj}),$$

the single-linkage amalgamation rule. Or, $\alpha_i = \alpha_j = \gamma = 0.5$ gives

$$d_{hk} = \max \, (d_{hi}, d_{hj}).$$

PROGRAMS

JOIN   successively joins closest rows of a matrix.

AMALG gives a rule for combining two rows to form a new row.

DIST   See program in Chapter 2.

```
      SUBROUTINE JOIN(A,M,N,NB,DM,RL)
C.......................................................20 MAY 1973
C.... JOIN FINDS CLOSEST PAIR OF ROWS AND AMALGAMATES THEM TO FORM A NEW ROW,
C.... CONTINUING UNTIL A SINGLE ROW REMAINS.   THE USER SHOULD PROVIDE DISTANCE
C.... AND AMALGAMATION ALGORITHMS, APPORPRIATE FOR HIS PARTICULAR DATA.
C.... M = NUMBER OF ROWS
C.... N = NUMBER OF COLUMNS
C.... A = M BY N BORDERED ARRAY
C.... NB = 3 BY M ARRAY DEFINING CLUSTER BOUNDARIES
C.... DM = 1 BY M SCRATCH ARRAY
C.... RL = 1 BY M ARRAY OF REORDERED ROW LABELS, SUITABLE FOR TREE DRAWING
C..........................................................................
      DIMENSION RL(M)
      DIMENSION A(M,N),NB(3,M),DM(M)
      DO 10 I=2,M
   10 NB(3,I)=I
      MM=M*(N-2)+1
C.... FIND CLOSEST PAIR
      DO 30 K=1,M
      IF(K.GT.M-2) GO TO 30
      DMIN=10.**20
      IMIN=2
      JMIN=3
      DO 21 I=2,M
      IF(NB(3,I).LT.0) GO TO 21
      DO 20 J=I,M
      IF(NB(3,J).LT.0) GO TO 20
      IF(J.EQ.I) GO TO 20
      Z=DIST(A(I,2),A(J,2),MM,M,2.)
      IF(Z.GE.DMIN) GO TO 20
      IMIN=I
      JMIN=J
      DMIN=Z
   20 CONTINUE
   21 CONTINUE
C.... FORM A NEW CLUSTER BY AMALGAMATING ROWS
      I=IMIN
      NB(1,K)=I
      II=NB(3,JMIN)
      DM(K)=DMIN
      NB(2,K)=II
      NB(3,JMIN)=-NB(3,I)
      NB(3,I)=II
      CALL AMALG(A(IMIN,2),A(JMIN,2),MM,M,2.)
   30 CONTINUE
C.... REORDER OBJECTS, ASSIGN LABELS
      J=M+1
      KC=M-2
      IF(M.EQ.2) RETURN
      L=NB(2,KC)
      DO 71 JJ=2,M
      J=M-JJ+2
      LL=-NB(3,L)
      NB(3,L)=J
      L=LL
   71 CONTINUE
      DO 72 K=1,KC
      L=NB(1,K)
      NB(1,K)=NB(3,L)
      L=NB(2,K)
   72 NB(2,K)=NB(3,L)
      DO 73 I=2,M
      J=NB(3,I)
   73 RL(J)=A(I,1)
C.... DEFINE LAST TWO CLUSTERS, AND CLUSTER DIAMETERS
      DMAX=0
      DO 40 I=1,KC
   40 IF(DM(I).GT.DMAX) DMAX=DM(I)
      WRITE(6,1) DMAX
    1 FORMAT(27H MAXIMUM JOINING DISTANCE =,F20.6)
      DO 50 I=1,KC
   50 NB(3,I)=(DM(I)/DMAX)*100
      DO 60 J=1,3
      DO 60 K=KC,M
   60 NB(J,K)=NB(J,KC)
      RETURN
      END
```

```
      SUBROUTINE AMALG(X,Y,MM,N,P)
C..............................................................20 MAY 1973
C.... AMALGAMATES X AND Y VECTORS TO BE EQUAL
C.... MM = LENGTH OF VECTOR
C.... M = SKIP FACTOR, TO BE USED WHEN X AND Y ARE ROWS OF A MATRIX( SEE DIST)
C.... X = FIRST VECTOR
C.... Y = SECOND VECTOR
C.... P = PARAMETER OF JOIN
C..............................................................................
      DIMENSION X(MM),Y(MM)
      DO 20 I=1,MM,N
      X(I)=(X(I)+Y(I))/2.
   20 Y(I)=X(I)
      RETURN
      END
```

CHAPTER 13

Minimum Mutation Methods

13.1 INTRODUCTION

From Imms (1957), "In the embryo of most insects, evident rudiments of paired abdominal appendages appear at some stage during development . . . . A variable number of these appendages may become transformed into organs that are functional during postembryonic life, while the remainder disappear. The most conspicuous of the persistent appendages are the cerci of the 11th segment, which exhibit wide diversity of form, and may even be transformed into forceps, as in the Japygidae, and the earwigs."

The cerci of each insect order is recorded as segmented, unsegmented, or absent, in Table 13.1. Given the tree connecting the insect orders (which is not well established), it would be desirable to predict for each cluster the cerci status of the most recent ancestor of the cluster. For example, the ancestral endopterygota probably had no cerci. A simple and plausible criterion for interpolating ancestral values is the minimum mutation requirement. Values are assigned to the ancestors so that a minimum number of mutations (or changes of value between an object and its most recent ancestor) occur. Rules for assigning values to minimize the mutations were given for a binary tree by Fitch (1971) and for a general tree, with optimality proofs, by Hartigan (1972). Prior structure sometimes exists on the values. For example, it might be specified that the only possible mutations of cerci are in the order segmented → unsegmented → absent. This case has been considered by Camin and Sokal (1965) and, with a more general partial order on the values, by Estabrooke (1968). With these prior structures, the interesting problem is construction of the tree, not the interpolation of values given the tree; the value of any ancestor is simply the most primitive value among the descendants.

13.2 MINIMUM MUTATION FITS

Preliminaries. A variable V has values $V(1)$, $V(2)$, . . . , $V(M)$ on objects $1, 2, . . . , M$. A tree of clusters $1, 2, . . . ,$ NC is given on the objects, specified by an *ancestor function T*. For each cluster I, $T(I)$, *the ancestor of I*, is the smallest cluster properly including I. The function T is defined for all I ($1 \leq I <$ NC), but it is not defined for the cluster NC, the cluster of all objects. The numbering of clusters is such that $I < T(I)$ for each I ($1 \leq I <$ NC).

During the algorithm, the quantity $V(I)$ is a subset of values corresponding to the Ith cluster. For $1 \leq I \leq M$, $V(I)$ will be the value of variable V for object I. At first,

233

Table 13.1   Presence of Cerci (Tail Appendages) in Insects

S, segmented cerci; A, cerci absent; U, unsegmented cerci (From Imms, 1957).

S	1. THYSANURA (Bristle tail)			36. APTERYGOTA
S	2. DIPLURA			
A	3. PROTURA			
A	4. COLLEMBOLA (Spring tail)			
S	5. EPHEMEROPTERA (May flies)	30. PALAEOPTERAN	34. EXOPTERYGOTA	37. PTERYGOTA
A	6. ODONODATA (Dragon flies)			
S	7. PLECOPTERA (Stoneflies)	31. ORTHOPTEROID		
S	8. GRYLLOBLATTODEA			
U	9. ORTHOPTERA (grasshoppers)			
U	10. PHASMIDA (Stick insect)			
U	11. DERMAPTERA (earwig)			
S	12. EMBIOPTERA			
S	13. DICTYOPTERA (cockroaches)			
U	14. ISOPTERA (termites)			
U	15. ZORAPTERA			
A	16. PSOCOPTERA (book lice)	32. HEMIPTEROID		
A	17. MALLOPHAGA (bird lice)			
A	18. SIPHUNCULATA (sucking lice)			
A	19. HEMIPTERA (plant bugs)			
A	20. THYSANOPTERA (thrips)			
A	21. NEUROPTERA (lacewings)	33. PANORPOID	35. ENDOPTERYGOTA	
A	22. MECOPTERA (scorpion flies)			
A	23. LEPIDOPTERA (butterflies)			
A	24. TRICHOPTERA (Caddis flies)			
A	25. DIPTERA (true flies)			
A	26. SIPHONAPTERA (fleas)			
A	27. HYMENOPTERA (ants)			
A	28. COLEOPTERA (beetles)			
A	29. STREPSIPTERA			

$V(I)$ is constructed to be the set of values at cluster I compatible with a minimum mutation fit to the tree below cluster I (the tree consisting of clusters included in I and of I itself). Later, $V(I)$ is the set of values at cluster I compatible with a minimum mutation fit to the whole tree. This terminology is illustrated in Table 13.2.

STEP 1.   For each I in turn ($M < I \leq$ NC) find the set of clusters $I(1), I(2), \ldots, I(J)$ such that $T[I(K)] = I$ ($1 \leq K \leq J$). Then $V(I)$ is the set of values of V which occur with maximum frequency in $V[I(1)], V[I(2)], \ldots, V[I(J)]$.

STEP 2.   For each I in turn ($M < I <$ NC), beginning with $I =$ NC $- 1$ and decreasing I by 1 at each step, if $V(I) \supset V[T(I)]$, set $V(I) = V[T(I)]$. If $V(I) \not\supset V[T(I)]$, find the set of clusters $I(1), I(2), \ldots, I(J)$ for which I is the ancestor. Then $V(I)$ is the set of values of V which occurs with maximum frequency or frequency one less than maximum in $V[I(1)], V[I(2)], \ldots, V[I(J)], V[T(I)]$.

Table 13.2 Illustration of Terminology Used in Minimum Mutation Fit

V

V(1) = A	1---I--I--I
V(2) = B	2---/ I I
V(3) = A	3--I I I
V(4) = C	4--/ --/ I
V(5) = A	5---I I
V(6) = A	6---/ ----/

CLUSTERS	7={1,2}	8={3,4}	9={5,6}	10={1,2,3,4}	11={1,2,3,4,5,6}
ANCESTORS	T(1)=T(2)=7	T(3)=T(4)=8	T(5)=T(6)=9	T(7)=T(8)=9	T(9)=T(10)=11
INITIAL V	V(7)={A,B}	V(8)={A,C}	V(9)={A}	V(10)={A}	V(11)={A}
FINAL V	V(7)={A}	V(8)={A}	V(9)={A}	V(10)={A}	V(11)={A}

NOTE 1. To obtain easily a single minimum mutation fit, in Step 2 change $V(I)$ to $V[T(I)]$ if $V(I) \supset V[T(I)]$, and otherwise change $V(I)$ by dropping all but one value.

NOTE 2. It is never necessary to change the values $V(I)$ originally given for the objects $1, 2, \ldots, M$.

13.3 APPLICATION OF MINIMUM MUTATION ALGORITHM TO CERCI IN INSECTS

The T values specifying tree structure may be recovered from Table 13.1. For example, for the four orders of *Apterygota*, $T(1) = 36$, $T(2) = 36$, $T(3) = 36$, and $T(4) = 36$.

STEP 1. The first cluster considered is 30, *Palaeopteran*. The set of clusters whose ancestor is *Palaeopteran* are 5 and 6, so $I(1) = 5$ and $I(2) = 6$. Since $V(5) = S$ and $V(6) = A$, the values S and A occur with maximum frequency. Therefore $V(30) = (S, A)$. The next cluster is 31, *Orthopteroid*, for which $I(1) = 7$, $I(2) = 8, \ldots,$ $I(9) = 15$. The corresponding V values are $S, S, U, U, U, S, S, U, U$. The value U occurs 5 times and the value S occurs only 4, so $V(31) = U$. Continuing, $V(32) = A$ and $V(33) = A$. A new situation arises with 34, *Exopterygota*, ancestor of 30, 31, 32. The corresponding V's are $\{S, A\}$, $\{U\}$, $\{A\}$. Among these V's, A is maximal, so $V(34) = A$.

At the conclusion of Step 1, $V(30) = (S, A)$, $V(31) = U$, $V(32) = A$, $V(33) = A$, $V(34) = A$, $V(35) = A$, $V(36) = (S, A)$, $V(37) = A$, and $V(38) = A$.

STEP 2. Running through the clusters in reverse order, cluster 37 is treated first. Since $T(37) = 38$, $V(37) = A \supset V(38) = A$, so $V(37) = A$ without change. For cluster 36, $V(36) = \{S, A\}$ includes $V(38) = A$, so $V(36) = V(38) = A$. The next interesting cluster is 31, $V(31) = U$. Since $T(31) = 34$, $V(31) = U \not\supset V(34) = A$, so the second option is necessary. Thus $V(31)$ becomes the set of values which are maximal or submaximal in $V(7)$, $V(8)$, ..., $V(15)$ and $V(34)$. The value U occurs five times, the value S four times, and the value A once. Thus $V(31) = U$.

Finally, after Step 2 is completed, $V(30) = A$, $V(31) = U$, $V(32) = A$, $V(33) = A$, $V(34) = A$, $V(35) = A$, $V(36) = A$, $V(37) = A$, and $V(38) = A$.

The minimum mutation fit thus gives value A to all insect ancestors, except the orthopteroids, which are assigned value U. It must be admitted that this violates the accepted theory, which is that cerci are primitive appendages, so that segmented $\rightarrow$ unsegmented $\rightarrow$ absent would be the expected order of evolution. [A slightly different tree would produce this as a possible sequence. For example, if the orthopteroid cluster were omitted, $V(34) = \{U, S\}$, $V(37) = \{U, S, A\}$, and $V(38) = \{S, A\}$. A minimum mutation sequence would then have all ancestors with segmented cerci, except *Endopterygota* (absent) and *Hemipteroid* (absent).]

The total number of data values is 29, but these are represented in the minimum mutation fit by nine symbols (the ancient A, the orthopteroid U, and the seven present-day objects differing from these).

13.4 SOME PROBABILITY THEORY FOR THE NUMBER OF MUTATIONS

If just a few mutations will explain a large number of values, the tree is validated. To make this precise, it is necessary to know how many mutations would be expected, if there were no relation between the tree and the data—that is, if the data values were assigned at random to the objects at the end of the tree.

Suppose there are K different data values, and M objects. The probability theory is easiest if the data values are assigned independently to the M objects, with the Ith object taking the Jth data value with probability $P(J)$. [This $P(J)$ might be estimated by the proportion of objects taking the Jth value.] For the Ith cluster, the set $V(I)$ will be defined, as in the algorithm of Section 13.2, as the set of values assigned to cluster I compatible with a minimum mutation fit to the tree below I. The number of mutations may be determined from the $V(I)$. Also $P[V(I) = A]$, where A is some subset of values of V, may be determined by recurrence relations from these probabilities for all the clusters whose ancestor is cluster I. In this way, the distribution of the number of mutations may be obtained for any given probabilities $\{P(J), J = 1, \ldots, K\}$ and any given tree. If there are many objects, this number of mutations is approximately normally distributed.

This procedure will be followed for the simple case when only two values are possible and when the tree is binary. Let the values be denoted by X and Y. Suppose that a cluster I is ancestor to clusters J and K. Then

$$
\begin{array}{llll}
V(I) = X & \text{if} & V(J) = X, & V(K) = X \\
& \text{or} & V(J) = XY, & V(K) = X \\
& \text{or} & V(J) = X, & V(K) = XY, \\
V(I) = Y & \text{if} & V(J) = Y, & V(K) = Y \\
& \text{or} & V(J) = XY, & V(K) = Y \\
& \text{or} & V(J) = Y, & V(K) = XY, \\
V(I) = XY & \text{if} & V(J) = X, & V(K) = Y \\
& \text{or} & V(J) = Y, & V(K) = X \\
& \text{or} & V(J) = XY, & V(K) = XY.
\end{array}
$$

A mutation will occur from I to J or from I to K, if and only if $V(J) = X$, $V(K) = Y$ or $V(K) = X$, $V(J) = Y$. Events in J and K are independent since they contain different objects. Therefore

$$P[V(I) = X] = P[V(J) = X]P[V(K) = X] + P[V(J) = XY]P(V(K) = X]$$
$$+ P[V(J) = X]P[V(K) = XY]$$

with similar expressions for $P[V(I) = Y]$ and $P[V(I) = XY]$. The probability that a mutation occurs from I to J or from I to K is $P[V(J) = X]P[V(K) = Y] + P[V(J) = Y]P[V(K) = X]$. Thus the expected number of mutations is

$$\sum \{P[V(J) = X]P[V(K) = Y] + P[V(J) = Y]P[V(K) = X]\}$$

summed over all clusters I. [Complex recurrence relations on the number of mutations in the tree below I when $V(I) = X$, when $V(I) = Y$, and when $V(I) = XY$ may be used to determine the distribution of mutations in the whole tree and the variance of this number of mutations. Since the number of mutations is approximately normal for many objects, the mean and variance are sufficient to determine the distribution approximately. For clusters far from the ends of the tree, it may be shown that $P[V(I) = X] = P[V(I) = Y] = P[V(I) = XY] = \frac{1}{3}$, approximately. This is not much use in determining the asymptotic distribution of mutations, since most mutations occur near the ends of the tree.]

For a small and simple tree, the probabilities $P[V(I) = X]$, $P[V(I) = Y]$, $P[V(I) = XY]$ and the expected number of mutations and the distribution of mutations are computed in Table 13.3. A typical recurrence relation occurs in computing $P[V(11) = X]$ by using probabilities for clusters 9 and 10:

$$P[V(11) = X] = P[V(9) = X]P[V(10) = X] + P[V(9) = X]P[V(10) = XY]$$
$$+ P[V(9) = XY]P[V(10) = X]$$

$$= \tfrac{1}{9} \times \tfrac{1}{9} + \tfrac{1}{9} \times \tfrac{8}{27} + \tfrac{1}{9} \times \tfrac{4}{9} = \tfrac{23}{243}.$$

The expected number of mutations is the sum of the probabilities of a mutation at each join. The actual distribution of mutations is computed here by enumeration over the 64 patterns of X's and Y's possible for the end objects.

The mathematics goes nicely when the variable takes only two values, X and Y, and when the tree is ternary (every cluster is an ancestor of just three clusters). In this case, the set $V(I)$ is either X or Y but cannot be XY. If I is ancestor to J, K, L,

$$P[V(I) = X] = P[V(J) = X]P[V(K) = X]P[V(L) = X]$$

$$\times \left(\frac{1}{P[V(I) = X]} + \frac{1}{P[V(J) = X]} + \frac{1}{P[V(K) = X]} - 2 \right).$$

13.5 REDUCED MUTATION TREE

Preliminaries. This algorithm successively amalgamates "closest" pairs of clusters, beginning with M clusters each consisting of a single object and stopping when a single cluster containing all objects remains. The tree structure at each stage is determined by an ancestor function T defined for every cluster. For clusters I that still remain to be amalgamated, $T(I) = 0$. For a cluster I that has been amalgamated, $T(I)$ denotes the cluster I was amalgamated into.

Table 13.3 Expected Number of Mutations

```
1.  --I--I--I

2.  --/  I  I

3.  --I  I  I

4.  --/--/  I

5.  --I    I

6.  --/-----/
```

CLUSTERS. 7={1,2}, 8={3,4}, 9={5,6}, 10={1,2,3,4}, 11={1,2,3,4,5,6}

X occurs with probability 1/3; Y occurs with probability 2/3 .

P[V(7)=X] = 1/9   P[V(7)=Y] = 4/9   P[V(7)=XY] = 4/9   P[MUTATION] = 4/9

P[V(8)=X] = 1/9   P[V(8)=Y] = 4/9   P[V(8)=XY] = 4/9   P[MUTATION] = 4/9

P[V(9)=X] = 1/9   P[V(8)=Y] = 4/9   P[V(8)=XY] = 4/9   P[MUTATION] = 4/9

P[V(10)=X] = 1/9   P[V(10)=Y] = 16/27   P[V(10)=XY] = 8/27   P[MUTATION] = 8/81

P[V(11)=X] = 23/243 P[V(11)=Y] = 160/243 P[V(11)=XY] = 60/243 P[MUTATION] = 28/243

EXPECTED NUMBER OF MUTATIONS = 376/243 = 1.547 .

P[MUTATIONS = 0] = 65/729

P[MUTATIONS = 1] = 264/729

P[MUTATIONS = 2] = 336/729

P[MUTATIONS = 3] = 64/729

Used in computing similarities between clusters, the set $V(I, J)$, defined for cluster I and variable J, is the set of values of variable J compatible with a minimum mutation fit of variable J to the tree of clusters included in cluster I (including I itself). After the tree is completed, the quantities $V(I, J)$ for all I determine a minimum mutation fit to the whole tree (this fit is not necessarily unique).

This algorithm is essentially a joining algorithm with distances and amalgamation rules specified on the sets $V(I, J)$.

STEP 1.   Initialize $V(I, J)$ $(1 \leq I \leq M, \ 1 \leq J \leq N)$ to be the value of the Jth variable on the Ith object. The number of clusters so far considered is $K = M$. Initialize $T(I) = 0$ $(1 \leq I \leq M)$.

STEP 2.   Compute the distance between clusters I and L as

$$1 - (\sum \{1 \leq J \leq N\} \#[V(I, J) \cap V(L, J)])(\sum \{1 \leq J \leq N\} \#[V(I, J) \cup V(L, J)]^{-1}.$$

Here $\#(S)$ denotes the number of values in the set S. For clusters which are single objects, this distance is just the proportion of mismatches between the two objects for the N variables. Let the pair of clusters I, L with $T(I) = 0$, $T(L) = 0$, which are closest in this distance, be IM, LM.

STEP 3. Increase K by 1; define $T(\text{IM}) = T(\text{LM}) = K$. Set

$$V(K, J) = V(\text{IM}, J) \cap V(\text{LM}, J) \qquad \text{if} \qquad V(\text{IM}, J) \cap V(\text{LM}, J) \neq \varnothing,$$

$$V(K, J) = V(\text{IM}, J) \cup V(\text{LM}, J) \qquad \text{if} \qquad V(\text{IM}, J) \cup V(\text{LM}, J) = \varnothing.$$

If $K < 2M - 1$, go to Step 2.

STEP 4. The tree is now complete. Identify a minimum mutation fit in each variable by dropping all but one value (arbitrarily selected) from $V(2M - 1, J)$; then for any I for which $V[T(I), J]$ contains a single value, replace $V(I, J)$ by $V[T(I), J]$ if $V(I, J) \supset V[T(I), J]$, and drop all but one value (arbitrarily selected) from $V(I, J)$ if $V(I, J) \not\supset V[T(I), J]$.

NOTE. The minimum mutation fit obtained in Step 4 is not necessarily unique, since an arbitrary choice of values in the sets $V(I, J)$ is sometimes necessary. The following technique does not yield a minimum mutation fit, but it does produce a unique specification of values on the tree from the $V(I, J)$ obtained in Step 3. For all I, J, if $V[T(I), J]$ consists of a single element and if $V(I, J) \supset V[T(I), J]$, set $V(I, J) = V[T(I), J]$. Now for each I ($1 \leq I < 2M - 1$) set $V(I, J) = \varnothing$ if $V(I, J) = V[T(I), J]$ or if $V(I, J)$ contains more than one element.

13.6 APPLICATION OF REDUCED MUTATION ALGORITHM TO AMINO ACID SEQUENCES

The data used are the amino acid sequences in the protein molecule cytochrome-c for a number of vertebrates, given in Table 13.4. Only the species man, monkey, chicken, duck, kangaroo, and rattlesnake will be considered. The sequence of joins is displayed in Table 13.5.

STEP 1. The initial values of $V(I, J)$ ($1 \leq I \leq 6$, $1 \leq J \leq 22$) are just the data values. The cluster number is $K = 6$. Initially $T(1) = 0$, $T(2) = 0$, $T(3) = 0$, $T(4) = 0$, $T(5) = 0$, and $T(6) = 0$, which indicates that these clusters are available for joining.

STEP 2. The distance between 1, human, and 2, monkey, is $1 - \frac{21}{23}$ since they fail to match only in position 102. The distance between 3, chicken, and 4, duck, is $1 - \frac{20}{24}$ since they fail to match in two positions. Over all 15 pairs of clusters, 1 and 2 are closest, so IM = 1, LM = 2.

STEP 3. Now $K = 7$, $T(1) = T(2) = 7$, and the values $V(7, J)$ must be defined. For $J = 1$ (position 3), $V(1, 1) = V$ and $V(2, 1) = V$. Thus $V(1, 1) \cap V(2, 1) \neq \varnothing$. Therefore $V(7, 1) = V(2, 1) \cap V(1, 1) = V$. And so it goes until $J = 20$ (position 102), where $V(1, 20) = T$, $V(2, 20) = A$, and $V(1, 20) \cap V(2, 20) = \varnothing$. Therefore $V(7, 20) = V(2, 20) \cup V(1, 20) = \{T, A\}$. Return to Step 2.

STEP 2 REPEATED. There is some novelty in computing distances between the new cluster 7, and the remaining clusters. For example, the distance between 7 and 3 is $1 - \frac{11}{34}$. Here there are 11 variables which match, including $V(7, 20) = \{A, T\}$ and $V(3, 20) = T$. The closest pair are chicken and duck, joined to make cluster 8. Then cluster 8 and kangaroo, joined to make cluster 9. Then clusters 9 and 7 joined to make cluster 10. And finally, the rattlesnake is admitted to the collection.

Table 13.4 Amino Acid Sequence in Cytochrome-*c* for Vertebrates

[From Dickerson, R. E. (1972). "The structure and history of an ancient protein," *Sci. Amer.* (No. 4), **222**, 58–72.] Each letter denotes an amino acid. Only positions which vary over the vertebrates are recorded.

	3	4	9	11	12	15	20	22	28	33	36	44	46	47	50	54	58	60	61	62	65	66	81	83	85	86	88	89	92	93	95	100	101	102	103	104
MAN--------	V	E	I	I	M	S	V	K	T	H	F	P	Y	S	A	N	I	G	E	D	M	E	I	V	I	K	K	E	A	D	I	K	A	T	N	E
MONKEY-----	V	E	I	I	M	S	V	K	T	H	F	P	Y	S	A	N	I	G	E	D	M	E	I	V	I	K	K	E	A	D	I	K	A	A	N	E
HORSE------	V	E	I	V	Q	A	V	K	T	H	F	P	F	T	D	N	T	K	E	E	M	E	I	V	I	K	K	T	E	D	I	K	A	T	N	E
DONKEY-----	V	E	I	V	Q	A	V	K	T	H	F	P	F	S	D	N	T	K	E	E	M	E	I	V	I	K	K	T	E	D	I	K	A	T	N	E
PIG--------	V	E	I	V	Q	A	V	K	T	H	F	P	F	S	D	N	T	G	E	E	M	E	I	V	I	K	K	G	E	D	I	K	A	T	N	E
DOG--------	V	E	I	V	Q	A	V	K	T	H	F	P	F	S	D	N	T	G	E	E	M	E	I	V	I	K	T	G	A	D	I	K	A	T	K	E
RABBIT-----	V	E	I	V	Q	A	V	K	T	H	F	V	F	S	D	N	T	G	E	D	M	E	I	V	I	K	K	D	A	D	I	K	A	T	N	E
WHALE------	V	E	I	V	Q	A	V	K	T	H	F	V	F	S	D	N	T	G	E	E	M	E	I	V	I	K	K	G	A	D	I	K	A	T	N	E
KANGAROO---	V	E	I	V	Q	A	V	K	T	N	F	P	F	T	D	N	I	G	E	D	M	E	I	V	I	K	K	G	A	D	I	K	A	T	N	E
CHICKEN----	I	E	I	V	Q	S	V	K	T	H	F	E	F	S	D	N	T	G	E	D	M	E	I	V	I	K	K	S	V	D	I	D	A	T	S	K
PIGEON-----	I	E	I	V	Q	S	V	K	T	H	F	E	F	S	D	N	T	G	E	D	M	E	I	V	I	K	K	A	A	D	I	Q	A	T	A	K
DUCK-------	V	E	I	V	Q	S	V	K	T	H	F	E	F	S	D	N	T	G	E	D	M	E	I	V	I	K	K	S	A	D	I	D	A	T	A	K
TURTLE-----	V	E	I	V	Q	A	V	K	T	N	I	E	F	S	E	N	T	G	E	E	M	E	I	V	I	K	K	A	A	D	I	D	A	T	S	K
RATTLESNAKE	V	E	I	V	Q	S	V	K	T	H	F	V	Y	S	A	N	I	G	E	D	M	E	V	T	L	S	K	K	T	N	I	E	K	T	A	A
BULLFROG---	V	E	I	T	M	A	C	K	V	Y	I	A	F	S	D	N	T	G	E	D	M	E	I	V	I	K	K	G	Q	D	I	S	A	C	S	K
TUNA-------	V	A	T	V	Q	A	V	N	V	W	F	E	Y	S	D	S	V	N	N	D	M	E	I	V	I	K	K	G	Q	D	V	S	A	T	S	-
DOGFISH----	V	E	V	V	Q	A	V	N	T	S	F	Q	F	S	D	S	T	Q	Q	E	R	I	I	V	L	K	K	S	Q	D	I	K	T	A	A	S

Table 13.5 Application of Reduced Mutation Algorithm to Amino Acid Sequences of Man, Monkey, Chicken, Duck, Kangaroo, Rattlesnake

(Only positions that vary are recorded.)

CLUSTER POSITION /	3	11	12	15	33	44	46	47	50	58	81	83	85	86	89	92	93	100	101	102	103	104
1. HUMAN (HN)	V	I	M	S	H	P	Y	S	A	I	I	V	I	K	E	A	D	K	A	T	N	E
2. MONKEY (MY)	V	I	M	S	H	P	Y	S	A	I	I	V	I	K	E	A	D	K	A	A	N	E
3. CHICKEN (CN)	I	V	Q	S	H	E	F	S	D	T	I	V	I	K	S	A	D	D	A	T	S	K
4. DUCK (DK)	V	V	Q	S	H	E	F	S	D	T	I	V	I	K	S	A	D	D	A	T	A	K
5. KANGAROO (KO)	V	V	Q	A	N	P	F	T	D	I	I	V	I	K	G	A	D	K	A	T	N	E
6. RATTLESNAKE (RE)	V	V	Q	S	H	V	Y	S	A	I	V	T	L	S	K	T	N	E	K	T	A	A

VALUES OF V(I,J)

	3	11	12	15	33	44	46	47	50	58	81	83	85	86	89	92	93	100	101	102	103	104
7. HN-MY	V	I	M	S	H	P	Y	S	A	I	I	V	I	K	E	A	D	K	A	TA	N	E
8. CN-DK	IV	V	Q	S	H	E	F	S	D	T	I	V	I	K	S	A	D	D	A	T	SA	K
9. CN-KO	V	V	Q	SA	NH	EP	F	ST	D	TI	I	V	I	K	SG	A	D	KD	A	T	NSA	EK
10. HN-KO	V	VI	QM	S	H	P	YF	S	DA	I	I	V	I	K	ESG	A	D	K	A	T	N	E
11. ALL	V	V	Q	S	H	PV	Y	S	A	I	IV	VT	IL	SK	ESGK	AT	DN	KE	AK	T	NA	EA

VALUES OF V(I,J) IN MINIMUM MUTATION FIT TO COMPUTED TREE

	3	11	12	15	33	44	46	47	50	58	81	83	85	86	89	92	93	100	101	102	103	104
11. ALL	V	V	Q	S	H	P	Y	S	A	I	I	V	I	K	E	A	D	K	A	T	N	E
10. HN-KO	V	V	Q	S	H	P	Y	S	A	I	I	V	I	K	E	A	D	K	A	T	N	E
9. CN-KO	V	V	Q	S	H	P	F	S	D	I	I	V	I	K	S	A	D	K	A	T	N	E
7. HN-MY	V	I	M	S	H	P	Y	S	A	I	I	V	I	K	E	A	D	K	A	T	N	E
8. CN-DK	V	V	Q	S	H	E	F	S	D	T	I	V	I	K	S	A	D	D	A	T	S	K

STEP 4.   To identify a minimum mutation fit to the constructed tree, begin with the full cluster, 11. Consider only the variable 15 (position 89). Then $V(11, 15) = $ ESGK. All but one value must be dropped, so set $V(11, 15) = E$. Now at $I = 10$, $V(10, 15) = $ ESG $\supset V(11, 15) = E$. Therefore, change $V(10, 15)$ to E. At $I = 9$, $Z(9, 15) = $ SG $\not\supset E$. The second option applies, so all but one value are dropped to set $V(9, 15) = S$. When this is done for all clusters and variables, the resulting values of $V(I, J)$ constitute a minimum mutation fit in every variable.

According to the Note in Section 13.5, a cluster takes a value for a given variable only if this value is uniquely determined in the minimum mutation fit. This procedure does not necessarily lead to a minimum mutation fit, but it does allow a unique summary representation of the data, given in Table 13.6.

Table 13.6   Representation of Amino Acid Data on Reduced Mutation Tree

```
 3 = V      12 = Q      33 = H      47 = S      58 = I
11 = V      15 = S      46 = Y      50 = A     102 = T

44 = P      85 = I      93 = D     103 = N              |  44 = V
81 = I      86 = K     100 = K     104 = E              |  81 = V
83 = V      92 = A     101 = A                          |  83 = T

11 = I      89 = E      46 = F      50 = D              | 85 = L
HUMAN      102 = A      44 = E      89 = S     15 = S    | 86 = S
            MONKEY      58 = T     100 = D     33 = H    | 89 = K
                                   104 = K     47 = T    | 92 = T
                         3 = I     103 = A     89 = G    | 93 = N
                       103 = S     DUCK    KANGAROO      | 100 = E
                       CHICKEN                           | 101 = K
                                                         | 103 = A
                                                         | 104 = A
                                                         RATTLESNAKE
```

13.7   THINGS TO DO

13.7.1   Using Minimum Mutation Techniques

The fitting algorithm is to be used with category data and a tree already computed. It results in a representation of the data with a minimum number of symbols. A typical use might be a summary representation of the votes of U.S. congressmen. An initial clustering of congressmen into blocs is necessary. Then each vote is reduced to a list of votes of the various blocs, with an individual congressman appearing only if he votes differently to the smallest bloc including him. (See Table 13.7 for such data and Table 13.8 for data on Indo-European languages.)

13.7.2   Uniqueness

The minimum mutation fit is not necessarily unique. There may be more than one value assigned to a node, which is consistent with a minimum mutation fit. However,

Table 13.7 Congressmen by Bills (90th Congress)

Code: 1, yes; 2, pair yes; 3, announced yes; 4, announced no; 5, pair no; 6, no; 7, general pair; 8, abstain; 9, absent; 0, sponsor absent.

```
SPONSOR BILLS
ASPINALL    1. Auth. Biscayne National Monument in Florida
PERKINS     2. Promote health and safety in building trades
PATMAN      3. Sr. extend 2 Yrs. auth. reg. interest and dividend rates
DINGELL     4. Rel. Dev. Fish Protein concentrate
PERKINS     5. Establish commission on Negro History and Culture
ASPINALL    6. Designate parts of Morris City, N.J., as Wilderness
UDALL       7. Provide overtime and standby pay for Trans. Dept.
EDWARDS     8. Amd. Bill for relief of sundry claimants
GROSS       9. Amd. Omnibus claims bill
GROSS      10. Strike Title 8 of omnibus claims bill
HALL       11. Strike Title 9 of omnibus claims bill
GROSS      12. Strike Title 10 of omnibus claims bill
HALL       13. Strike Title 11 of omnibus claims bill
TALCOTT    14. Strike Title 14 of omnibus claims bill
POAGE      15. Take FD and AG ACT AMD SPKRS TBLE AGREE S CONF
```

	1	2	3	4	5	6	7	8	9	10	11	12	13	14	15
ABERNETHY	1	1	6	1	1	6	1	1	1	6	1	6	1	1	1
ALBERT	1	1	1	1	1	1	1	1	6	6	6	6	6	6	6
WIDNALL	1	1	1	1	7	1	7	7	9	9	9	9	9	9	9
ANDREWS	1	1	6	1	1	6	6	1	1	6	1	1	6	1	1
ARENDS	1	6	6	1	6	1	1	1	1	1	1	6	1	6	1
WILSON	1	5	7	1	7	1	7	1	9	9	9	9	9	9	9
ASHMORE	7	7	5	1	1	7	7	7	9	9	9	9	9	9	9
ASPINALL	1	1	1	1	1	1	1	6	6	6	6	6	6	6	6
AYRES	1	1	1	1	1	1	1	6	6	6	6	6	6	1	1
BARING	8	1	6	7	7	7	1	1	1	9	1	1	1	1	1
ANDERSON	1	1	6	1	1	1	1	1	1	6	6	6	1	9	1
ASHBROOK	1	5	5	7	7	7	7	7	1	1	1	1	1	1	1
BATTIN	5	6	6	1	6	1	1	1	9	6	1	1	1	1	1
BELL	1	1	1	1	1	1	1	1	1	9	9	9	9	9	9
CLANCY	1	6	6	1	6	1	1	1	1	6	1	1	6	1	1
DAVIS	1	1	1	1	1	6	6	1	1	6	6	6	6	6	6
DOLE	1	6	6	1	1	1	1	1	1	6	1	1	6	6	1

Table 13.8 Indo-European Languages

A subset of data belonging to Professor Dyen, Linguistics Department, Yale University.

WORD

LANGUAGE	All	Bad	Belly	Black	Bone	Day	Die	Drink
Russian	vse	ploxoj	zivot	cernyj	kost	den	umirat	pit
French	tout	mauvais	ventre	noir	os	jour	mourir	boire
Spanish	todo	mal	vientre	negro	hueso	dia	morir	beber
Italian	tutto		ventre	nero	osso	giorno	morire	bere
German	alle	schlect	bauch	schwarz	knochen	tag	sterben	trinken
Swedish	all	dalig	buk	svart	ben	dag	do	dricka
English	all	bad	belly	black	bone	day	die	drink
Welsh C	pawb	drwg	bola	du	asgwrn	dydd	marw	yfed
Greek	olos	kakos	kilya	mavros	kokalo	mera	petheno	pino
Irish	vile	olc	bolg	dubh	chaimh	la	doluidh	olaim
Persian	hame	bad	shekam	siah	ostokhan	ruz	mordan	nushidan
Bengali	sob	bad	pei	kalo	har	din	mora	khaoa
Hindi	sob	khorab	pei	kala	hoddi	din	morna	pina

LANGUAGE	Ear	Eat	Egg	Eye	Father	Fish	Five	Foot
Russian	uxo	est	jojco	glaz	otec	ryba	pjat	noga
French	oreille	manger	oeuf	oeil	pere	poisson	cinq	pied
Spanish	oreja	comedere	heuvo	ojo	padre	pez	cinco	pie
Italian	orecchio	mangiare	novo	occhio	padre	pesce	cinque	pie
German	ohr	essen	ei	auge	vater	fische	funf	fuss
Swedish	ora	ata	agg	oga	fader	fisk	fem	fot
English	ear	eat	egg	eye	father	fish	five	foot
Welsh C	clust	bwyta	wy	llygad	tad	pisgodyn	pump	troed
Greek	afti	troo	avgho	mati	pateras	psari	pende	podhi
Irish	cluas	ithim	ubh	suil	athair	iasc	cuigear	cos
Persian	gush	khordan	tokhm	chashm	pedar	mahi	panz	pa
Bengali	kan	khaoa	onda	cok	baba	mac	pac	pa
Hindi	kan	khana	onda	akh	bap	mocchi	pac	per

the number of mutations in the fit, in the part of the tree descended from a node I, is equal to the minimum number of mutations when only this part of the tree is considered or to one more than this minimum.

13.7.3 Probability Models

A mutation occurs from X at node I to Y at node J, one of the descendants of I, with probability $P(X, Y)$. Assume that $P(X, X) = p$ for all values X and that $P(X, Y) = r$ for all $X \neq Y$. Show that, if $p \geq r$, then a minimum mutation fit and a maximum likelihood fit coincide.

13.7.4 Dittoing

The sequence 1 0 1 0 0 1 0 0 1 0 0 0 1 0 0 1 1 1 1 1 0 may be reduced to
1 0 1 0 . 1 0 . 1 0 . . 1 0 . 1 0,
where a dot signifies that a character is identical to the previous character. Such identities are the stuff of clustering. Another coding is

1 0 0 0 1 C 0 1 1 0 0 1 1 0 0 2 1 0 0 1 1 4 0,

with the even positions counting the number of times the previous value is to be repeated. This can produce large gains in storage and understanding if the neighboring members of the data sequence are highly correlated.

In this technique, which is well known in communications theory, each value is connected to the previous one for prediction purposes.

1 0 1 0 0 1 0 0.

One generalization is to allow the *I*th object to be linked to the *J*th object, where *J* is greater than *I* but *J* is not necessarily $I + 1$. (It is convenient to number the objects from the right.)

1 0 1 0 0 1 0 0.

In this case, $J = \text{MT}(I) > I$ is the predictor for the *I*th object. Of course, the predictor array MT defines a tree. If the positions that make predictions are initially empty, the assignment of values to positions is the minimum mutation problem.

Another natural generalization allows transformations of values; for example, if neighboring values were negatively correlated, a one would be predicted to follow a zero. Such transformations may also be used with a tree prediction structure.

13.7.5 Unrooted Trees

The minimum mutation fit assigns values to the nodes of the tree to minimize the number of changes between neighboring nodes. For many variables, the error associated with a tree is $\sum \rho(I, J)$, where *I* and *J* are neighboring nodes and ρ is the number of mismatches, summed over all variables, between nodes *I* and *J*. This measure of error is the same whichever node is the root. Thus the "best" minimum mutation tree will have no preferred root; the search is for an unrooted tree.

13.7.6* Real Variables

For a real variable, values *V* must be assigned to each node (or cluster) *I* to minimize $\sum \rho[V(I), V(J)]$, summed over neighboring nodes *I* and *J*. First, choosing $\rho(x, y) = (x - y)^2$ leads to optimal $V(I)$ which are solutions of certain linear equations. These solutions may be obtained, in minimum mutation style, as follows. Associate with each cluster *I* a pair $A(I), B(I)$ such that $V(I) = A(I) V[T(I)] + B(I)$. For the objects at the ends of the tree, set $A(I) = 0$, $B(I) = V(I)$. Define $A(I)$ and $B(I)$ iteratively, if *I* is the ancestor of $J(1), \ldots, J(K)$, by

$$A(I) = (K + 1 - \sum \{1 \le L \le K\} A(J(L)))^{-1},$$
$$B(I) = A(I) \sum \{1 \le L \le K\} B(J(L)).$$

At the root, compute $V(I) = B(I)/[1 - A(I)]$ and compute each V value successively by using the equation

$$V(I) = A(I)V[T(I)] + B(I) .$$

Choosing $\rho(x, y) = |x - y|$ permits a solution following the minimum mutation method, and the two problems coincide for 0–1 variables. Associate with each cluster I an interval $[U(I), V(I)]$ within which the optimal value lies. For the objects at the end of the tree, $U(I) = V(I)$. If cluster I is the ancestor of $J(1), \ldots, J(K)$, define $[U(I), V(I)]$ from $[U[J(1)], V[J(1)]], \ldots, [U[J(K)], V[J(K)]]$ to be the interval of values X for which the number of U values greater than X equals the number of V values less than X. (It is true, but not obvious, that these maximal X values form an interval. If $U = V$ always, X will be the median.) Now, moving from the root toward the ends, choose any value in the interval $[U(I), V(I)]$ at the root. Then, for each J, choose the value in $[U(J), V(J)]$ which is closest to the fitted value for the ancestor of J.

13.7.7 Iteration

In the real case with $\rho(x, y) = (x - y)^2$ [or $\rho(x, y) = |x - y|$], it is simple to program an iterative procedure for computing the nodal values. Each nodal value is replaced by the mean (or median) of the neighboring nodal values. This works especially quickly for the median. The iteration converges to an optimal solution by a standard convexity argument, although this solution may not be unique in the case of the median.

A similar iterative procedure does not necessarily converge in the category-variables matching-distance case.

REFERENCES

CAMIN, J. H., and SOKAL, R. R. (1965). "A method for deducing branching sequences in phylogeny." *Evolution* **19**, 311–326. Each character is assumed to evolve through a known sequence of values. The goodness of a tree, constructed on objects with known values, is measured by the number of mutations in all characters. Some approximate methods of constructing the tree are applied to data on modern and ancient horses.

CAVALLI-SFORZA, L. L., and EDWARDS, A. W. F. (1967). "Phylogenetic analysis—models and estimation procedures." *Amer. J. Human Genetics* **19**, 233–257. Some distance models are discussed in which the distance between any two objects is the sum of the distances of the links in the unique chain connecting the two objects. A minimum-distance tree connecting all objects is one for which the total of the link distances is a minimum. This is a continuous-variable version of the minimum-mutation tree. No solution is cheaply available to the complete problem, but, given the tree structure, the optimal link distances and positions of ancestral nodes may be computed.

ESTABROOKE, G. F. (1968). "A general solution in partial orders for the Camin-Sokal model in phylogeny." *J. Theor. Biol.* **21**, 421–438. Instead of an order on the values taken by each character, each character evolves through a specified tree of values. The tree is a partial order with the requirement that any set of values have a unique most recent ancestor.

FITCH, W. M. (1971). "Toward defining the course of evolution: minimum change for a specific tree typology." *Systematic Zool.* **20**, 406–416. The correct rules for finding minimum-mutation fits to a given binary tree are stated (without proof).

FARRIS, J. S. (1970). "Methods for computing Wagner trees." *Systematic Zool.* **19**, 83–92. Consider objects as points in N-dimensional space and the ancestors also as points. For a given tree, the ancestors are positioned to minimize the sum of distances between each object or ancestor and its ancestor. If the distance

$$D(\mathbf{X}, \mathbf{Y}) = \sum \{1 \geq J \geq N\} |X(J) - Y(J)|$$

is used, recurrence relations similar to those in Section 13.2 determine the optimal positions of the ancestors.

This procedure applies to category variables taking only two values.

An algorithm for constructing the tree begins with the closest two objects and adds the other objects one at a time, at each stage adding that object which least increases the sum of link distances of the tree.

HARTIGAN, J. A. (1972). "Minimum mutation fits to a given tree." *Biometrics* **29** 53–65. Rules are proved for finding all the minimum-mutation fits to an arbitrary given tree.

IMMS, A. D. (1957). *A General Textbook of Entomology*, Methuen, London.

PROGRAMS

MMFIT   minimum-mutation fit to a given tree.

LINK   constructs tree by adding objects in succession to minimize sum of link distances.

MIDDLE   amalgamates three objects to form a single object.

DOT   represents an array in a dot matrix, using minimum-mutation fits.

```
      SUBROUTINE MMFIT(X,M,MT,K,XM,NVAL)
C.........................................................................20 MAY 1973
C.... FINDS MINIMUM MUTATION FIT OF VARIABLE X TO TREE MT .
C.... INPUT VALUES OF X GIVEN IN INDICES I,I .LE. M, OUTPUT VALUES  I, I .GT. M.
C.... M = NUMBER OF OBJECTS, FIRST OBJECT IGNORED
C.... K = NUMBER OF NODES OF TREE, K.GT.M.
C.... X = 1 BY K ARRAY
C          ON INPUT X(I) IS REAL BUT TAKES INTEGER VALUES, 1.LE.X(I).LE.NVAL.
C          ON OUTPUT, X CONTAINS VALUES FITTED TO NODES
C.... MT = 1 BY K TREE ARRAY, MT(I).GT.I EXCEPT AT I=K.
C.... XM = K BY NVAL, SCRATCH ARRAY
C.... NVAL = MAXIMUM VALUE OF X(I)
C.........................................................................
      DIMENSION X(K),XM(K,NVAL),MT(K)
C.... CHECK RANGE OF X(I)
      DO 20 I=2,M
      NC=0
      IF(X(I).GE.1.AND.X(I).LE.NVAL) GO TO 20
      WRITE(6,1) I
      NC=1
   20 CONTINUE
    1 FORMAT(I5,33HTH VALUE OF VARIABLE OUT OF RANGE)
      IF(NC.NE.0) RETURN
C.... OPTIMAL ASSIGNMENT TO NODES IGNORING REST OF TREE
      DO 29 I=M,K
   29 IF(I.GT.M) X(I)=0.
      DO 30 I=2,K
      DO 30 J=1,NVAL
   30 XM(I,J)=0
      DO 31 I=2,M
      J=X(I)
   31 XM(I,J)=1
      DO 32 I=2,K
      XMAX=0
      DO 33 J=1,NVAL
   33 IF(XM(I,J).GT.XMAX) XMAX=XM(I,J)
      DO 34 J=1,NVAL
      IF(XM(I,J).LT.XMAX) XM(I,J)=0
   34 IF(XM(I,J).NE.0) XM(I,J)=1
      IF(I.EQ.K) GO TO 32
      II=MT(I)
      DO 35 J=1,NVAL
   35 XM(II,J)=XM(II,J)+XM(I,J)
   32 CONTINUE
C.... FIND OPTIMAL ASSIGNMENT OVER WHOLE TREE
      DO 40 I=2,K
      II=K-I+2
      IT=MT(II)
      DO 40 J=1,NVAL
   40 IF((X(II).EQ.0.OR.X(IT).EQ.J).AND.XM(II,J).EQ.1.) X(II)=J
      RETURN
      END
```

```
      SUBROUTINE LINK(A,K,N,MT,DD,P)
C.......................................................................20 MAY 1973
C.... K = 2*NUMBER OF OBJECTS
C.... N = NUMBER OF COLUMNS
C.... A = K BY N BORDERED ARRAY
C.... MT = 1 BY K TREE ARRAY, MT(I) NOT NECESSARILY GREATER THAN I.
C.... DD = 1 BY K ARRAY, DD(I)=DISTANCE OF OBJECT I AT AMALGAMATION
C.... P = PARAMETER SPECIFYING TYPE OF LINKAGE
C          P=2.,MEANS
C          P=1.,MEDIANS
C          P=0.,MATCHING
C................................................................................
      DIMENSION MT(K),A(K,N)
         DIMENSION DD(K)
      DATA CL/4HCLUS/
      M=K/2
      DO 10 I=M,K
      IF(I.EQ.M) GO TO 10
      A(I,1)=CL
   10 CONTINUE
      DO 19 I=2,K
   19 MT(I)=0
C.... INITIALIZE TREE
      MM=M*(N-2)+1
      DM=10.**10
      IM=2
      JM=3
      DO 20 I=2,M
      DO 20 J=I,M
      IF(J.EQ.I) GO TO 20
      MM=K*(N-2)+1
      CALL MIDDLE(A(I,2),A(J,2),A(I,2),A(K,2),D,MM,K,P)
      CALL MIDDLE(A(I,2),A(J,2),A(J,2),A(K,2),D1,MM,K,P)
      D3=D+D1
      IF(D3.GE.DM) GO TO 20
      IM=I
      JM=J
      DM=D3
   20 CONTINUE
      MT(JM)=JM
      MT(IM)=JM
C.... FIND DISTANCES TO INITIAL LINK
      DO 25 L=2,M
      IF(MT(L).GT.0) GO TO 25
      CALL MIDDLE(A(IM,2),A(JM,2),A(L,2),A(K,2),DD(L),MM,K,P)
      MT(L)=-IM
   25 CONTINUE
C.... CONSTRUCT CLUSTERS
      KK=K-3
      DO 30 I=M,KK
      IF(I.EQ.M) GO TO 30
      DM=10.**10
      I1=2
      I2=2
      I3=3
C.... FIND BEST OBJECT TO ADD TO TREE
      DO 50 L=2,M
      IF(MT(L).GT.0) GO TO 50
      IF(DD(L).GE.DM) GO TO 50
      I1=-MT(L)
      I2=MT(I1)
      I3=L
      DM=DD(L)
   50 CONTINUE
      CALL MIDDLE(A(I1,2),A(I2,2),A(I3,2),A(I,2),D,MM,K,P)
      MT(I3)=I
      MT(I)=I2
      MT(I1)=I
C.... UPDATE DISTANCE ARRAY
      DO 60 L=2,M
      IF(MT(L).GT.0) GO TO 60
      CALL MIDDLE(A(I1,2),A(I,2),A(L,2),A(K,2),D1,MM,K,P)
      CALL MIDDLE(A(I2,2),A(I,2),A(L,2),A(K,2),D2,MM,K,P)
      CALL MIDDLE(A(I3,2),A(I,2),A(L,2),A(K,2),D3,MM,K,P)
      IF(I1.EQ.-MT(L)) DD(L)=10.**10
      IF(D1.LT.DD(L)) MT(L)=-I1
      IF(D1.LT.DD(L)) DD(L)=D1
      IF(D2.LT.DD(L)) MT(L)=-I
      IF(D2.LT.DD(L)) DD(L)=D2
      IF(D3.LT.DD(L)) MT(L)=-I3
      IF(D3.LT.DD(L)) DD(L)=D3
   60 CONTINUE
   30 CONTINUE
      RETURN
      END
```

```
      SUBROUTINE MIDDLE(X,Y,Z,U,D,M,N,P)
C.........................................................20 MAY 1973
C.... COMPUTES MIDDLE OF THREE VECTORS X,Y,Z, AND DISTANCE FROM MIDDLE
C.... X = M BY 1 VECTOR
C.... Y = M BY 1 VECTOR
C.... Z = M BY 1 VECTOR
C.... U = M BY 1 VECTOR
C.... D = DISTANCE OF X,Y,Z, FROM U
C.... M = NUMBER OF ELEMENTS IN VECTOR
C.... N = SKIP FACTOR, FOR USE WITH ROWS OF MATRIX. SEE DIST.
C.... P = PARAMETER SPECIFYING DISTANCE MEASURE
C          P = 2., MEANS
C          P = 1., MEDIANS
C          P = 0., MATCHING
C.........................................................
      DIMENSION X(M),Y(M),Z(M),U(M)
      DO 20 I=1,M,N
      IF(P.EQ.2) GO TO 21
      IF((X(I)-Y(I))*(X(I)-Z(I)).LE.0.) U(I)=X(I)
      IF((Y(I)-X(I))*(Y(I)-Z(I)).LE.0.) U(I)=Y(I)
      IF((Z(I)-X(I))*(Z(I)-Y(I)).LE.0.) U(I)=Z(I)
      GO TO 20
   21 U(I)=(X(I)+Y(I)+Z(I))/3.
   20 CONTINUE
      D=0
      DD=0
      DO 30 I=1,M,N
      DD=DD+1
      IF(P.EQ.0) GO TO 31
      D=D+(ABS(X(I)-U(I)))**P+(ABS(Y(I)-U(I)))**P+(ABS(Z(I)-U(I)))**P
      GO TO 30
   31 IF(X(I).NE.U(I)) D=D+1
      IF(Y(I).NE.U(I)) D=D+1
      IF(Z(I).NE.U(I)) D=D+1
   30 CONTINUE
      D=D/DD
      RETURN
      END
```

```
      SUBROUTINE DOT(A,M,N,MT,K,XM,NVAL)
C............................................................................20 MAY 1973
C.... REPRESENTS AN ARRAY A, IN A DOT MATRIX, USING THE TREE MT.
C.... THE VALUE AT CASE I IS DOTTED IF IT AGREES WITH THE VALUE AT THE SMALLEST
C     CLUSTER INCLUDING I.
C.... A = M BY N BORDERED ARRAY, TAKING INTEGER VALUES BETWEEN 1 AND NVAL.LE.35.
C.... M = NUMBER OF ROWS
C.... N = NUMBER OF COLUMNS
C.... MT = 1 BY K TREE ARRAY, MT(I).GT.I EXCEPT AT I=K.  K.LE.100.
C.... K = NUMBER OF NODES OF TREE, K.GT.M.
C.... XM = K BY NVAL SCRATCH ARRAY
C............................................................................
      DIMENSION A(M,N),MT(K),XM(K,NVAL)
      DIMENSION X(100)
      DIMENSION AA(36)
      DATA AA/1H.,1H1,1H2,1H3,1H4,1H5,1H6,1H7,1H8,1H9,1HA,1HB,
     *1HC,1HD,1HE,1HF,1HG,1HH,1HI,1HJ,1HK,1HL,1HM,1HN,
     *1HO,1HP,1HQ,1HR,1HS,1HT,1HU,1HV,1HW,1HX,1HY,1HZ/
C.... CHECK RANGE OF A
      DO 20 I=2,M
      DO 20 J=2,N
   20 IF(A(I,J).LT.1.OR.A(I,J).GT.NVAL) WRITE(6,1) I,J
    1 FORMAT(I5,2H ,,I5,27H TH DATA VALUE OUT OF RANGE     )
C.... COUNT NUMBER IN EACH CLUSTER
      DO 70 I=2,K
      XM(I,1)=0
   70 IF(I.LE.M) XM(I,1)=1.
      DO 71 I=2,K
      IF(I.EQ.K) GO TO 71
      J=MT(I)
      XM(J,1)=XM(J,1)+XM(I,1)
   71 CONTINUE
      WRITE(6,3)
    3 FORMAT(5HOTREE)
C.... PRINT TREE, A LINE AT A TIME.
      DO 80 I=2,M
      J=K
      XM(J,1)=XM(J,1)-1
      X(K)=AA(25)
      DO 81 L=2,K
      LL=K-L+2
      IF(LL.EQ.K) GO TO 81
      IF(XM(LL,1).GT.0) XM(LL,1)=-XM(LL,1)
      X(LL)=AA(1)
      IF(MT(LL).NE.J) GO TO 81
      IF(XM(LL,1).EQ.0) GO TO 81
      J=LL
      XM(LL,1)=-XM(LL,1)-1.
      X(J)=AA(25)
   81 CONTINUE
      WRITE(6,7) J,A(J,1),(X(L),L=2,K)
   80 CONTINUE
C.... GO THROUGH VARIABLES, ONE AT A TIME
      WRITE(6,4)((I,J=1,10),I=1,9)
    4 FORMAT(22X,18(5I1,1X))
      WRITE(6,5)((I,I=1,9),J=1,10)
    5  FORMAT(9X,10(2H 0,4I1,1X,5I1))
      WRITE(6,6)
    6 FORMAT(10X,120(1H-))
      DO 50 JJ=2,N
      DO 30 I=2,M
   30 X(I)=A(I,JJ)
      CALL MMFIT(X,M,MT,K,XM,NVAL)
      DO 41 I=2,K
      IF(I.EQ.K) GO TO 40
      II=MT(I)
      IF(X(I).EQ.X(II)) X(I)=0.
   40 CONTINUE
      J=X(I)+1.
      IF(J.LT.1) J=1
      IF(J.GT.35) J=36
      X(I)=AA(J)
   41 CONTINUE
      WRITE(6,7) JJ,A(1,JJ),(X(I),I=2,K)
    7 FORMAT(I4,A5,3X,20(3A1,1X,2A1))
   50 CONTINUE
      RETURN
      END
```

CHAPTER 14

Direct Splitting

14.1 INTRODUCTION

For Table 14.1, the percentage Republican vote for president in the Southern States in 1900–1968, it is natural to seek clusters of states and also clusters of years. For example, Kentucky, Delaware, and Maryland have similar voting patterns over all years, and also 1932, 1936, 1940, 1944, the Roosevelt years, have similar votes in all states.

In Table 14.2, the years have been clustered using euclidean distance, averaging squared differences over states. For example, $D(32, 40) = \{[(2 - 1)^2 + (4 - 3)^2 + (7 - 11)^2 + (40 - 42)^2 + (36 - 41)^2 + (35 - 48)^2]/6\}^{1/2}$. The states have been similarly clustered using euclidean distance, averaging squared differences over years. The clusters of years and states are plausible. For example, KY, MD, MO are very close together, SC, LA, MS are not quite so close to each other, and there are large distances between the two groups.

Yet the clusters are somewhat removed from the original data, since they refer to distances rather than data. It would be much more useful to describe the clusters of states in terms of their behavior in different years and similarly to describe the clusters of years in terms of their behavior in different states.

What is needed is a model for clusters that expresses interaction between the clusters of states and clusters of years. For a general data matrix, in which all values in the matrix are expressed on the same scale, such a model is given in Table 14.3. In this model, there are three types of clusters: the row clusters, the column clusters, and the data clusters. Each data cluster is the subset of data values indexed by the rows in a row cluster and the columns in a column cluster. The data clusters partition the data matrix, whereas the row and column clusters each form a tree. Within each cluster, the data values are identical. The data matrix could thus be reduced by specifying a single value for each data cluster, as in the second matrix in Table 14.3.

A direct splitting algorithm, for identifying clusters according to the above model was first given by Hartigan (1972). Procedures for clustering variables and then clustering cases (in order, rather than simultaneously) are given in Tryon and Bailey (1970). In Sonquist (1971), a single target variable is predicted by splitting the sample into two clusters according to a known binary variable, splitting one of these two clusters by a further binary variable, and so on. At each stage, that split is chosen which most reduces prediction error. Procedures of this type are applicable to two-way (or many-way) clustering models when all the margins are of size 2. For example the

Table 14.1 Republican Vote for President

Southern states by twentieth century years, in percentages (From Peterson, S. (1963) *A Statistical History of the American Presidential Elections*, Ungar, New York).

STATES	00	04	08	12	16	20	24	28	32	36	40	44	48	52	56	60	64	68
ALABAMA (AL)	35	21	24	8	22	31	27	48	14	13	14	18	19	35	39	42	70	14
ARKANSAS (AR)	35	40	37	20	28	39	29	39	13	18	21	30	21	44	46	43	44	31
DELAWARE (DE)	54	54	52	33	50	56	58	65	51	43	45	45	50	52	55	49	39	45
FLORIDA (FL)	19	21	22	8	18	31	28	57	25	24	26	30	34	55	57	52	48	41
GEORGIA (GA)	29	18	31	4	7	29	18	43	8	13	15	18	18	30	33	37	54	30
KENTUCKY (KY)	49	47	48	25	47	49	49	59	40	40	42	45	41	50	54	54	36	44
LOUISIANA (LA)	21	10	12	5	7	31	20	24	7	11	14	19	17	47	53	29	57	23
MARYLAND (MD)	52	49	49	24	45	55	45	57	36	37	41	48	49	55	60	46	35	42
MISSISSIPPI (MS)	10	5	7	2	5	14	8	18	4	3	4	6	3	40	24	25	87	14
MISSOURI (MO)	46	50	49	30	47	55	50	56	35	38	48	48	42	51	50	50	36	45
NORTH CAR. (NC)	45	40	46	12	42	43	55	29	29	27	26	33	33	46	49	48	44	40
SOUTH CAR. (SC)	7	5	6	1	2	4	2	9	2	1	4	4	4	49	25	49	59	39
TENNESSEE (TN)	45	43	46	24	43	51	44	54	32	31	33	39	37	50	49	53	44	38
TEXAS (TX)	31	22	22	9	17	24	20	52	11	12	19	17	25	53	55	49	37	40
VIRGINIA (VA)	44	37	38	17	32	38	33	54	30	29	32	37	41	56	55	52	46	43
WEST VIRGINIA (WV)	54	55	53	21	49	55	49	58	44	39	43	45	42	48	47	54	32	40

model in Table 14.4, is correctly analyzed by this technique. In extending the technique to marginal variables taking many values, the clusters of marginal values obtained may possibly overlap, so that no simple representation of the data matrix is possible, with a reordering based on the marginal clusters.

The two-way splitting algorithm given here combines the discipline on marginal clusters in Table 14.3 with the generality of the Sonquist procedure (it is applicable to many marginal variables). The outcome of the algorithm is a partition of the response variable, specified by a hierarchial clustering of each of the marginal variables.

14.2 BINARY SPLITTING ALGORITHM

Preliminaries. This algorithm is the basic component of the one-way and two-way splitting algorithms. It may be studied independently as a method of dividing a set of cases into two clusters. It is assumed that data values $A(I, J)$ $(1 \leq I \leq M, 1 \leq J \leq N)$ are given, where I denotes a case and J denotes a variable. Weights $W(I, J)$ are given for the Ith case and Jth variable, where $W(I, J)$ is the number of observations combined in the data value $A(I, J)$. The two clusters of cases to be constructed are $C(1)$ and $C(2)$. The weighted averages of the Jth variable over the two clusters are $B(1, J)$ and $B(2, J)$. Thus

$$B(1, J) = (\sum \{I \in C(1)\} \, W(I, J)A(I, J))(\sum \{I \in C(1)\} \, W(I, J))^{-1}.$$

The error of a partition is based on the difference between the weighted averages,

$$e[P(M, 2)] = \left(-\sum \{1 \leq J \leq N\} \, [B(1, J) - B(2, J)]^2\right)\left(\frac{1}{D(1, J)} + \frac{1}{D(2, J)}\right)^{-1},$$

where

$$D(L, J) = \sum \{I \in C(L)\} \, W(I, J).$$

The algorithm finds the case whose removal from the rest most decreases the error and begins a cluster with this case. Next, that case is transferred to the new cluster which most reduces the error (the error may increase, but the case is transferred nevertheless). This procedure continues until all cases have been transferred. The split that has the smallest error during the M transfers is the final split. The error after the Kth transfer is $E(K)$.

Table 14.2 Marginal Trees on Percentage Republican in Presidential Elections

YEARS

		1932	1936	1940	1960	1964	1968
STATES	SC	2	1	4	49	59	39
	MS	4	3	4	25	87	14
	LA	7	11	14	29	57	23
	KY	40	40	42	54	36	44
	MD	36	37	41	46	35	42
	MO	35	38	48	50	36	45

DISTANCES BETWEEN YEARS

(ROOT OF AVERAGE SQUARED DIFFERENCE, OVER STATES)

32					
3	40				
6	5	36			
25	23	24	60		
18	18	18	8	68	
50	45	47	30	34	64

DISTANCES BETWEEN STATES (OVER YEARS)

SC					
12	LA				
18	14	MS			
23	26	36	KY		
28	21	35	4	MD	
28	25	36	4	4	MO

STEP 1.   Initially,

$$D(1, J) = \sum \{1 \leq I \leq M\} W(I, J),$$

$$B(1, J) = \sum \{1 \leq I \leq M\} \frac{W(I, J)A(I, J)}{D(1, J)},$$

and

$$D(2, J) = B(2, J) = 0.$$

Set $K = 0.$ $E(0) = 0.$

Table 14.3   Direct Partition Model

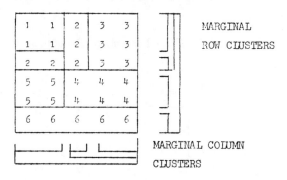

DATA
CLUSTER

MARGINAL
ROW CLUSTERS

MARGINAL COLUMN
CLUSTERS

Table 14.4   The Sonquist Technique

	C = 1	C = 2
R = 1	1	2
R = 2	1	3

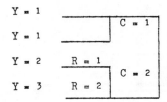

SPLIT FIRST ON COLUMNS,
SPLIT SECOND ON ROWS
WITHIN FIRST COLUMN.

STEP 2.   The decrease in error due to transferring case I from cluster 1 to cluster 2 is

$$EE = \sum \{1 \leq J \leq N\} \, [A(I, J) - B(1, J)]^2 \, \frac{D(1, J)W(I, J)}{D(1, J) - W(I, J)}$$

$$- \sum \{1 \leq J \leq N\} \, [A(I, J) - B(2, J)]^2 \, \frac{D(2, J)W(I, J)}{D(2, J) + W(I, J)} \, .$$

The case I, not equal to $I(1), \ldots, I(K)$, is transferred, which maximizes EE.

STEP 3. Increase K by 1 and set $I(K) = I$, $E(K) = E(K - 1) -$ EE. Update cluster means and cluster weights:

$$D(1, J) = D(1, J) - W(I, J),$$

$$B(1, J) = B(1, J) + W(I, J) \frac{B(1, J) - A(I, J)}{D(1, J)} \, ,$$

$$D(2, J) = D(2, J + W(I, J),$$

$$B(2, J) = B(2, J) - W(I, J) \frac{B(2, J) - A(I, J)}{D(2, J)} \, .$$

If K is less than M, return to Step 2.

STEP 4.   Set $C(2) = \{I(1), I(2), \ldots, I(K)\}$, where $E(K)$ is a minimum over all K $(1 \leq K \leq M)$.

14.3 APPLICATION OF BINARY SPLITTING ALGORITHM TO VOTING DATA WITH MISSING VALUES

Table 14.5 was derived from Table 14.2 by setting each value missing with probability 0.5. In terms of the algorithm in Section 14.2, $W(I, J) = 1$ if the Ith year is present for the Jth state, and $W(I, J) = 0$ if the value is missing. Note that years are treated as objects or cases and states are the variables.

STEP 1.   Initially, $D(1, J)$ is the sum of weights for variable J, and $B(1, J)$ is the average of values present for variable J. For example, for $J = 1$, SC, there are four years present with values 2, 1, 59, and 39, so $D(1, 1) = 4$, $B(1, 1) = (2 + 1 + 59 + 39)/4 = 25$. The other values of $D(1, J)$, $D(2, J)$, $B(1, J)$, and $B(2, J)$ are given in Table 14.5 beneath the line $K = 0$.

STEP 2.   The object whose transfer to the second cluster most reduces the error is 1964, the fifth year, $I = 5$.

STEP 3.   Increase K by 1, so $K = 1$, $I(1) = 5$, $E(1) = -6041$. Cluster means and cluster weights are updated. For example,

$$D(1, 1) = D(1, 1) - W(5, 1) = 4 - 1 = 3,$$
$$B(1, 1) = 25 + (25 - 59)/3 = 14.$$

Since $K = 1 \leq 6$, return to Step 2.

STEP 2 REPEATED.   The next object to be transferred is 1960, $I = 4$. And then, successively, 1968, 1932, 1940, and 1936. Note that after 1936 is transferred the error

Table 14.5 Binary Splitting Algorithm on Small Voting Matrix with Missing Values (50%, at Random)

YEARS

	(1) 1932	(2) 1936	(3) 1940	(4) 1960	(5) 1964	(6) 1968
SC	2	1			59	39
MS		3	4	25	87	14
LA		11		29		
KY	40		42	54		44
MD	36	37				
MO	35	38	48	50		

$K = 0$ $\quad E = 0$

	B(1,J)	D(1,J)	B(2,J)	D(2,J)
SC	25	4	0	0
MS	27	5	0	0
LA	20	2	0	0
KY	45	4	0	0
MD	37	2	0	0
MO	43	4	0	0

$K = 1$ $\quad I(1) = 5$ $\quad E = -6041$

B(1,J)	D(1,J)	B(2,J)	D(2,J)
14	3	59	1
12	4	87	1
20	2	0	0
45	4	0	0
37	2	0	0
43	4	0	0

$K = 2$ $\quad I(2) = 4$ $\quad E = -4679$

B(1,J)	D(1,J)	B(2,J)	D(2,J)
14	3	59	1
7	3	56	2
11	1	29	1
42	3	54	1
37	2	0	0
40	3	50	1

$K = 3$ $\quad I(3) = 6$ $\quad E = -4270$

B(1,J)	D(1,J)	B(2,J)	D(2,J)
1	2	49	2
3	2	42	3
11	1	29	1
41	2	49	2
37	2	0	0
40	3	50	1

$K = 4$ $\quad I(4) = 1$ $\quad E = -2674$

B(1,J)	D(1,J)	B(2,J)	D(2,J)
1	1	33	3
3	2	42	3
11	1	29	1
42	1	46	3
37	1	36	1
43	2	43	2

$K = 5$ $\quad I(5) = 3$ $\quad E = -1626$

B(1,J)	D(1,J)	B(2,J)	D(2,J)
1	1	33	3
3	1	33	4
11	1	29	1
0	0	45	4
37	1	36	1
38	1	44	3

should again be zero. This requirement checks the hand calculations. (There is a small rounding error in Table 14.5, so that the final error is 30 rather than zero.)

STEP 4.   The smallest error occurs at $K = 1$, so the second cluster consists of 1964 alone.

This algorithm is a refinement of the K-means algorithm, takes $O(M^2)$ computations, and so is appropriate only for moderate data sets. It is superior to the K-means algorithm in being more likely to reach the globally optimal partition. It is guaranteed to reach the global optimum for objects in one dimension.

14.4 ONE-WAY SPLITTING ALGORITHM

Preliminaries. A data matrix $A(I, J)$ $(1 \leq I \leq M, 1 \leq J \leq N)$ is given and a threshold T. Clusters are to be constructed so that all variables have a within-cluster variance of no greater than T. (Different thresholds for different variables are accommodated by scaling the variables.) During the algorithm, the data matrix is reordered so that each cluster is a contiguous sequence of objects. The clusters $1, 2, 3, \ldots, KC$ are constructed with cluster I determined by $\mathrm{MIN}(I)$, $\mathrm{MAX}(I)$ the two clusters it splits into. For a minimal cluster I, $\mathrm{MIN}(I)$ and $\mathrm{MAX}(I)$ are defined to be the first and last objects in cluster I. Also, after the algorithm is complete, define $V(I)$ to be the set of variables which meet threshold within I but not within any larger cluster. The algorithm proceeds by progressively splitting clusters. At the Kth step, there will be K clusters $I(1), I(2), \ldots, I(K)$ partitioning the objects, the minimal clusters in the set $1, 2, 3, \ldots, 2K - 1$. During the algorithm, $V[I(J)]$ is the set of variables that do not meet threshold for any larger cluster. A binary split is performed on some $I(J)$, using only the variables in $V[I(J)]$ that exceed threshold. The two new clusters, $2K$ and $2K + 1$, have $V(2K) = V(2K + 1)$ defined to be the set of variables in $V[I(J))$, which exceed threshold in $I(J)$. Then $V[I(J)]$ is changed to be the variables in $V[I(J)]$ which do not exceed threshold in $I(J)$, and the splitting continues. The splitting stops when every $V[I(J)]$ contains only variables which meet threshold in $I(J)$.

The basic idea is that only variables which exceed threshold within a given cluster are used in splitting that cluster.

STEP 1.   Define $\mathrm{MIN}(1) = 1$, $\mathrm{MAX}(1) = M$. Set $K = 1$, $I(1) = 1$, $L = 1$. Define $V(1) = \{1, 2, \ldots, N\}$.

STEP 2.   Try splitting $I(L)$. Compute the variance of each variable J $\{J \in V[I(L)]\}$ over the cases in the cluster $I(L)$. If none of these variances exceed T, go to Step 4. Otherwise, go to Step 3.

STEP 3.   Let VV be the set of variables whose variance, over the cases in $I(L)$, is greater than T. Apply the binary splitting algorithm to the cases I, $\mathrm{MIN}[I(L)] \leq I \leq \mathrm{MAX}[I(L)]$, over the variables in VV. Reorder the cases so that the first cluster consists of objects IL, IL $+ 1, \ldots,$ II and the second consists of objects II $+ 1, \ldots,$ IU. Define two new clusters $K + 1$ and $K + 2$ with $\mathrm{MIN}(K + 1) = $ IL, $\mathrm{MAX}(K + 1) = $ II, $\mathrm{MIN}(K + 2) = $ II $+ 1$, $\mathrm{MAX}(K + 2) = $ IU.   Define $\mathrm{MAX}[I(L)] = K + 2$, $\mathrm{MIN}[I(L)] = K + 1$. Define $V(K + 1) = V(K + 2) = $ VV, and change $V[I(L)]$ to $V[I(L)] - $ VV. Set $I(L) = K + 1$, $I(K + 1) = K + 2$, increase K by 1, and return to Step 2.

STEP 4.   Increase L by 1 and, if $L \leq K$, go to Step 2. Otherwise, stop.

14.5 ONE-WAY SPLITTING ALGORITHM APPLIED TO
REPUBLICAN PERCENTAGES

Either states or years may be treated as variables. In this problem years will be variables. The threshold variance is 25. As in other clustering techniques, a small threshold will give too many small clusters and a large threshold will give too few large clusters. A threshold of 25 corresponds to a standard deviation of 5, which seems a reasonable error between similar states.

STEP 1. $MIN(1) = 1$, $MAX(1) = 6$. $K = 1$, $I(1) = 1$, $L = 1$. $V(1) = \{1, 2, 3, 4, 5, 6\}$.

STEP 2.   Try splitting $I(1) = 1$. All variables have a variance exceeding 25, obviously.

STEP 3.   The set of variables to be used in the split, VV, is $\{1, 2, 3, 4, 5, 6\}$. The binary splitting algorithm produces two clusters, $\{SC, MS, LA\}$ and $\{KY, MD, MO\}$. New clusters 2 and 3 are defined with $MIN(2) = 1$, $MAX(2) = 3$ and $MIN(3) = 4$, $MAX(3) = 6$.   Redefine $MIN(1) = 2$, $MAX(1) = 3$.   Define $V(2) = V(3) = \{1, 2, 3, 4, 5, 6\}$, set $V(1) = \varnothing$, $I(1) = 2$, $I(2) = 3$, $K = 2$, and return to Step 2.

STEP 2 REPEATED. Try splitting $I(1) = 2$. All variables except $J = 1$ have a variance exceeding 25.

STEP 2 REPEATED.   Then VV $= \{2, 3, 4, 5, 6\}$, and the binary splitting algorithm produces clusters $\{MS\}$ and $\{SC, LA\}$. New clusters 4 and 5 are defined with $MIN(4) = 1$, $MAX(4) = 1$, $MIN(5) = 2$, $MAX(5) = 3$, $MIN(2) = 4$, $MIN(2) = 5$. Define $V(4) = V(5) = \{2, 3, 4, 5, 6\}$, set $V(2) = \{1\}$, $I(1) = 4$, $I(3) = 5$, $K = 3$, and return to Step 2.

STEP 2 REPEATED.   Since $I(1) = 4$ contains only one object, the variances for all variables are zero; go to Step 4.

STEP 4.   Set $L = 2$. Since $L \leq 3$, go to Step 2.

STEP 2 REPEATED.   The cluster $I(2) = 3$ consists of objects KY, MD, MO. All variables have a variance less than 25. Go to Step 4.

STEP 4 REPEATED.   Set $L = 3$, and go to Step 2.

STEP 2 REPEATED.   The cluster $I(3) = 5$ consists of objects SC, LA. The relevant variables $V(5)$ are $\{2, 3, 4, 5, 6\}$. Of these, all but $J = 5$ exceed threshold. The cluster splits into two single-object clusters $\{SC\}$ $\{LA\}$. These two clusters are 6 and 7, with $MIN(6) = 2$, $MAX(6) = 2$, $MIN(7) = 3$, $MAX(7) = 3$, $MIN(5) = 6$, $MAX(5) = 7$. Define $V(6) = V(7) = \{2, 3, 4, 6\}$, $V(5) = 5$, $I(3) = 6$, $I(4) = 7$, $K = 4$; return to Step 2.

No further splits take place. The history of the splits is represented in Table 14.6. A method of using the splitting information to summarize the table is given in Table 14.7. This method is analogous to the representation of category data by using minimum-mutation fits. The novelty of the one-way splitting algorithm lies in the association with each cluster of the variables approximately constant over the cluster.

Table 14.6 One-Way Splitting Algorithm Applied to Republican Percentages

	32	36	40	60	64	68
SC	2	1	4	49	59	39
MS	4	3	4	25	87	14
IA	7	11	14	29	57	23
KY	40	40	42	54	36	44
MD	36	37	41	46	35	42
MO	35	38	48	50	36	45

FIRST SPLIT

	32	36	40	60	64	68
SC	2	1	4	49	59	39
MS	4	3	4	25	87	14
IA	7	11	14	29	57	23
KY	40	40	42	54	36	44
MD	36	37	41	46	35	42
MO	35	38	48	50	36	45

(outlined values ignored in later steps)

SECOND SPLIT

	32	36	40	60	64	68
MS	4	3	4	25	87	14
SC	2	1	4	49	59	39
IA	7	11	14	29	57	23
KY	40	40	42	54	36	44
MD	36	37	41	56	35	42
MO	35	38	48	50	36	45

THIRD SPLIT

	32	36	40	60	64	68
MS	4	3	4	25	87	14
SC	2	1	4	49	59	39
IA	7	11	14	29	57	23
KY	40	40	42	54	36	44
MD	36	37	41	46	35	42
MO	35	38	48	50	36	45

Table 14.7 Representation of Data Using One-Way Splitting Algorithm

VARIABLE 1 = 1932

VARIABLE 2 = 1936

VARIABLE 3 = 1940

VARIABLE 4 = 1960

VARIABLE 5 = 1964

VARIABLE 6 = 1968

1 = 37 3 = 44 5 = 36	1 = 4		
2 = 39 4 = 50 6 = 44	5 = 58		2 = 3
MO MD KY	2 = 11	2 = 1	3 = 4
	3 = 14	3 = 4	4 = 25
	4 = 29	4 = 49	5 = 87
	6 = 23	6 = 39	6 = 14
	IA	SC	MS

14.6 TWO-WAY SPLITTING ALGORITHM

Preliminaries. A model for which the two-way splitting algorithm is appropriate is given in Table 14.3. There is a data matrix $A(I, J)$ ($1 \leq I \leq M, 1 \leq J \leq N$) with entries comparable among both rows and columns. The algorithm computes row clusters $1, 2, \ldots, KR$, column clusters $1, 2, \ldots, KC$ and data clusters $1, 2, 3, \ldots,$ KD. The data cluster K is the submatrix of data values with rows in the row cluster IR(K) and with columns in the column cluster IC(K). Each data point lies in just one data cluster, so these clusters partition the matrix. The row clusters $1, 2, \ldots, KR$ form a tree. The structure of the tree is described by a pair MINR(I), MAXR(I), which specify the two largest clusters properly included in I, except for minimal clusters, when they specify minus the first and last row in cluster I. The matrix is reordered during the algorithm so that row clusters are contiguous sequences of rows. Similarly, for columns, the column clusters are $1, 2, \ldots, KC$, forming a tree described by the pair MINC(I), MAXC(I), which for the minimal clusters are minus the first and last columns in cluster I. Minus values of MINC and MAXC thus identify minimal clusters.

During the algorithm, the quantity VR(K) is the average variance of columns in the Kth data cluster, and the quantity VC(K) is the average variance of rows in the Kth data cluster. The quantity VR(K) is set zero if all column variances are less than a threshold T, and the quantity VC(K) is set zero if all row variances are less than T. At the conclusion of the algorithm, all VR(K) and VC(K) are zero.

At each step an average variance is computed for the Ith row cluster by averaging over all VR(K) for which VR$(K) \neq 0$, IR$(K) = I$. Similarly, an average variance is computed for the Ith column cluster by averaging over all VC(K) for which VC$(K) \neq 0$, IC$(K) = I$. That row or column cluster which has a maximum average variance is split by using the binary splitting algorithm. If the Ith row cluster is split, the data used will be the data clusters K such that IR$(K) = I$, VR$(K) \neq 0$. After the split, each of these clusters will be replaced by two clusters, and two new row clusters will exist. New values of VR and VC are computed within the new clusters, and the next step is performed.

The terminology is illustrated in Table 14.8.

STEP 1.   Initially KD $= 1$, MINR$(1) = -1$, MAXR$(1) = -M$, MINC$(1) = -1$, MAXC$(1) = -N$. Also KR $= 1$, KC $= 1$, IR$(1) = 1$, IC$(1) = 1$. The quantity VR(1) is the average variance of columns,

$$\text{VR}(1) = \sum \{1 \leq J \leq N\} \{1 \leq I \leq M\} \frac{[A(I, J) - \text{AVE}(J)]^2}{(M-1)N}$$

where AVE$(J) = \sum \{1 \leq I \leq M\} A(I, J)/M$. If all column variances are less than T, set VR$(1) = 0$. Similarly define VC(1), the average variance of rows.

STEP 2.   For each L ($1 \leq L \leq$ KR), define the quantities RSUM(L) and RDF(L) as follows. Initially set RSUM$(L) = $ RDF$(L) = 0$ ($1 \leq L \leq$ KR). For all K ($1 \leq K \leq$ KD), if VR$(K) \neq 0$, increase RSUM[IR(K)] by VR(K)\{MAXC[IC(K)] $-$ MINC[IC(K)] $+ 1$\} and increase RDF[IR(K)] by MAXC[IC(K)] $-$ MINC[IC(K)] $+$ 1. The quantity RSUM(L)/RDF(L) is then the average variance of the Lth row cluster over all data clusters with row margin equal to L and variances exceeding threshold.

Table 14.8 Two-Way Splitting Terminology

ROW CLUSTERS   : 1 = {1} ,   2 = {2} ,   3 = {1,2} ,   4 = {3,4} .

COLUMN CLUSTERS: 1 = {1,2} ,   2 = {3} ,   3 = {1,2,3} ,   4 = {4,5} .

DATA CLUSTERS   : 1 = {(1,1), (1,2), (1,3), (2,1), (2,2), (2,3)} ,

2 = {(1,4), (1,5)} ,   3 = {(2,4), (2,5)} ,   4 = {(3,1), (3,2), (4,1), 4,2)} ,

5 = {(3,3), (4,3)} ,   6 = {3,4), (3,5), (4,4), (4,5)} .

DATA CLUSTERS → ROW CLUSTERS   : IR(1) = 3 ,   IR(2) = 1 ,   IR(3) = 3 ,   IR(4) = 4 ,

IR(5) = 4 ,   IR(6) = 4 .

DATA CLUSTERS → COLUMN CLUSTERS: IC(1) = 3 ,   IC(2) = 4 ,   IC(3) = 4

IC(4) = 1 ,   IC(5) = 2 ,   IC(6) = 4 .

ROW TREE STRUCTURE   : MINR(1) = 1 ,   MAXR(1) = 1

MINR(2) = 2 ,   MAXR(2) = 2

MINR(3) = 1 ,   MAXR(3) = 2

MINR(4) = 3 ,   MAXR(4) = 4

COLUMN TREE STRUCTURE: MINC(1) = 1 ,   MAXC(1) = 2

MINC(2) = 3 ,   MAXC(2) = 3

MINC(3) = 1 ,   MAXC(3) = 2

MINC(4) = 4 ,   MAXC(4) = 5 .

STEP 3.   Similarly, compute for the Lth column cluster the average variance of rows in data clusters with column margin equal to L and row variances exceeding threshold. Let this average be $\text{CSUM}(L)/\text{CDF}(L)$.

STEP 4.   If the maximum value over all row and column clusters of $\text{RSUM}(L)/\text{RDF}(L)$ and $\text{CSUM}(L)/\text{CDF}(L)$ is zero, stop. Otherwise, choose that row or column cluster with maximum variance—say, the row cluster, L. Necessarily, L will be a minimal row cluster. Consider the data matrix obtained by combining data clusters K for which $\text{IR}(K) = L$, $\text{VR}(K) \neq 0$. The binary splitting algorithm is applied to the rows of this matrix to yield row clusters [after reordering of the rows $\text{MINR}(L)$, $\text{MINR}(L) + 1, \ldots, \text{MAXR}(L)$] $\{I1, I1 + 1, \ldots, I2\}$ and $\{I2 + 1, I2 + 2, \ldots, I3\}$.

STEP 5.   Now the clusters are updated as a result of the split. First, two new row clusters $KR + 1$ and $KR + 2$ are defined, with $MINR(KR + 1) = -I1$, $MINR(KR + 2) = -I2 + 1$, $MAXR(KR + 1) = -I2$, $MAXR(KR + 2) = -I3$. Also redefine $MINR(L) = KR + 1$, $MAXR(L) = KR + 2$. Each cluster K $(1 \leq K \leq KD)$ with $IR(K) = L$, $VR(K) \neq 0$ is itself destroyed, and generates two new clusters. Define $IR(K) = KR + 1$, increase KD by one, and define $IR(KD) = KR + 2$, $IC(KD) = IC(K)$. Also compute new average cluster variances VR and VC within the new clusters K and KD. Increase KR by 2. Return to Step 2. (Step 5 is analogous if a column cluster is split.)

14.7   TWO-WAY SPLITTING ALGORITHM APPLIED TO REPUBLICAN VOTE FOR PRESIDENT

(See Table 14.9 to follow the sequence of splits.)
   Choose a threshold variance, $T = 25$.

STEP 1.   Initialize the number of clusters, $KD = 1$, and the cluster boundaries $MINR(1) = -1, MAXR(1) = -6, MINC(1) = -1, MAXC(1) = -6$. The number

Table 14.9   Sequence of Splits of Republican Vote for President

INITIAL DATA

	32	36	40	60	64	68
SC	2	1	4	49	59	39
M	4	3	4	25	87	14
IA	7	11	14	29	57	23
KY	40	40	42	54	36	44
MD	36	37	41	46	35	42
MO	35	38	48	50	36	45

FIRST SPLIT

	32	36	40	60	68	64
SC	2	1	4	49	39	59
M	4	3	4	25	14	87
IA	7	11	14	29	23	57
KY	40	40	42	54	44	36
MD	36	37	41	46	42	35
MO	35	38	48	50	45	36

SECOND SPLIT

	32	36	40	60	68	64
SC	2	1	4	49	39	59
M	4	3	4	25	14	87
IA	7	11	14	29	23	57
KY	40	40	42	54	44	36
MD	36	37	41	46	42	35
MO	35	38	48	50	45	36

THIRD SPLIT

	32	36	40	60	64	68
M	4	3	4	25	14	87
SC	2	1	4	49	39	59
IA	7	11	14	29	23	57
KY	40	40	42	54	44	36
MD	36	37	41	46	42	35
MO	35	38	48	50	45	36

FOURTH SPLIT

	32	36	40	60	68	64
M	4	3	4	25	14	87
SC	2	1	4	49	39	59
IA	7	11	14	29	23	57
KY	40	40	42	54	44	36
MD	36	37	41	46	42	35
MO	35	38	48	50	45	36

FIFTH SPLIT

	32	36	40	60	68	64
M	4	3	4	25	14	87
SC	2	1	4	49	39	59
IA	7	11	14	29	23	57
KY	40	40	42	54	44	36
MD	36	37	41	46	42	35
MO	35	38	48	50	45	36

SIXTH SPLIT

	32	36	40	60	68	64
M	4	3	4	25	14	87
SC	2	1	4	49	39	59
IA	7	11	14	29	23	57
KY	40	40	42	54	44	36
MD	36	37	41	46	42	35
MO	35	38	48	50	45	36

SEVENTH SPLIT

	32	36	40	60	68	64
M	4	3	4	25	14	87
SC	2	1	4	49	39	59
IA	7	11	14	29	23	57
KY	40	40	42	54	44	36
MD	36	37	41	46	42	35
MO	35	38	48	50	45	36

FINAL (= SEVENTH) SPLIT

	32	36	40	60	68	64
M	4	3	4	25	14	87
SC	2	1	4	49	39	59
IA	7	11	14	29	23	57
KY	40	40	42	54	44	36
MD	36	37	41	46	42	35
MO	35	38	48	50	45	36

ROW CLUSTERS   (MS)((SC)(IA))(KY MD MO)
COL CLUSTERS   ((32 36)(40))((60)(68))(64)

of initial row clusters is KR = 1, and the number of initial column clusters is KC = 1. The row cluster corresponding to data cluster 1 is IR(1) = 1. The column cluster corresponding to data cluster 1 is IC(1) = 1.

STEP 2. The quantity RSUM(1) is the sum of all above-threshold variances within columns in data clusters whose marginal row cluster is 1. Thus RSUM(1) = 1827. The quantity RDF(1) is the number of such variances, RDF(1) = 6. Then RSUM(1)/RDF(1) = 304.

STEP 3. For column cluster 1, CSUM(1)/CDF(1) = 365.

STEP 4. The maximum value of RSUM(L)/RDF(L) and CSUM(L)/CDF(L) is 365. The cluster to be split is thus column cluster 1. The data clusters involved in the split consist of the single data cluster, 1. The binary splitting algorithm splits the columns into years {1932, 1936, 1940, 1960, 1968} and {1964}. After reordering, the clusters are {1, 2, 3, 4, 5} and {6} in the order 1932, 1936, 1940, 1960, 1968, 1964.

STEP 5. Two new column clusters 2, 3 are constructed with MINC(2) = −1, MAXC(2) = −5, MINC(3) = −6, MAXC(3) = −6. Redefine MINC(1) = 2, MAXC(1) = 3, denoting that column cluster 1 has divided into column clusters 2 and 3. The data cluster 1 is destroyed and replaced by two new data clusters. For the new data cluster 1, IR(1) = 1, IC(1) = 2, so that the new marginal column cluster is 2. For the new data cluster 2, IR(2) = 1, IC(2) = 3. Finally the average variances of rows and columns are computed in the new clusters: VR(1) = 281, VR(2) = 350, and VC(1) = 133, VC(2) = 0. The second-row variance VC(2) is zero because the rows in cluster 2 consist of a single element. Increase KC to 3 and KD to 2. Return to Step 2.

STEP 2 REPEATED. There is a single row cluster with RSUM(1)/RDF(1) = 304.

STEP 3 REPEATED. There are three column clusters, 1, 2, 3. Initially CSUM(L) = CDF(L) = 0 (1 ≤ L ≤ 3). For all K (1 ≤ K ≤ 2), increase CSUM[IC(K)] by VC(K){MAXR[IR(K)] − MINR[IR(K)] + 1} and increase CDF[IC(K)] by MAX[IR(K)] − MINR[IR(K)] + 1. That is, with K = 1, increase CSUM(2) by 133 × 5 and increase CDF(2) by 5, and with K = 2, since VC(2) = 0, make no change. The optimal value of CSUM(L)/CDF(L) is therefore 133 for L = 2.

STEP 4. The row or column cluster with maximum average column or row variance is row cluster 1, for which the average is 304. The binary splitting algorithm yields row clusters {SC, MS, LA} and {KY, MD, MO} or, without changing the ordering, rows {1, 2, 3} and {4, 5, 6}.

STEP 5. The clusters are updated by first adding row clusters 2,3 with MIN(2) = −1, MAXR(2) = −3, MINR(3) = −4, MAXR(3) = −6. The row cluster 1 splits into MINR(1) = 2 and MAXR(1) = 3. Each data cluster K (1 ≤ K ≤ 2) is destroyed and generates two new clusters. The cluster 1 has IR(1) = 2, IC(1) = 2. Increase KD to 3 with IR(3) = 3, IC(3) = 2. Compute VR(1) = 116, VC(1) = 268, VR(3) = 0, VC(3) = 31. Note that VR(2) = 0 because all columns in the third data cluster have variance less than the threshold, 25. In later row splits, the third data cluster will be ignored. For K = 2, IR(2) = 2, IC(2) = 3. Increase KD to 4 with IR(4) = 3, IC(4) = 4. Compute VR(2) = 281, VC(2) = 0, VR(4) = 0, VC(4) = 0. Increase KR to 3. Return to Step 2.

Table 14.10 Reduced Representation of Republican Vote Based on Two-Way Split

	32	36	40	60	68	64
MS	4			25	14	87
SC	2			49	39	58
LA	11		26			
KY	38		44	50	43	36
MD						
MO						

STEP 2 REPEATED. There are three row clusters, for which $\text{RSUM}(L) = 0$, $\text{RDF}(L) = 0$, initially. There are contributions from the four data clusters: for $K = 1$, $\text{IR}(1) = 2$, $\text{VR}(1) = 116$, so $\text{RSUM}(2)$ becomes 698, $\text{RDF}(2)$ becomes 5; for $K = 2$, $\text{IR}(2) = 2$, $\text{VR}(2) = 281$, so $\text{RSUM}(2)$ becomes 979, $\text{RDF}(2)$ becomes 6; for $K = 3$ and $K = 4$, $\text{VR} = 0$, so no change in RSUM occurs. The maximum value of $\text{RSUM}(L)/\text{RDF}(L) = 163$ at $L = 2$.

STEP 3 REPEATED. The maximum value of $\text{CSUM}(L)/\text{CDF}(L) = 149$ at $L = 2$.

STEP 4 REPEATED. Split row cluster 2. The clusters involved in the split are those with $\text{IR}(K) = 2$—that is, $K = 1$, $K = 2$. The split is into rows {MS} and {SC, LA}, or, after reordering the rows, {1} and {2, 3}.

STEP 5 REPEATED. New row clusters 4 and 5 are defined, with $\text{MINR}(4) = -1$, $\text{MAXR}(4) = -1$, $\text{MINR}(5) = -2$, $\text{MAXR}(5) = -3$. Redefine $\text{MINR}(2) = 4$, $\text{MAXR}(2) = 5$. Each of the clusters 1 and 2 is destroyed and replaced by two new clusters. Define $\text{IR}(1) = 4$, $\text{IC}(1) = 2$, $\text{IR}(5) = 5$, $\text{IC}(5) = 1$, increasing KD to 5. The new average variances are $\text{VR}(1) = 1$, $\text{VC}(1) = 90$, $\text{VR}(5) = 82$, $\text{VC}(5) = 85$. For cluster 2, define $\text{IR}(2) = 4$, $\text{IC}(2) = 3$ and $\text{IR}(6) = 5$, $\text{IC}(6) = 3$, increasing KD to 6. The new average variances are all zero. Increase KR to 5 and return to Step 2.

The splitting continues in this way until all within-cluster variances of rows and columns are within threshold. The complete sequence of splits for this data set is given in Table 14.9, and a reduced representation of the data on the fifteen final clusters is given in Table 14.10. The basic conclusions are that KY, MD, and MO are similar in all years. The years 1932, 1936, and 1940 are similar in each of the states MS, SC, and LA, but {1932, 1936} differ from {1940} in the states KY, MD, and MO.

14.8 THINGS TO DO

14.8.1 Running the Splitting Algorithm

The binary splitting algorithm is a more rigorous version of the K-means algorithm, applicable to the same types of data. Use, for example, the mammal's milk or the New Haven school data. The one-way splitting algorithm treats variables and cases differently, ending with a hierarchical clustering of cases in which each cluster is characterized by certain variables being approximately constant within the cluster. The variables may be measured on different scales, but they should be rescaled before application of the algorithm so that an error of one unit means the same for each variable. Acidosis patients (Table 14.11) are suggested as a trial data set.

Table 14.11   Acidosis Patients

FIRST VARIABLE = PH IN CEREBROSPINAL FLUID (nanomol/litre)

SECOND VARIABLE = PH IN BLOOD (nanomol/litre)

THIRD VARIABLE = HCO3 IN CEREBROSPINAL FLUID (millimoles/litre)

FOURTH VARIABLE = HCO3 IN BLOOD (millimoles/litre)

FIFTH VARIABLE = CO2 IN CEREBROSPINAL FLUID (mm mercury)

SIXTH VARIABLE = CO2 IN BLOOD (mm mercury)

39.8	38.0	22.2	23.2	38.8	36.5
53.7	37.2	18.7	18.5	45.1	28.3
47.3	39.8	23.3	22.1	48.2	36.4
41.7	37.6	22.8	22.3	41.6	34.6
44.7	38.5	24.8	24.4	48.5	38.8
47.9	39.8	22.0	23.3	46.2	38.5
48.4	36.7	21.0	21.3	44.5	32.6
48.4	35.1	23.9	24.0	50.6	35.0
48.4	45.7	18.6	14.9	39.4	28.8
41.7	81.3	9.8	4.2	17.8	12.9
46.2	42.7	15.5	15.0	31.3	26.8
48.4	42.2	19.6	18.7	41.6	32.6
49.6	55.0	14.6	11.3	31.8	25.7
47.3	59.4	10.4	7.5	21.5	18.6
42.7	49.0	15.3	9.5	26.9	19.0
38.5	47.9	13.7	9.4	23.0	18.7
46.2	36.1	23.2	27.3	46.9	39.8
51.3	39.6	23.1	26.8	52.3	43.3
49.0	40.1	18.9	20.0	40.4	32.4
46.0	41.4	18.9	20.0	44.8	40.1
50.9	40.8	23.3	25.5	52.0	42.3
50.0	41.4	24.6	25.2	53.8	42.3
49.0	39.5	24.5	26.4	52.4	42.3
49.4	40.9	22.9	28.5	53.1	47.2
47.2	38.5	27.2	26.4	54.9	40.9
47.7	38.0	26.2	29.2	53.5	44.6
49.0	41.9	27.6	28.9	58.0	49.0
53.1	39.4	26.2	33.0	59.9	52.2
52.5	53.8	29.4	32.0	66.2	56.9
51.3	41.7	28.4	31.1	62.6	52.4
52.7	43.8	30.4	34.6	69.0	61.4
48.2	38.6	29.4	34.0	60.6	52.6
42.7	37.2	20.7	18.6	38.6	29.2
44.2	33.9	20.7	19.1	40.3	26.8
43.6	35.5	21.9	21.8	41.7	32.2
49.0	33.9	22.4	23.3	46.9	30.6
54.9	33.8	22.9	34.8	45.1	32.1
46.6	37.3	22.5	24.9	44.8	37.4
47.5	36.4	22.3	22.9	45.4	33.3
44.3	32.8	22.8	24.8	42.8	32.5

Table 14.12 Profitability of Sectors of U.S. Economy

Profit as a percentage of stockholders' equity.

```
 2 ALL MANUFACTURING CORPORATIONS (EXCEPT NEWSPAPERS)
 3 TOTAL CURABLE
 4 MOTOR VEHICLES AND EQUIPMENT.
 5 ELECTRICAL MACHINERY, EQUIPMENT, AND SUPPLIES.
 6 MACHINERY (EXCEPT ELECTRICAL).
 7 FABRICATED METAL PRODUCTS
 8 PRIMARY IRON AND STEEL INDUSTRY
 9 PRIMARY NON-FERROUS METAL INDUSTRIES.
10 STONE, CLAY, AND GLASS PRODUCTS
11 FURNITURE AND FIXTURES.
12 LUM ER AND WOOD PRODUCTS (EXCEPT FURNITURE).
13 INSTRUMENTS AND RELATED PRODUCTS.
14 MISCELLANEOUS MANUFACTURING (INCLUDING ORDNANCE).
 2 TOTAL NON-DURABLE
 3 FOOD AND KINDRED PRODUCTS
 4 TOBACCO MANUFACTURES.
 5 TEXTILE MILL PRODUCTS
 6 APPAREL AND RELATED PRODUCTS
 7 PAPER AND ALLIED PRODUCTS.
 8 PRINTING AND PUBLISHING (EXCEPT NEWSPAPERS).
 9 CHEMICALS AND ALLIED PRODUCTS.
10 PETROLEUM REFINING.
11 RUBBER AND MISCELLANEOUS PLASTIC PRODUCTS.
12 LEATHER AND LEATHER PRODUCTS.
```

		59	60	61	62	63	64	65	66	67	68
2	ALL	10.	9.	9.	10.	10.	12.	13.	13.	12.	12.
3	TOTA	10.	9.	8.	10.	10.	12.	14.	14.	12.	12.
4	MOTO	14.	14.	11.	16.	17.	17.	20.	16.	12.	15.
5	ELEC	13.	10.	9.	10.	10.	11.	14.	15.	13.	12.
6	MACH	10.	8.	8.	9.	10.	13.	14.	15.	13.	12.
7	FABR	8.	6.	6.	8.	8.	10.	13.	15.	13.	12.
8	PRIM	8.	7.	6.	5.	7.	9.	10.	10.	8.	8.
9	PRIM	8.	7.	7.	8.	8.	10.	12.	15.	11.	11.
10	STON	13.	10.	9.	9.	9.	10.	10.	10.	8.	9.
11	FUR	9.	7.	5.	8.	8.	10.	13.	14.	12.	12.
12	LUM	9.	4.	4.	6.	8.	10.	10.	10.	9.	15.
13	INS	13.	12.	11.	12.	12.	14.	18.	21.	18.	17.
14	MIS	9.	9.	10.	9.	9.	10.	11.	15.	13.	12.
2	TOTA	10.	10.	10.	10.	10.	12.	12.	13.	12.	12.
3	FOOD	9.	9.	9.	9.	9.	10.	11.	11.	11.	11.
4	TOBA	13.	13.	14.	13.	13.	13.	14.	14.	14.	14.
5	TEXT	8.	6.	5.	6.	6.	9.	11.	10.	8.	9.
6	APPA	9.	8.	7.	9.	8.	12.	13.	13.	12.	13.
7	PAPE	10.	9.	8.	8.	8.	9.	9.	11.	9.	10.
8	PRIN	11.	11.	9.	10.	9.	13.	14.	16.	13.	13.
9	CHEM	14.	12.	12.	12.	12.	14.	15.	15.	13.	13.
10	PET	10.	10.	10.	10.	11.	11.	12.	12.	13.	12.
11	RUB	11.	9.	9.	10.	9.	11.	12.	12.	10.	12.
12	LEA	9.	6.	4.	7.	7.	11.	12.	13.	12.	13.

Table 14.13   Connecticut Votes for President

From Scammon, R. M. (1965). *America at the Polls*, University of Pittsburgh, Pittsburgh.

	1920			1924			1928		
	Rep.	Dem.	Other	Rep.	Dem.	Other	Rep.	Dem.	Other
FAIRFIELD	55251	24761	3101	58041	18815	10791	71410	55491	1047
HARTFORD	54046	30287	4646	61381	28139	9622	75997	65789	1169
LITCHFIELD	14405	6938	504	15499	6645	3129	19157	10766	138
MIDDLESEX	8447	4170	331	9383	4009	995	11205	7380	115
NEW HAVEN	65938	37977	5559	69164	36247	15108	80952	82657	1439
NEW LONDON	17422	9209	889	18205	8615	2386	21378	16299	292
TOLLAND	5135	2308	322	5161	2239	885	6502	4256	126
WINDHAM	8594	5071	207	9488	5475	974	10040	9447	66

	1932			1936			1940		
	72238	64367	8092	67846	87329	8088	91190	93688	829
	72611	72322	5220	65652	103450	7216	88155	114336	462
	18682	13469	660	18850	17468	875	22956	19537	49
	10770	9286	344	10925	12294	359	13447	13044	39
	79019	86826	8296	76614	117308	10689	103100	126072	517
	19721	19576	858	21367	24999	1337	23389	28286	98
	5857	4985	455	5965	6676	488	7503	7669	25
	9522	10801	206	11466	12605	857	12079	14989	43

	1944			1948			1952		
	103693	99181	2423	118636	90767	7669	167278	106403	1814
	95224	127841	1153	105262	124874	5157	150332	146551	832
	24019	19212	248	26848	18628	823	35735	20163	107
	14315	13551	176	16119	14609	537	22157	15722	74
	108883	123450	1811	120769	121591	6633	165917	136476	1149
	24153	29304	285	27416	29425	973	38148	31374	148
	8208	7721	117	9012	7970	347	13466	9425	52
	12032	14886	104	13692	15433	328	17979	15535	74

	1956			1960			1964		
	199841	84890	116	167778	146442	6	125576	194782	261
	175894	126923	25	136459	195403	2	88811	240071	622
	40029	17226	6	34043	29062	3	20834	40172	62
	25496	13851	8	22045	22158	1	14697	30517	71
	191215	112208	36	136852	188685	2	97656	218743	171
	43453	27317	11	38070	40625	1	24391	54551	49
	15880	9111	1	15386	14575	0	9951	22195	58
	20029	13553	2	15180	20105	0	9080	25233	19

The two-way splitting algorithm produces more beautiful pictures, but it treats variables and cases symmetrically. Thus the data must be prescaled so that all data values are comparable. (This requirement restricts the range of data types to which the algorithm may be applied.) Suitable data sets are the profitability of various sectors of the economy (Table 14.12) and the votes for President in Connecticut (Table 14.13). The votes should be converted to percentage Republican.

14.8.2 Missing Values

The two-way splitting algorithm permits an estimation of missing values once the clustering is complete. A missing value is estimated by the mean of the smallest data cluster containing it.

14.8.3 Error Analysis

An informal error analysis is provided by repeating the algorithm on the data several times, at each run declaring each value missing with probability 0.5. Clusters that are stable or reliable will reappear (with perhaps a little change) in most of the runs. This technique prevents finding and interpreting excessive numbers of clusters.

14.8.4 Range

Instead of using variance to measure spread within a cluster, it is plausible to use the range or the mean deviation from the median. These measures will be less sensitive to a few large values, which might dominate the choice of the first few splits, even though they will be isolated in small pockets later. Both measures are more convenient than the variance for small hand calculations.

14.8.5 Binary Splits and Range

Divide a set of M points in N dimensions into two clusters to minimize the sum of the ranges within clusters summed over all variables. Show that the division conforms to order for at least one variable, so that each value of the variable in one cluster is less than or equal to each value of the variable in the other cluster. Using the range in the one-way splitting algorithm thus allows identification of a few variables for each split which conform to the split.

14.8.6 Experiments

It would be interesting to know whether the binary split algorithm often produces better clusters than the K-means algorithm. This should be tried for various numbers of cases, numbers of variables, and data types. Various normal mixtures would be plausible trial data.

For the two-way splitting algorithm, begin with an 8×6 data matrix divided into four data clusters. In each cluster the value is constant plus normal noise for each data value. See how well the data clusters are recovered for various choices of constants and noise variance.

14.8.7 One-Way Model

The one-way splitting algorithm partitions the data matrix into data clusters in which values are constant, but the corresponding marginal column cluster always consists of a single variable. This is necessary because values are not directly comparable between variables.

REFERENCES

FISHER, W. D. (1969). *Clustering and Aggregation in Economics*, Johns Hopkins Book, Baltimore, Md. Input-output matrices in economics are data matrices in which both rows and columns refer to industrial sectors. In reducing such a matrix, it is necessary to cluster rows and columns simultaneously; thus, if rows 1 and 2 are joined, columns 1 and 2 must also be joined. Fisher studies an algorithm for this called the "lockstep" progressive merger procedure.

GOOD, I. J. (1965). "Categorization of classifications." *Mathematics and computer science in biology and medicine*, H.M. Stationery Office, London. Good suggests removing row and column means from a two-way table and then generating a two-way split (into four data clusters) of the residuals by maximizing $\sum r_i c_j A_{ij}$, where $\{r_i\}$ is a row vector of ± 1's, corresponding to row clusters, $\{c_j\}$ is a column vector of ± 1's corresponding to column clusters, and $\{A_{ij}\}$ is the residual matrix. This technique is to be repeated on each of the clusters obtained.

HARTIGAN, J. A. (1972). "Direct clustering of a data matrix." *J. Am. Stat. Assoc.* **67**, 123–129. An algorithm similar to the two-way splitting algorithm is proposed. At each stage in the algorithm, the matrix is divided into a number of clusters, each of which is a submatrix corresponding to a row cluster and a column cluster. A single one of these clusters is split, by rows or by columns, at each stage in the algorithm.

SONQUIST, J. A. (1971). *Multivariate Model Building: The Validation of a Search Strategy*. Institute for Social Research, The University of Michigan, Ann Arbor, Mich. The automatic interaction detection technique, quoting from p. 20: "The technique is a step-wise application of a one-way analysis of variance model. Its objective is to partition the sample into a series of non-overlapping subgroups whose means explain more of the variation in the dependent variable than any other such set of subgroups." A set of objects is given on which are defined a dependent variable Y and a number of independent variables $X_1, X_2, \ldots, X_p$. During the algorithm, a partition $1, 2, 3, \ldots, K$ of the objects is constructed by successively splitting.

STEP 1. Begin with an initial partition consisting of the set of all objects, $K = 1$.

STEP 2. If all members of the partition have a sum of squared deviations of Y less than P, stop. If $K > R$, stop.

STEP 3. Choose an arbitrary member of the partition not tried before—say, I. For each variable X_J $(1 \leq J \leq P)$ consider splits of I into two clusters, in the first of which $X_J \in A_J$ and in the second of which $X_J \notin A_J$. (The possible forms of the set A_J depend on the type of variable A_J. If X_J is a category variable, A_J is any subset of the possible values of X_J; if X_J is an ordered variable, A_J is of form $\{X_J \mid X_J \leq C_J\}$.) The split with largest between-cluster sum of squares over all X_J and over all A_J is executed, provided this between-cluster sum of squares exceeds Q. Return to Step 2.

This algorithm has been applied to a large number of problems, especially in the social sciences. The book contains a number of real examples, some careful studies validating the technique (using random data), and comparisons with other techniques.

TRYON, R. C., and BAILEY, D. E. (1970). *Cluster Analysis*, McGraw-Hill, New York. Variables are first clustered in a factor analytic approach, and a few factors are obtained. By using these factors as variables in the euclidean distance calculations, objects are clustered by algorithms of the K-means type.

PROGRAMS

SPLIT splits rows into two clusters to approximately minimize the within-cluster sum of squares averaged across the columns.

SPLIT1   splits rows until all variables have sufficiently small variance within all clusters.
OUT1   prints output from SPLIT1.
SPLIT2   partitions data matrix into blocks within which all row and column variances are sufficiently small.
CSPLIT   splits columns in SPLIT2.
RSPLIT   splits rows in SPLIT2.

```
      SUBROUTINE SPLIT1(A,W,M,N,NB,K,KC,TH,NC,XX)
C.................................................................20 MAY 1973
C.... SPLITS ARRAY A INTO TWO ROW CLUSTERS, THEN SPLITS EACH OF THESE,
C.... CONTINUING UNTIL EVERY VARIABLE HAS VARIANCE LESS THAN THE THRESHOLD TH
C.... WITHIN EACH CLUSTER.
C.... USES SUBROUTINE SPLIT.  AFTER RUNNING SPLIT1, USE OUT1 FOR PRINTING BLOCKS
C.... A = M BY N BORDERED DATA MATRIX,PERMUTED ON OUTPUT
C.... W = M BY N BORDERED WEIGHT MATRIX ( SET W(I,J)=0 FOR MISSING VALUES)
C.... NB = 4 BY K BLOCK ARRAY, EACH BLOCK IS A ROW CLUSTER BY A VARIABLE
C            NB(1,K)=FIRST ROW IN BLOCK
C            NB(2,K)=LAST ROW IN BLOCK
C            NB(3,K)=NB(4,K)=COLUMN IN BLOCK
C.... M = NUMBER OF ROWS
C.... N = NUMBER OF COLUMNS
C.... K = NUMBER OF BLOCKS,NEVER EXCEEDS M*N.
C.... KC = ACTUAL NUMBER OF BLOCKS
C.... NC = 2 BY M SCRATCH ARRAY
C.... XX = 4 BY N SCRATCH ARRAY
C.... TH = THRESHOLD VARIANCE FOR VARIABLES WITHIN CLUSTERS
C..............................................................................
      DIMENSION A(M,N),W(M,N),NB(4,K),NC(2,M)
      DIMENSION XX(4,N)
C.... INITIALIZE CLUSTER OF ALL ROWS
      NC(1,1)=2
      NC(2,1)=M
      KR=0
      KC=0
   50 KR=KR+1
      IF(KR.EQ.0) RETURN
      SP=0
      IL=NC(1,KR)
      IU=NC(2,KR)
C.... IDENTIFY VARIABLES WITHIN THRESHOLD WITHIN CLUSTER, DEFINE BLOCKS.
      DO 20 J=2,N
      W(1,J)=1
      S1=0
      S2=0
      S3=0
      DO 21 I=IL,IU
      IF(W(I,J).EQ.0) GO TO 21
      S1=S1+W(I,J)
      S2=S2+W(I,J)*A(I,J)
      S3=S3+W(I,J)*A(I,J)**2
   21 CONTINUE
         IF(S1.EQ.0) GO TO 19
      S2=S2/S1
      S3=S3/S1-S2**2
      IF(S3.GT.TH) SP=1
      IF(S3.GT.TH) GO TO 20
      KC=KC+1
      NB(1,KC)=IL
      NB(2,KC)=IU
      NB(3,KC)=J
      NB(4,KC)=J
      DO 22 I=IL,IU
   22 W(I,J)=0
   19 W(1,J)=0
   20 CONTINUE
C.... SPLIT CLUSTER KR IF NECESSARY
      IF(SP.EQ.0) KR=KR-2
      IF(SP.EQ.0) GO TO 50
      CALL SPLIT(W,A,M,N,IL,IU,IM,XX,DM)
      NC(2,KR+1)=NC(2,KR)
      NC(2,KR)=IM
      NC(1,KR+1)=IM+1
      GO TO 50
      END
```

```
      SUBROUTINE SPLIT(W,A,M,N,IL,IU,IM,XX,DM)
C...................................................................20 MAY 1973
C.... USED IN SPLIT1 TO SPLIT A ROW CLUSTER ON SELECTED VARIABLES
C.... M = NUMBER OF ROWS
C.... N = NUMBER OF COLUMNS
C.... A = M BY N BORDERED ARRAY
C.... XX = 4 BY N MEANS MATRIX
C.... DM = REDUCTION IN WITHIN CLUSTER SUM OF SQUARE
C.... W = M BY N WEIGHT MATRIX
C.... W(I,J)=0 MEANS A(I,J) WILL BE IGNORED
C            W(I,1)=0, I NOT ASSIGNED(SPECIFIED ON INPUT)
C            W(I,1)=1, I IN FIRST CLUSTER, SPECIFIED ON OUTPUT
C            W(I,1)=2, I IN SECOND CLUSTER, SPECIFIED ON OUTPUT
C...................................................................
      DIMENSION W(M,N),A(M,N),XX(4,N)
C.... FIND MEAN OF ALL CASES
      TH=10.**(-6)
      DO 30 J=2,N
      XX(1,J)=0
      XX(3,J)=0
      XX(2,J)=TH
      XX(4,J)=TH
   30 CONTINUE
      DO 32 J=2,N
      IF(W(1,J).EQ.0) GO TO 32
      DO 31 I=IL,IU
      XX(1,J)=XX(1,J)+A(I,J)*W(I,J)
   31 XX(2,J)=XX(2,J)+W(1,J)
   33 XX(1,J)=XX(1,J)/XX(2,J)
   32 CONTINUE
      DM=0
      DD=0
      DO 50 IC=IL,IU
      II=IU-IC+IL
      DMAX=-10.**10
      IMAX=II
C.... EFFECT OF MOVING ITH CASE
      DO 51 I=IL,II
      D=0
      DO 57 J=2,N
      IF(W(1,J).EQ.0) GO TO 57
      IF(XX(2,J).EQ.W(I,J)) XX(2,J)=W(I,J)+TH
      D=D+W(I,J)*XX(2,J)*(A(I,J)-XX(1,J))**2/(XX(2,J)-W(I,J))
      D=D-W(I,J)*XX(4,J)*(A(I,J)-XX(3,J))**2/(XX(4,J)+W(I,J))
   57 CONTINUE
      IF(D.LE.DMAX) GO TO 51
      IMAX=I
      DMAX=D
   51 CONTINUE
      DD=DD+DMAX
      IF(DD.GT.DM) IM=II-1
      IF(DD.GT.DM) DM=DD
C.... UPDATE MEANS OF TWO CLUSTERS
      I=IMAX
      W(IMAX,1)=2
      DO 55 J=2,N
      IF(W(1,J).EQ.0) GO TO 55
      XX(2,J)=XX(2,J)-W(I,J)
      IF(XX(2,J).LT.TH) XX(2,J)=TH
      XX(1,J)=XX(1,J)+(XX(1,J)-A(I,J))*W(I,J)/XX(2,J)
      XX(4,J)=XX(4,J)+W(I,J)
      XX(3,J)=XX(3,J)-(XX(3,J)-A(I,J))*W(I,J)/XX(4,J)
   55 CONTINUE
C.... INTERCHANGE SELECTED ROW WITH LAST FEASIBLE ROW
      DO 56 J=1,N
      C=A(I,J)
      A(I,J)=A(II,J)
      A(II,J)=C
      C=W(I,J)
      W(I,J)=W(II,J)
   56 W(II,J)=C
   50 CONTINUE
C.... DEFINE CLUSTERS
      RETURN
      END
```

```
      SUBROUTINE OUT1(A,M,N,NB,KC)
C...........................................................20 MAY 1973
C.... USE AFTER SPLIT1.
C.... PRINTS OUT MATRIX A WHEN BLOCKS NB EACH CONTAIN ONLY ONE COLUMN
C.... A = M BY N BORDERED ARRAY
C.... M = NUMBER OF ROWS
C.... N = NUMBER OF COLUMNS
C.... NB = 4 BY KC BLOCK ARRAY
C          NB(1,K)=FIRST ROW
C          NB(2,K)=LAST ROW
C          NB(3,K)=NB(4,K)=COLUMN
C.... KC = NUMBER OF BLOCKS
C..........................................................................
      DIMENSION A(M,N),NB(4,KC)
      DIMENSION AA(20)
      DATA BLANK,DASH/4H    ,4H----/
      WRITE(6,1)A(1,1)
    1 FORMAT(18H ONE WAY SPLIT OF ,A4)
      NN=(N-1)/20+1
      DO 70 LL=1,NN
      JL=(LL-1)*20+1
      JU=LL*20
      IF(JU.GT.N) JU=N
      WRITE(6,2)(A(1,J),J=JL,JU)
      WRITE(6,2)(DASH,J=JL,JU)
    2 FORMAT(2X,20(A4,2X))
      DO 60 I=2,M
      WRITE(6,3)(A(I,J),J=JL,JU)
    3 FORMAT(1X,A4,1X,20F6.2/(7X,20F6.2))
      DO 50 L=1,20
   50 AA(L)=BLANK
      NC=0
      DO 30 K=1,KC
      L=0
      IF(NB(2,K).EQ.I) L=NB(3,K)
      IF(L.GE.JL.AND.L.LE.JU) AA(L-JL+1)=DASH
   30 IF(L.NE.0) NC=1
      IF(NC.EQ.0) GO TO 60
      WRITE(6,2)(AA(L),L=1,20)
   60 CONTINUE
      WRITE(6,5)
    5 FORMAT(1H1)
   70 CONTINUE
      RETURN
      END
```

```
      SUBROUTINE SPLIT2(A,M,N,NB,KD,KA,XR,XC,TH,IORD)
C..................................................................20 MAY 1973
C.... SPLITS ARRAY A BY ROWS AND COLUMNS SIMULTANEOUSLY.  ALL DATA VALUES SHOULD
C.... BE COMPARABLE.  MISSING VALUES ARE REPRESENTED BY A(I,J)=99999.
C.... USES RSPLIT,CSPLIT
C.... A = M BY N BORDERED ARRAY
C.... M = NUMBER OF ROWS
C.... N = NUMBER OF COLUMNS
C.... KD = MAXIMUM NUMBER OF BLOCKS(NEVER EXCEEDS M*N)
C.... KA = COMPUTED NUMBER OF BLOCKS
C.... NB = 4 BY KD BLOCK ARRAY GIVING FIRST AND LAST ROWS AND COLUMNS
C.... XR = 9 BY M SCRATCH ARRAY
C          XR(1,I) = FIRST ROW IN ROW CLUSTER I
C          XR(2,I) = LAST ROW IN ROW CLUSTER I
C          XR(3,I) = REDUCTION IN SSQ DUE TO SPLITTING
C          XR(4,I) = LAST ROW IN FIRST CLUSTER OF SPLIT OF I
C          XR(5,I) = 1 IF ROW IS INCLUDED IN PRESENT COLUMN SPLIT
C          XR(6,I) = NUMBER OF COLUMNS IN ITH ROW OF PRESENT COLUMN SPLIT
C          XR(7,I) = MEAN OF ITH ROW, FIRST COLUMN CLUSTER
C          XR(8,I)= NUMBER OF COLUMNS, SECOND CLUSTER
C          XR(9,I) = MEAN OF ITH ROW, SECOND CLUSTER
C.... XC = 9 BY N SCRATCH ARRAY,USED LIKE XR
C.... TH = THRESHOLD(EACH ROW AND COLUMN VARIANCE WITHIN A BLOCK  LESS THAN)
C          IORD = 0 ROWS AND COLUMNS PERMUTED
C          IORD = 1 LEAVE ROW ORDER
C          IORD = 2 LEAVE COLUMN ORDER
C          IORD = 3 LEAVE ROW AND COLUMN ORDER
C..................................................................................
      DIMENSION A(M,N),NB(4,KD),XR(9,M),XC(9,N)
C.... INITIALIZE BLOCKS AND ROW AND COLUMN CLUSTERS
      XR(1,1)=2
      XR(2,1)=M
      XC(1,1)=2
      XC(2,1)=N
      KR=1
      KC=1
      KA=1
      NB(1,1)=2
      NB(2,1)=M
      NB(3,1)=2
      NB(4,1)=N
      IR=1
      IC=1
      K=KD
      CALL RSPLIT(A,M,N,NB,K,KA,XR,XC,TH,IR,IORD)
      CALL CSPLIT(A,M,N,NB,K,KA,XR,XC,TH,IC,IORD)
   70 CONTINUE
C.... FIND BEST ROW OR COLUMN SPLIT
      IB=1
      XB=0
      DO 60 I=1,KR
      IF(XR(3,I).LE.XB) GO TO 60
      XB=XR(3,I)
      IB=I
   60 CONTINUE
      DO 61 J=1,KC
      IF(XC(3,J).LE.XB) GO TO 61
      XB=XC(3,J)
      IB=J+M
   61 CONTINUE
      IF(XB.EQ.0) GO TO 80
C.... SPLIT ROW CLUSTER
      KKC=KA
      IF(IB.GT.M) GO TO 64
      IL=XR(1,IB)
      IU=XR(2,IB)
      IM=XR(4,IB)
      DO 62 K=1,KA
      IF(NB(1,K).NE.IL.OR.NB(2,K).NE.IU) GO TO 62
      KKC=KKC+1
      NB(1,KKC)=IM+1
      NB(2,KKC)=NB(2,K)
      NB(2,K)=IM
      NB(3,KKC)=NB(3,K)
      NB(4,KKC)=NB(4,K)
   62 CONTINUE
      KA=KKC
      XR(2,IB)=IM
```

274

```
         KR=KR+1
         XR(1,KR)=IM+1
         XR(2,KR)=IU
         CALL RSPLIT(A,M,N,NB,KD,KA,XR,XC,TH,IB,IORD)
         CALL RSPLIT(A,M,N,NB,KD,KA,XR,XC,TH,KR,IORD)
         GO TO 70
      64 CONTINUE
C.... SPLIT COLUMN CLUSTER
         JB=IB-M
         JL=XC(1,JB)
         JU=XC(2,JB)
         JM=XC(4,JB)
         DO 65 K=1,KA
         IF(NB(3,K).NE.JL.OR.NB(4,K).NE.JU) GO TO 65
         KKC=KKC+1
         NB(3,KKC)=JM+1
         NB(4,KKC)=NB(4,K)
         NB(4,K)=JM
          NB(1,KKC)=NB(1,K)
         NB(2,KKC)=NB(2,K)
      65 CONTINUE
         KA=KKC
         XC(2,JB)=JM
         KC=KC+1
         XC(1,KC)=JM+1
         XC(2,KC)=JU
         CALL CSPLIT(A,M,N,NB,KD,KA,XR,XC,TH,KC,IORD)
         CALL CSPLIT(A,M,N,NB,KD,KA,XR,XC,TH,JB,IORD)
         GO TO 70
      80 CONTINUE
         DO 81 K=1,KA
         DO 81 J=1,4
      81 IF(NB(J,K).LT.0) NB(J,K)=-NB(J,K)
         RETURN
         END
```

```
      SUBROUTINE CSPLIT(A,M,N,NB,KD,KA,XR,XC,TH,IR,IORD)
C.............................................................20 MAY 1973
C.... SPLITS COLUMNS, FOR ARGUMENT DESCRIPTION SEE SPLIT2.
C...........................................................................
      DIMENSION A(M,N),NB(4,KD),XR(9,M),XC(9,N)
      XM=99999.
      DO 23 I=2,M
   23 XR(5,I)=0
C.... LOOK FOR BLOCKS WITHIN THRESHOLD
      JL=XC(1,IR)
      JU=XC(2,IR)
      DO 20 K=1,KA
      IF(NB(3,K).NE.JL.OR.NB(4,K).NE.JU) GO TO 20
      IL=NB(1,K)
      IF(IL.LT.0) IL=-IL
      IU=NB(2,K)
C.... COMPUTE VARIANCES
      NC=0
      DO 21 I=IL,IU
      S1=0
      S2=0
      S3=0
      DO 22 J=JL,JU
      IF(A(I,J).EQ.XM) GO TO 22
      S1=S1+1
      S2=S2+A(I,J)
      S3=S3+A(I,J)**2
   22 CONTINUE
      XR(6,I)=S1
      IF(S1.NE.0) XR(7,I)=S2/S1
      IF(S1.NE.0) S3=S3/S1-(S2/S1)**2
      IF(S3.GT.TH) XR(5,I)=1
      IF(S3.GT.TH) NC=1
   21 CONTINUE
      IF(NC.EQ.0) NB(3,K)=-NB(3,K)
   20 CONTINUE
C.... FIND BEST COLUMN SPLIT
      DO 30 I=2,M
      XR(8,I)=0
   30 XR(9,I)=0
      DM=0
      XC(3,IR)=0
      XC(4,IR)=JL
      DO 31 J=JL,JU
      IF(J.EQ.JU) GO TO 31
      JJ=JU-J+JL
      JD=JJ
      DD=-10.**30
      DO 32 L=JL,JJ
      IF(IORD.GE.2.AND.L.NE.JJ) GO TO 32
      DL=0
      DO 33 I=2,M
      IF(XR(5,I).EQ.0) GO TO 33
      IF(A(I,L).EQ.XM) GO TO 33
      DL=DL+(A(I,L)-XR(7,I))**2*(XR(6,I)+1)/XR(6,I)
      DL=DL-(A(I,L)-XR(9,I))**2*XR(8,I)/(XR(8,I)+1.)
   33 CONTINUE
      IF(DL.LE.DD) GO TO 32
      DD=DL
      JD=L
   32 CONTINUE
   34 CONTINUE
C.... INTERCHANGE JD AND JJ
      DO 35 I=1,M
      C=A(I,JJ)
      A(I,JJ)=A(I,JD)
   35 A(I,JD)=C
C.... UPDATE MEANS
      DO 36 I=2,M
      IF(XR(5,I).EQ.0) GO TO 36
      IF(A(I,JJ).EQ.XM) GO TO 36
      XR(6,I)=XR(6,I)-1
      IF(XR(6,I).NE.0) XR(7,I)=XR(7,I)+(XR(7,I)-A(I,JJ))/XR(6,I)
      XR(8,I)=XR(8,I)+1
      XR(9,I)=XR(9,I)-(XR(9,I)-A(I,JJ))/XR(8,I)
   36 CONTINUE
      DM=DM+DD
      IF(DM.LT.XC(3,IR)) GO TO 31
      XC(3,IR)=DM
      XC(4,IR)=JJ-1
   31 CONTINUE
      RETURN
      END
```

276

```
      SUBROUTINE RSPLIT(A,M,N,NB,KD,KA,XR,XC,TH,IR,IORD)
C........................................................................20 MAY 1973
C.... FIND OPTIMAL ROW SPIT, FOR ARGUMENT DESCRIPTION SEE SPLIT2
C........................................................................
      DIMENSION A(M,N),NB(4,KD),XR(9,M),XC(9,N)
      XM=99999.
      DO 23 J=2,N
   23 XC(5,J)=0
C.... LOOK FOR BLOCKS WITHIN THRESHOLD
      IL=XR(1,IR)
      IU=XR(2,IR)
      DO 20 K=1,KA
      IF(NB(1,K).NE.IL.OR.NB(2,K).NE.IU) GO TO 20
      JL=NB(3,K)
      JU=NB(4,K)
      IF(JL.LT.0) JL=-JL
C.... COMPUTE VARIANCES
      NC=0
      DO 21 J=JL,JU
      S1=0
      S2=0
      S3=0
      DO 22 I=IL,IU
      IF(A(I,J).EQ.XM) GO TO 22
      S1=S1+1
      S2=S2+A(I,J)
      S3=S3+A(I,J)**2
   22 CONTINUE
      XC(6,J)=S1
      IF(S1.NE.0) S3=S3/S1-(S2/S1)**2
      IF(S1.NE.0) XC(7,J)=S2/S1
      IF(S3.GT.TH) XC(5,J)=1
      IF(S3.GT.TH) NC=1
   21 CONTINUE
      IF(NC.EQ.0) NB(1,K)=-NB(1,K)
   20 CONTINUE
C.... FIND BEST ROW SPLIT
      DO 30 J=2,N
      XC(8,J)=0
   30 XC(9,J)=0
      DM=0
      XR(3,IR)=0
      XR(4,IR)=IL
      DO 31 I=IL,IU
      IF(I.EQ.IU) GO TO 31
      II=IU-I+IL
      ID=II
      DD=-10.**30
      DO 32 L=IL,II
      IF(IORD.EQ.1.OR.IORD.EQ.3.AND.L.NE.II) GO TO 32
      DL=0
      DO 33 J=2,N
      IF(XC(5,J).EQ.0) GO TO 33
      IF(A(L,J).EQ.XM) GO TO 33
      DL=DL+(A(L,J)-XC(7,J))**2*(XC(6,J)+1)/XC(6,J)
      DL=DL-(A(L,J)-XC(9,J))**2*XC(8,J)/(XC(8,J)+1)
   33 CONTINUE
      IF(DL..E.DD) GO TO 32
      DD=DL
      ID=L
   32 CONTINUE
   37 CONTINUE
C.... INTERCHANGE ID AND II
      DO 35 J=1,N
      C=A(II,J)
      A(II,J)=A(ID,J)
   35 A(ID,J)=C
C.... UPDATE MEANS
      DO 36 J=2,N
      IF(XC(5,J).EQ.0) GO TO 36
      IF(A(II,J).EQ.XM) GO TO 36
      XC(6,J)=XC(6,J)-1.
      IF(XC(6,J).NE.0) XC(7,J)=XC(7,J)+(XC(7,J)-A(II,J))/XC(6,J)
      XC(8,J)=XC(8,J)+1
      XC(9,J)=XC(9,J)-(XC(9,J)-A(II,J))/XC(8,J)
   36 CONTINUE
      DM=DM+DD
      IF(DM.LT.XR(3,IR)) GO TO 31
      XR(3,IR)=DM
      XR(4,IR)=II-1
   31 CONTINUE
      RETURN
      END
```

CHAPTER 15

Direct Joining

15.1 INTRODUCTION

In Table 15.1, a number of species of the yeast candida are distinguished by their production of acid with various carbohydrates. It is desired to cluster the species and to characterize each cluster by its oxidation-fermentation behavior. It is also desirable to cluster the carbohydrates (if, for example, two carbohydrates always produced the same reaction, it would be unnecessary to use both). It is therefore necessary to construct clusters simultaneously on the species and on the carbohydrates.

The three-tree cluster model postulates clusters of rows, clusters of columns, and clusters of the data values themselves. Each of the three types of clusters forms trees. (The change from the clustering model used in Chapter 14 is that here the data clusters form a tree rather than a partition.) This model is exemplified in Table 15.3. The responses within a data cluster C are all equal, once values corresponding to clusters properly included in C are removed. This model for two-way tables can obviously be generalized to many-way tables.

The candida data are of a specialized type, in which a comparable response is induced on all combinations of candida and carbohydrates. This wide homogeneity of the response simplifies the simultaneous clustering of candida and carbohydrates. The more general type of data structure is the cases-by-variables structure, in which responses are comparable between cases for the same variable but not (without some prescaling) between variables for the same case. There is a correspondence between the two data structures shown in Table 15.2. In general, consider an N-way table with a response variable Y. The Jth margin of the table has a number of possible classes, and so it is a category variable—say, $V(J)$. Each entry in the table is identified by a response Y and the values of the margins, $V(1), V(2), \ldots, V(N)$. Each entry in the table is a case from a data matrix with $N + 1$ variables, corresponding to the response variable Y and the $N + 1$ marginal variables.

The many-tree clustering model on a data matrix with N marginal category variables and a response variable Y constructs clusters in each of the marginal variables and clusters on the cases themselves, so that (i) the response variable Y is equal within clusters of cases (after proper subclusters are removed), and (ii) each cluster of cases is the set of cases corresponding to a product of clusters of values in the marginal

278

Table 15.1 Oxidation-Fermentation Patterns in Species of Candida

[From Hall, C. T., Webb, C. D., and Papageorge, C. (1972).] Use of an oxidation-fermentation medium in the identification of yeasts, *HSMHA Health Rep.* **87,** 172–176.] The oxidation-fermentation of various carbohydrates is used in differentiating various strains of yeast in the species *Candida*.

```
CARBOHYDRATES:  1 = GLUCOSE, 2 = MALTOSE, 3 = SUCROSE, 4 = LACTOSE,

5 = GALACTOSE, 6 = MELIBIOSE, 7 = CELLOBIOSE, 8 = INOSITOL, 9 = XYLOSE,

10 = RAFFINOSE, 11 = TREHALOSE, 12 = DULCITOL
```

	1	2	3	4	5	6	7	8	9	10	11	12
1. C. ALBICANS	+	+	+	-	+	-	-	-	-	-	+	-
2. C. TROPICALIS	+	+	+	-	+	-	+	-	+	-	+	-
3. C. KRUSEI	+	-	-	-	-	-	-	-	-	-	-	-
4. C. PARAPSILOSIS	+	+	+	-	+	-	-	-	+	-	+	-
5. C. GUILLERMONDII	+	+	+	-	+	+	+	-	+	+	+	+
6. C. STELLATOIDEA	+	+	-	-	+	-	-	-	+	-	+	-
7. C. PSEUDOTROPICALIS	+	-	+	+	+	-	+	-	+	+	-	-
8. C. VINI	+	-	-	-	-	-	-	-	-	-	-	-

(+ means oxidative production of acid; - means no acid production).

variables. For example, if variable 1 takes values A, B, C and variable 2 takes values X, W, Z, U, then a data cluster would be the set of cases corresponding to the clusters (A, B) and (X, Z, U)—that is, to the set of cases $\{V(1) = A,\ V(2) = X\}$, $\{V(1) = A, V(2) = Z\}$, $\{V(1) = A, V(2) = U\}$, $\{V(1) = B, V(2) = X\}$, $\{V(1) = B, V(2) = Z\}$, $\{V(1) = B, V(2) = U\}$.

It will be seen in the above model that the clusters are evaluated by the response variable Y, but they are constrained by the marginal variables $V(1), \ldots, V(N)$. In regression, Y would be the dependent variable and $\{V(J), 1 \leq J \leq N\}$ would be the independent variables. The above showed that an N-way table may be treated as a data matrix with $N + 1$ variables. Conversely, a data matrix may be treated as an N-way table.

Specifically, consider an arbitrary data matrix of category variables. Each variable is a partition of the cases, so the basic problem is the construction of an overall tree of cases that conforms as best as possible to the original partitions. Some combinations of values of the category variables will appear in the data matrix, and some will not. Make an enlarged data matrix with one case for each combination of values and a response for each case equal to the number of times that combination of values appeared in the original data. This new data matrix is equivalent to an N-way table with responses equal to the number of times a particular combination of values appears. The outcome of the many-way clustering will be trees on each of the marginal variables and a clustering of cases.

The algorithm to perform this construction is a joining algorithm that joins two values of one of the category variables at each step.

Table 15.2   Candida Data in Cases-by-Variables Form

Note that there are four columns of cases.

ACID PRODUCTION	YEAST SPECIES	CARBO HYDRATE	AP	YS	CH	AP	YS	CH	AP	YS	CH
+	1	1	+	3	1	+	5	1	+	7	1
+	1	2	-	3	2	+	5	2	-	7	2
+	1	3	-	3	3	+	5	3	+	7	3
-	1	4	-	3	4	-	5	4	+	7	4
+	1	5	-	3	5	+	5	5	+	7	5
-	1	6	-	3	6	+	5	6	-	7	6
-	1	7	-	3	7	+	5	7	+	7	7
-	1	8	-	3	8	-	5	8	-	7	8
-	1	9	-	3	9	+	5	9	+	7	9
-	1	10	-	3	10	+	5	10	+	7	10
+	1	11	-	3	11	+	5	11	-	7	11
-	1	12	-	3	12	+	5	12	-	7	12
+	2	1	+	4	1	+	6	1	+	8	1
+	2	2	+	4	2	+	6	2	-	8	2
+	2	3	+	4	3	-	6	3	-	8	3
-	2	4	-	4	4	-	6	4	-	8	4
+	2	5	+	4	5	+	6	5	-	8	5
-	2	6	-	4	6	-	6	6	-	8	6
+	2	7	-	4	7	-	6	7	-	8	7
-	2	8	-	4	8	-	6	8	-	8	8
+	2	9	+	4	9	+	6	9	-	8	9
-	2	10	-	4	10	-	6	10	-	8	10
+	2	11	+	4	11	+	6	11	-	8	11
-	2	12	-	4	12	-	6	12	-	8	12

15.2   TWO-WAY JOINING ALGORITHM

Preliminaries. The data matrix $\{A(I, J),\ 1 \leq I \leq M,\ 1 \leq J \leq N\}$ consists of zeroes or ones. The outcome of the clustering is a tree of clusters of the data entries, each of which is the set of data in the matrix corresponding to a cluster of rows and a cluster of columns. Denoting the data clusters by $1, 2, \ldots,$ KD, let IR(K) be the row cluster corresponding to the Kth data cluster and let IC(K) be the corresponding column cluster The row clusters are $1, 2, \ldots,$ KR, and the tree structure is specified by the function JR(I), which is the smallest cluster properly including I. Similarly, the column clusters are $1, 2, \ldots,$ KC and the column tree structure is specified by JC(I), the smallest column cluster properly including column cluster I.

The algorithm at each step joins that pair of rows or columns that are closest in a certain measure of distance. The closest pair of rows, for example, are joined to make

Table 15.3 Three-Tree Model

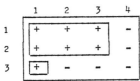

CARBOHYDRATES

		1	2	3	4
CANDIDA	1	+	+	+	−
	2	+	+	+	−
	3	+	−	−	−

ACID	CANDIDA	CARBOHYDRATES	
+	1	1	
+	1	2	DATA
+	1	3	CLUSTERS
+	2	1	
+	2	2	
+	2	3	
+	3	1	
−	3	2	
−	3	3	
−	1	4	
−	2	4	
−	3	4	

CANDIDA CLUSTERS	CARBOHYDRATE CLUSTERS
1	1
2	2
3	3
	4

a new row by using a certain amalgamation rule, possible data clusters are identified wherever the two rows do not match, and the algorithm proceeds to the next step.

STEP 1.   Initially KD = 0, KR = M, KC = N, JR(I) = 0 ($1 \leq I \leq M$), and JC(I) = 0 ($1 \leq I \leq N$).

STEP 2.   For each pair of row clusters ($1 \leq I, J \leq$ KR) with JR(I) = JR(J) = 0, compute the distance $(n_{01} + n_{10})/(n_{00} + n_{11} + n_{01} + n_{10})$, where n_{XY} is the number of columns K, with JC(K) = 0, and with $A(I, K) = X$ and $A(J, K) = Y$. Similarly compute the distance between each pair of column clusters ($1 \leq I, J \leq$ KC) with JC(I) = JC(J) = 0.

STEP 3.   Join the closest pair of rows or columns—say the closest pair of rows, I, J. Increase KR to KR + 1, define JR(I) = JR(J) = KR and JR(KR) = 0. For each column K, JC(K) = 0, define $A($KR$, K)$ to be $A(I, K) \cap A(J, K)$, if $A(I, K) \cap A(J, K) \neq \varnothing$, and to be $A(I, K) \cup A(J, K)$ otherwise. If $A(I, K) \cap A(J, K) = \varnothing$, define IR(KD + 1) = I, IC(KD + 1) = K and IR(KD + 2) = J, IC(KD + 2) =

K, which thus generates two new data clusters. Increase KD by 2. Return to Step 2, unless there is only one row, $(1 \leq I \leq \text{KR})$ with $\text{JR}(I) = 0$ and one column $(1 \leq I \leq \text{KC})$ with $\text{JC}(I) = 0$.

STEP 4.   Reorder the data matrix in rows and columns so that row and column clusters consist of contiguous sequences of rows and columns.

STEP 5.   Consider the data values $\{A(I, J), 1 \leq I \leq M, 1 \leq J \leq N\}$ with respect to the data clusters $1 \leq K \leq \text{KD}$. For each cluster K, define $A(K)$ to be the set of values (either $\{0\}$, $\{1\}$, or $\{0, 1\}$) which occur maximally in $A(J)$, where J runs over the maximal proper subclusters of K. Define $A(K) = A(I, J)$ on clusters consisting of a single data value.

STEP 6.   For the data cluster K of all data values, if $A(K) = \{0, 1\}$, set $A(K) = \{1\}$. Eliminate the cluster K if the minimal cluster L including K has $A(L) \subset A(K)$. The final number of clusters is $(\text{KD} + 1)/2$.

15.3   APPLICATION OF TWO-WAY JOINING ALGORITHM TO CANDIDA

The value in the data matrix will be $+$ or $-$ rather than 0 or 1. (See Tables 15.4 and 15.5 for the final results.)

STEP 1.   Initially specify the number of clusters, $\text{KD} = 0$, the number of row clusters, $\text{KR} = 8$, and the number of column clusters, $\text{KC} = 12$. The structure is initialised by $\text{JR}(I) = 0$, $1 \leq I \leq 6$ and $\text{JC}(I) = 0$ $(1 \leq I \leq 12)$.

STEP 2.   Distances are computed between each pair of row clusters I, J with $\text{JR}(I) = \text{JR}(J) = 0$. For example, with $I = 1$, $J = 2$, $n_{++} = 5$, $n_{+-} = 0$, $n_{-+} = 2$, $n_{--} = 5$. These numbers, of course, add to 12 which is the number of columns. The distance is $\frac{2}{12}$. Similarly, distances are computed for each pair of columns.

STEP 3.   There are some pairs of rows and columns that are identical, and these will be joined first. The algorithm does not specify the order of joining. Taking first columns 6 and 12, increase KC to 13, define $\text{JC}(6) = 13$, $\text{JC}(12) = 13$, $\text{JC}(13) = 0$. Define $A(K, 13) = A(K, 6)$ for $1 \leq K \leq 8$, since $A(K, 6) \cap A(K, 12) = A(K, 6)$ always. Because of this identity, no data clusters are generated. Return to Step 2 and join identical columns 11 and 2 to make column cluster 14, and then join identical rows 3 and 8 to make row cluster 9. Return to Step 2.

STEP 2 REPEATED.   Compute distances between row clusters I, J $(1 \leq I, J \leq 9)$, excluding I, J equal to 3 and 8 since $\text{JR}(3) = \text{JR}(8) = 9$. The distance between $I = 1$ and $J = 2$ uses the counts $n_{++} = 4$, $n_{+-} = 0$, $n_{-+} = 2$, $n_{--} = 4$, so the distance is $\frac{2}{10}$, whereas it was $\frac{2}{12}$ before the joining of the two identical column pairs. This is the mechanism for interaction between the row clustering and the column clustering. For example, two identical (or very similar) columns will have reduced weight in the clustering of rows because they will be joined early on. The closest pair of rows or columns is the pair of rows $I = 4$, $J = 6$, whose distance is $\frac{1}{10}$.

STEP 3.   Join row 4 and row 6. Increase KR to 10, define $\text{JR}(4) = \text{JR}(6) = 10$, $\text{JR}(10) = 0$. For column 1, $A(4, 1) = +$, $A(6, 1) = +$, so $A(4, 1) \cap A(6, 1) = +$, and therefore $A(10, 1) = +$. Column 2 is not considered because $\text{JC}(2) = 14$. For

Table 15.4 Preliminary Outcomes of Joining Algorithm

CARBOHYDRATES

	4	8	6	12	11	2	7	10	9	5	3	1
7	[+]	[-]	-	-	-	-	+	+	+	+	+	+
8	-	-	-	-	-	-	-	-	-	-	-	+
3	-	-	-	-	-	-	-	-	-	-	-	+
1	-	-	-	-	+	+	-	-	-	+	+	+
6	-	-	-	-	+	+	-	-	+	+	[-]	+
4	-	-	-	-	+	+	-	-	+	+	[+]	+
2	-	-	-	-	+	+	+	-	+	+	+	+
5	-	-	+	+	+	+	+	+	+	+	+	+

ROW CLUSTERS					COL CLUSTERS					DATA CLUSTERS					
I	JR(I)	I	JR(I)		I	JC(I)	I	JC(I)		K	IR(K)	IC(K)	K	JR(K)	JC(K)
1	12	9	13		1	19	13	20		1	4	3	13	13	17
2	11	10	11		2	14	14	20		2	6	3	14	13	19
3	9	11	12		3	15	15	19		3	2	7	15	5	20
4	10	12	13		4	18	16	17		4	10	7	16	14	20
5	15	13	14		5	15	17	21		5	1	9	17	15	21
6	10	14	15		6	13	18	22		6	11	9	18	15	22
7	14				7	16	19	21		7	7	4	19	15	23
8	9				8	18	20	22		8	7	8			
					9	17	21	23		9	9	1			
					10	16	22	23		10	9	15			
					11	14				11	12	13			
					12	13				12	12	14			

column 3, $A(4, 3) = +$, $A(6, 3) = -$, so $A(10, 1) = \pm$. Also construct two new data clusters with IR(1) = 4, IC(1) = 3, IR(2) = 6, IC(2) = 3. Increase KD to 2. Return to Step 2.

STEP 2 REPEATED. In distance calculations between row clusters, the value $\pm$ is ignored entirely. For example, between row 1 and row 10, $n_{++} = 3$, $n_{+-} = 0$, $n_{-+} = 1$, $n_{--} = 5$, with the third column ignored entirely. The algorithm continues until a single row cluster and a single column cluster remain. The final clusters of data, rows, and columns are given in Table 15.4. Each data cluster corresponds to a mismatch in two joined row or column clusters.

STEP 4. The data matrix has been reordered in Tables 15.4 and 15.5, so that data clusters are represented as contiguous blocks.

Table 15.5   Final Data Clusters from Joining Candida

CARBOHYDRATES

CANDIDA		4	8	6	12	11	2	7	20	9	5	3	1
CANDIDA	7	+	-	-	-	-	-	+	+	+	+	+	+
	8	-	-	-	-	-	-	-	-	-	-	-	+
	3	-	-	-	-	-	-	-	-	-	-	-	+
	1	-	-	-	-	+	+	-	-	-	+	+	+
	6	-	-	-	-	+	+	-	-	+	+	-	+
	4	-	-	-	-	+	+	-	-	+	+	+	+
	2	-	-	-	-	+	+	+	-	+	+	+	+
	5	-	-	+	+	+	+	+	+	+	+	+	+

STEP 5. In Table 15.4, there are nineteen data clusters within which data values are constant. Many of these clusters are unnecessary; for example, the cluster at (7, 8) may be dropped because this value is already implied by the larger cluster including it. These next steps prune away the unnecessary clusters by using the same techniques as in the minimum-mutation fit.

For $K = 4$, the subclusters are (4, 7) and (6, 7), for each of which $A = -$; thus $A(4) = -$. For $K = 13$, the subclusters are $K = 3, 4, 1, 11$ and 11 single points. The value $-$ occurs 13 times, and the value $+$ occurs twice. Thus $A(13) = -$.

STEP 6. In Step 6, unnecessary data clusters are pruned away. The largest cluster has $A(19) = \pm$; set $A(19) = +$. Since $A(18) = + \supset A(19)$, eliminate cluster 18. Similarly, $A(13) = + \supset A(19)$, and so cluster 13 is eliminated. The final data clusters appear in Table 15.5, where it will be seen that only ten data clusters are necessary; this reduction in the number of clusters to $(KD + 1)/2$ always occurs.

An overall criterion for this type of data is the representation of the data values exactly in a minimum number of clusters. It is still necessary to have a search procedure through the immense number of possible structures, and the joining algorithm is such a search procedure.

15.4   GENERALIZATIONS OF TWO-WAY JOINING ALGORITHM

The algorithm may obviously be generalized to many-way tables. At each step, one of the N marginal variables is selected, and two of its values are amalgamated. Measures of distance and amalgamation rules are obtained analogously to the two-way case.

A second direction of generalization is the response variable. In the two-way joining algorithm, the response is 0-1, but it could be a category variable with more than two categories, it could be an ordered variable, it could be an integer (as in contingency tables), or it could be on an interval scale. Different measures of distance and amalgamation rules will be necessary for each of these variable types.

For an interval scale, each data point should be regarded as a range of values (Y_1, Y_2); before any clustering takes place, $Y_1 = Y_2$ for every data point. A threshold T is given. Two data points, (Y_1, Y_2) and (Z_1, Z_2), are compatible if $\max (Y_2, Z_2) - \min (Y_1, Z_1) \leq T$. The distance between two vectors of data points will be the number of compatible pairs. Two data points (Y_1, Y_2) and (Z_1, Z_2) amalgamate to $(\min (Y_1, Z_1), \max (Y_2, Z_2))$ if compatible, and otherwise the amalgamated value is missing.

15.5 SIGNIFICANCE TESTS FOR OUTCOMES OF TWO-WAY JOINING ALGORITHM

Let M be the number of rows, let N be the number of columns, and let K ones and $MN-K$ zeroes be assigned to the data matrix $\{A(I, J), 1 \leq I \leq M, 1 \leq J \leq N\}$ at random. What is the distribution of KD, the number of data clusters?

Some asymptotic results are available. If $M \to \infty$, $K \to \infty$, but N remains fixed, then $KD = 2^N$. If $M \to \infty$ but K and N remain fixed, then $KD - 1$ equals the number of occupied cells when K balls are distributed at random among N cells. If $M \to \infty$, $N \to \infty$, but K remains fixed, $KD = K + 1$.

Some empirical results are given in Table 15.6 for relatively small matrices. Only means and variances are given, as the distributions are approximately normal as M, N, $K^{1/2}$ approach ∞ at comparable rates. Some general conclusions are that the

Table 15.6 The Number of Data Clusters when a Given Number of Ones are Assigned at Random to an M by N 0–1 Matrix. (Means and variance in 100 repetitions.)

M = 6 N = 2 ‖ 1's =	1	2	3	4	5	6
MEAN, VARIANCE =	2.00, 0.00	2.42, .24	2.81, .15	3.19, .35	3.43, .29	3.67, .32

M = 4 N = 3 ‖ 1's =	1	2	3	4	5	6
MEAN, VARIANCE =	2.00, .00	2.60, .24	3.07, .17	3.77, .66	3.94, .46	4.03, .45

M = 12 N = 2 ‖ 1's =	1	3	5	7	9	11
MEAN, VARIANCE =	2.00, .00	2.83, .14	3.20, .24	3.56, .27	3.86, .12	3.88, .11

M = 8 N = 3 ‖ 1's =	1	3	5	7	9	11
MEAN, VARIANCE =	2.00, .00	3.05, .23	4.01, .47	4.86, .84	5.51, .95	5.73, .94

M = 6 N = 4 ‖ 1's =	1	3	5	7	9	11
MEAN, VARIANCE =	2.00, .00	3.23, .28	4.25, .79	5.38, .92	6.10, .97	6.04, .94

M = 16 N = 3 ‖ 1's =	2	6	10	14	18	22
MEAN, VARIANCE =	2.68, .22	4.14, .42	5.40, 1.06	6.75, .69	7.20, .60	7.27, .62

M = 12 N = 4 ‖ 1's =	2	6	10	14	18	22
MEAN, VARIANCE =	2.76, .18	4.75, .73	6.39, .90	7.98, 1.40	9.13, 1.43	9.46, 1.57

M = 8 N = 6 ‖ 1's =	2	6	10	14	18	22
MEAN, VARIANCE =	2.75, .19	5.17, .90	7.42, 1.28	9.21, 1.76	10.10, 2.07	10.43, 1.81

M = 12 N = 8 ‖ 1's =	2	10	18	26		
MEAN, VARIANCE =	2.87, .11	7.86, 1.04	11.88, 2.23	15.57, 2.63		

M = 16 N = 6 ‖ 1's =	2	10	18			
MEAN, VARIANCE =	2.83, .14	7.38, .84	10.82, 2.07			

means and variance increase for M and N fixed, as K^2 increases toward $\frac{1}{2}MN$. For MN fixed, means and variances increase as M increases toward $\sqrt{MN}$. To illustrate the case M large, N and K small, consider $M = 16$, $N = 3$, $K = 2$. Then $KD - 1$ is the number of occupied cells in distributing two balls at random in three cells, so $KD = 2$ with probability $\frac{1}{3}$, $KD = 3$ with probability $\frac{2}{3}$; KD has mean 2.67, and variance 0.22.

15.6 DIRECT JOINING ALGORITHM FOR VARIABLES ON DIFFERENT SCALES

Preliminaries. This algorithm simultaneously clusters variables and cases when the variables are measured on different scales. The joining operation for variables is somewhat more complicated than that for the two-way algorithm, where all data values are comparable. At each stage in the algorithm, every variable is transformed to have the same first and third quartiles, and previous two-way joining rules are used on the transformed data.

As in the algorithm of Section 15.2, there is the data matrix $\{A(I, J), 1 \leq I \leq M, 1 \leq J \leq N\}$, there are the data clusters $1, 2, \ldots, KD$ with $IR(K)$, $IC(K)$ denoting the corresponding row and column clusters, there are the row clusters $1, 2, \ldots, KR$ with $JR(I)$ denoting the smallest cluster properly including I, and there are the column clusters $1, 2, \ldots, KC$ with $JC(I)$ denoting the smallest cluster properly including I. There is also the linear transformation $B(J)$, $C(J)$ of the Jth variable, such that $\{B(J) A(I, J) + C(J), 1 \leq I \leq M, 1 \leq J \leq N\}$ is a homogeneous data matrix. These linear transformations are discovered during the execution of the algorithm.

STEP 1.   Initially $KD = 0$, $KR = M$, $KC = N$, $JR(I) = 0$ $(1 \leq I \leq M)$ and $JC(I) = 0$ $(1 \leq I \leq N)$. As the first standardization of the variables, choose $B(J)$, $C(J)$ so that $B(J) A(I, J) + C(J)$ has the first quartile equal to -1 and the third quartile equal to $+1$.

STEP 2.   For each pair of row clusters $[1 \leq I, J \leq KR, JR(I) = JR(J) = 0]$, the distance is the proportion of column clusters K $[1 \leq K \leq KC, JC(K) = 0]$, such that $|A(I, K) - A(J, K)| \geq 1$. (The threshold 1 is chosen rather arbitrarily.) Similarly, for each pair of column clusters $[1 \leq I, J \leq KC, JC(I) = JC(J) = 0]$, the distance is the proportion of row clusters K $[1 \leq K \leq KR, JR(K) = 0]$, such that $|A(K, J) - A(K, I)| \geq 1$. If no pairs of row or column clusters satisfying the conditions exist, go to Step 7 with $K = 1$.

STEP 3.   If a pair of column clusters are closest, go to Step 5. If a pair of row clusters I, J are closest, they are amalgamated to form a new cluster $KR + 1$ with $JR(I) = JR(J) = KR + 1$, $JR(KR + 1) = 0$. Increase KR by 1. For each K $[1 \leq K \leq KC, JC(K) = 0]$, define $A(KR, K)$ to be $[A(I, K) + A(J, K)]/2$ if $|A(I, K) - A(J, K)| < 1$. Otherwise, define $A(KR, K)$ to be $A(I, K)$ or $A(J, K)$, whichever is smallest in absolute value, increase KD by 1, and set $IC(KD) = K$, $IR(KD) = I$ or J, according as $A(I, K)$ or $A(J, K)$ is largest in absolute value.

STEP 4.   Redefine $B(J)$, $C(J)$ for each column cluster J $[1 \leq J \leq KC, JC(J) = 0]$, such that the values $\{A(I, J), 1 \leq I \leq KR, JR(I) = 0\}$ have the first quartile equal to -1 and the third quartile equal to $+1$. Return to Step 2.

STEP 5. Suppose the column clusters I, J are closest. Amalgamate them to form a new column cluster $KC + 1$ with $JC(I) = JC(J) = KC + 1$, $JC(KC + 1) = 0$. Increase KC by 1. For each row cluster K $[1 \le K \le KR, JR(K) = 0]$, define $A(K, KC) = [A(K, I) + A(K, J)]/2$ if $|A(K, I) - A(K, J)| < 1$. Otherwise, set $A(K, KC) = A(K, I)$ or $A(K, J)$, whichever is smaller in absolute value, increase KD by 1, and set $IR(KD) = K$, $IC(KD) = I$ or J according as $A(K, I)$ or $A(K, J)$ is greater in absolute value.

STEP 6. Define $B(KC)$, $C(KC)$ so that $A(I, KC)B(KC) + C(KC)$ has the first quartile equal to -1 and the third quartile equal to $+1$. Return to Step 2.

STEP 7. If $K > N$, stop. Otherwise, change $B(K)$, $C(K)$ as follows. Let $L = K$.

STEP 8. Let $L = JC(L)$. If $JC(L) = 0$, increase K by 1 and return to Step 7. Change $B(K)$ to $B(L) B(K)$, and change $C(K)$ to $C(L) + C(K) B(L)$. Repeat Step 8.

This algorithm is applied to percentages of farmland devoted to various crops in counties in Ohio. The initial data are given in Table 15.7, and the final data in Table 15.8. There is a distinctive cluster of five counties which are high on hay and relatively low in other crops. The use of first and third quartiles for scaling the variables is somewhat arbitrary, and more careful scaling techniques will be discussed in Chapter 16.

Table 15.7 Ohio Croplands

The percent of total harvested cropland allocated to various crops in selected Ohio counties, *U.S. Census of Agriculture* (1949).

	CORN	MIXED	SMALL GRAINS	WHEAT	OATS	BARLEY	SOYBEAN	HAY
ADAMS	42.41		0.21	22.47	1.07	0.37	0.62	27.80
ALLEN	34.43		0.13	23.76	18.35	0.11	12.18	15.31
ASHTABULA	22.88		0.24	13.52	15.67	0.02	1.30	38.89
ATHENS	26.61		0.18	8.89	3.42	0.05	0.71	53.91
DELAWARE	33.52		0.13	17.60	11.33	0.16	11.82	22.69
CLINTON	48.45		0.24	29.50	3.10	0.25	2.72	9.85
GALLIA	31.38		0.83	13.07	2.03	0.60	0.71	44.07
GEAUGA	23.04		0.21	12.68	17.44	0.11	0.41	37.80
HANCOCK	36.13		0.12	24.64	16.56	0.13	13.91	16.46
HIGHLAND	4?.33		0.11	31.57	1.59	0.05	1.46	16.10
MEIGS	28.20		0.28	14.08	3.06	0.18	0.67	46.71
PORTAGE	26.67		0.11	19.13	18.67	0.03	0.69	27.33
PUTNAM	30.97		0.13	24.16	15.28	0.13	14.10	12.61
WARREN	43.23		0.09	24.97	3.20	0.24	4.68	18.72
WASHINGTON	25.08		0.08	13.43	1.96	0.66	1.06	50.27

From U.S. census of agriculture, 1949. (There is some prejudice for clustering counties contiguously.)

Table 15.8   Application of Direct Joining Algorithm to Ohio Croplands

	CORN	SOYBEAN	HAY	OATS	WHEAT
ADAMS------	42.41	.62	27.80	1.07	22.47
WARREN-----	43.23	4.68	18.72	3.20	24.97
PORTAGE----	26.67	.69	27.33	18.67	19.13
GALLIA-----	31.38	.71	44.07	2.03	13.07
WASHINGTON-	25.08	1.06	50.27	1.96	13.43
MEIGS------	28.20	.67	46.71	3.06	14.08
ASHTABULA--	22.88	1.30	38.89	15.67	13.52
GEAUGA-----	23.04	.41	37.80	17.44	12.68
DELAWARE---	33.52	11.82	22.69	11.33	17.60
PUTNAM-----	30.97	14.10	12.61	15.28	24.16
ALLEN------	34.43	12.18	15.31	18.35	23.76
HANCOCK----	36.13	13.91	16.46	16.56	24.64
ATHENS-----	22.61	0.71	53.91	3.42	8.89
HIGHLAND---	57.83	1.46	16.10	1.59	31.57
CLINTON----	48.45	2.72	9.85	3.10	29.50

15.7   THINGS TO DO

15.7.1   Running the Two-Way Joining Algorithm

The most important restriction in this algorithm is the requirement that data values be comparable across both rows and columns. Since variables are frequently measured on different scales, they must be scaled, often rather arbitrarily, before the algorithm may be applied.

The general algorithm requires specification of a within-cluster threshold. No absolute guidelines are available for the choice of the threshold. For 0–1 data or category responses in general, the threshold should be zero. Then all values within blocks will be identical. For continuous data, reasonable thresholds are in the range of $\frac{1}{20}-\frac{1}{2}$ of the standard deviation of all data values. Tables 15.9 and 15.10 contain data sets on patterns of food consumption and on varieties of languages spoken in various European countries.

15.7.2   Two-Way Clustering Models

The basic component of the two-way clustering models is the block, or submatrix of the data, in which some simple model must be obeyed by the data. For example, all values must be equal or all values must lie within a certain range.

Table 15.9   European Food

(From *A Survey of Europe Today*, The Reader's Digest Association Ltd., 25 Berkeley Square, London.) Percentage of all households with various foods in house at time of questionnaire. Foods by countries.

		WG	IT	FR	NS	BM	LG	GB	PL	AA	SD	SW	DK	NY	FD	SP	ID
GC	ground coffee	90	82	88	96	94	97	27	72	55	73	97	96	92	98	70	13
IC	instant coffee	49	10	42	62	38	61	86	26	31	72	13	17	17	12	40	52
TB	tea or tea bags	88	60	63	98	48	86	99	77	61	85	93	92	83	84	40	99
SS	sugarless sweet.	19	2	4	32	11	28	22	2	15	25	31	35	13	20	-	11
BP	packaged biscuits	57	55	76	62	74	79	91	22	29	31	-	66	62	64	62	80
SP	soup (packages)	51	41	53	67	37	73	55	34	33	69	43	32	51	27	43	75
ST	soup (tinned)	19	3	11	43	25	12	76	1	1	10	43	17	4	10	2	18
IP	instant potatoes	21	2	23	7	9	7	17	5	5	17	39	11	17	8	14	2
FF	frozen fish	27	4	11	14	13	26	20	20	15	19	54	51	30	18	23	5
VF	frozen vegetables	21	2	5	14	12	23	24	3	11	15	45	42	15	12	7	3
AF	fresh apples	81	67	87	83	76	85	76	22	49	79	56	81	61	50	59	57
OF	fresh oranges	75	71	84	89	76	94	68	51	42	70	78	72	72	57	77	52
FT	tinned fruit	44	9	40	61	42	83	89	8	14	46	53	50	34	22	30	46
JS	jam (shop)	71	46	45	81	57	20	91	16	41	61	75	64	51	37	38	89
CG	garlic clove	22	80	88	15	29	91	11	89	51	64	9	11	11	15	86	5
BR	butter	91	66	94	31	84	94	95	65	51	82	68	92	63	96	44	97
ME	margarine	85	24	47	97	80	94	94	78	72	48	32	91	94	94	51	25
OO	olive, corn oil	74	94	36	13	83	84	57	92	28	61	48	30	28	17	91	31
YT	yoghurt	30	5	57	53	20	31	11	6	13	48	2	11	2	-	16	3
CD	crispbread	26	18	3	15	5	24	28	9	11	30	93	34	62	64	13	9

The clustering structure assumed here is that the family of blocks form a tree, the family of row clusters (one to each block) forms a tree, and the family of column clusters forms a tree. The tree properties for row and column clusters guarantee, after some permutation of rows and columns, that each block consists of a set of contiguous data points.

For any family of blocks, a plausible model is

$$A(I, J) = \sum \{1 \leq K \leq L\}\, R(I, K) C(K, J) B(K),$$

where

$R(I, K) = 1$ if row I lies in block K,

$R(I, K) = 0$ if row I does not lie in block K,

$C(J, K) = 1$ if column J lies in block K,

$C(J, K) = 0$ if column J does not lie in block K,

$B(K)$ is the value associated with block K.

This is a factor analysis model that may be fitted in a stepwise algorithm, which at each step identifies the new block and the new block value which most reduces deviation between the data and the model. See also 17.8.2.

Table 15.10   Languages Spoken in Europe

(From *A Survey of Europe Today*, The Reader's Digest Association Ltd., 25 Berkeley Square, London.) Percentages of persons claiming to speak the language "enough to make yourself understood."

LANGUAGES			
FI	FINNISH	DU	DUTCH
SW	SWEDISH	FL	FLEMISH
DA	DANISH	FR	FRENCH
NO	NORWEGIAN	IT	ITALIAN
EN	ENGLISH	SP	SPANISH
GE	GERMAN	PO	PORTUGUESE

COUNTRY	FI	SW	DA	NO	EN	GE	DU	FL	FR	IT	SP	PO
WEST GERMANY.	0	0	0	0	21	100	2	1	10	2	1	0
ITALY........	0	0	0	0	5	3	0	0	11	100	1	0
FRANCE.......	0	2	3	0	10	7	1	1	100	12	7	1
NETHERLANDS..	0	0	0	0	41	47	100	100	16	2	2	0
BELGIUM......	0	0	0	0	14	15	0	59	44	2	1	0
LUXEMBOURG...	0	0	0	0	31	100	4	1	92	10	0	0
GREAT BRITAIN	0	0	0	0	100	7	0	0	15	3	2	0
PORTUGAL.....	0	0	0	0	9	0	0	0	10	1	2	100
AUSTRIA......	0	0	0	0	18	100	1	1	4	2	1	0
SWITZERLAND..	0	0	0	0	21	83	1	2	64	23	3	1
SWEDEN.......	5	100	10	11	43	25	0	0	6	1	1	0
DENMARK......	0	22	100	20	38	36	1	1	10	3	1	0
NORWAY.......	0	25	19	100	34	19	0	0	4	1	0	1
FINLAND......	100	23	0	0	12	11	0	0	2	1	0	0
SPAIN........	0	0	0	0	5	1	0	0	11	2	100	0
IRELAND......	0	0	0	0	100	1	0	0	2	0	0	ϵ

For example,

$$\begin{bmatrix} 3 & 5 & 2 & 3 \\ 8 & 5 & 5 & 3 \\ 5 & 2 & 5 & 0 \end{bmatrix} = 3\begin{bmatrix} 1 & 1 & 0 & 1 \\ 1 & 1 & 0 & 1 \\ 0 & 0 & 0 & 0 \end{bmatrix} + 2\begin{bmatrix} 0 & 1 & 1 & 0 \\ 0 & 1 & 1 & 0 \\ 0 & 1 & 1 & 0 \end{bmatrix} + 5\begin{bmatrix} 0 & 0 & 0 & 0 \\ 1 & 0 & 1 & 0 \\ 1 & 0 & 1 & 0 \end{bmatrix}$$

$$= 3\begin{bmatrix} 1 \\ 1 \\ 0 \end{bmatrix}\begin{bmatrix} 1 & 1 & 0 & 1 \end{bmatrix} + 2\begin{bmatrix} 1 \\ 1 \\ 1 \end{bmatrix}\begin{bmatrix} 0 & 1 & 1 & 0 \end{bmatrix} + 5\begin{bmatrix} 0 \\ 1 \\ 1 \end{bmatrix}\begin{bmatrix} 1 & 0 & 1 & 0 \end{bmatrix}$$

$$= \begin{bmatrix} 3 & 2 & 0 \\ 3 & 2 & 5 \\ 0 & 2 & 5 \end{bmatrix}\begin{bmatrix} 1 & 1 & 0 & 1 \\ 0 & 1 & 1 & 0 \\ 1 & 0 & 1 & 0 \end{bmatrix}$$

The overlapping clusters that arise from this procedure are difficult to present or interpret, and the actual data values within blocks do not obey any simple rule because there may be contributions from many other blocks. If the blocks form a tree, all values within any one block, not lying in a smaller block, will be equal. The requirement that blocks form a tree considerably complicates the fitting procedure.

15.7.3 Distances, Amalgamation, and Blocks

The basic ingredients of a joining algorithm are the methods of computing distances, of amalgamating, and of computing blocks. These operate during the process of joining two rows or two columns.

For example, from Table 15.1,

$$
\begin{array}{lccc}
\text{Row 4} & +\ +\ +\ - & +\ -\ -\ - & +\ -\ +\ - \\
\text{Row 6} & +\ +\ -\ - & +\ -\ -\ - & +\ -\ +\ -.
\end{array}
$$

The distance between the two rows is the proportion of mismatches, $\frac{1}{12}$. These rows will be joined before any more distant pairs of rows or columns. The amalgamation rule represents a mismatch as $\pm$, so that the new row is

$$
\text{Row (4, 6)}\qquad +\ +\ \pm\ - \qquad +\ -\ -\ - \qquad +\ -\ +\ -.
$$

At this stage, either (row 4, col 3) or (row 6, col 3) might be a block, and in the 0–1 two-way joining algorithm the decision is postponed until all rows and columns have been joined. Thus, during the algorithm, more blocks and more corresponding unspecified values such as $\pm$ are retained than is really necessary.

A better, though more complicated, method decides the blocks and the amalgamated values at each step during the running of the algorithm. In the above example, row 1 is found most similar to 4 and 6. Since row 1 takes the value $+$ in column 3, the correct value there is set at $+$, and (row 6, col 3) is declared a block. (At this stage, the block consists of a single element, but in later stages each row will have been amalgamated from a number of rows, each column form a number of columns, and the blocks will correspond to sets of rows and columns.)

15.7.4 Range

For continuous-response variables, it is not practical to require exact equality within blocks, but it is plausible to require that all values within a block have a range less than a specified threshold T. During the algorithm, each data value will be a range $(Y1, Y2)$, and, indeed, the original data values are frequently faithfully represented this way.

The distance between two vectors of such values is the sum of combined ranges in each position, with this combined range set back to the threshold value T if it exceeds T. The amalgamation rule is the combined range in positions where the combined range is within threshold and a more complicated procedure otherwise, which is to be described.

Thus, for a threshold of 10,

$$
\begin{array}{lccccc}
\text{Row 1} & (1, 3) & (2, 5) & (2, 3) & (5, 6) & (5, 5) \\
\text{Row 2} & (5, 6) & (7, 10) & (2, 4) & (7, 16) & (18, 19).
\end{array}
$$

The distance is $(6 - 1) + (10 - 2) + (4 - 2) + (16 - 5)^* + (19 - 5)^* = 35,$

where the asterisked values are set back to 10. The amalgamated values are

Row (1, 2)	(1, 6)	(2, 10)	(2, 4)	(5, 6)	(6, 18)
Row 3	(2, 5)	(7, 12)	(4, 7)	(5, 12)	(6, 18),

where row 3 is the row closest to (1, 2). Consider position 4, where (5, 6) and (7, 16) are out of threshold. Since (5, 6) is within threshold of (5, 12), the amalgamated value is (5, 6) and (7, 16) would become a block. Another case is position 6, where both (5, 5) and (18, 19) are out of threshold with (6, 18). In this case (5, 5) and (18, 19) are blocks and the amalgamated value is (6, 18).

15.7.5 Computational Expenses

Since the data matrix is destroyed during the operation of a joining algorithm, it is necessary to set aside space for saving the original data. The basic storage costs are thus $2MN$. Storage is not often important because the time costs are so heavy for large arrays. The number of multiplications, nearly all occurring in repeated distance calculations, is proportional to M^2N^2.

15.7.6 Identical Rows and Columns

If identical, or nearly identical, pairs of rows exist, they will be quickly joined and have the same weight as a single row in later calculations. The joining algorithms are thus insensitive to accidents of sampling which overrepresent some types of cases or some types of properties. This overrepresentation is one of the considerations necessary in weighting variables for distance calculations; the weighting takes place automatically in two-way joining algorithms.

15.7.7 Minimum Number of Blocks

Usually it will be desired to represent the data in a minimum total number of blocks, and the joining algorithms join rows or columns to least increase the number of blocks. An exact minimizing technique seems impracticable except for very small data sets. The blocks in a minimum structure will not be unique in general. Consider

$$
\begin{array}{cccc}
+ & + & - & - \\
+ & + & + & + \\
+ & - & + & -
\end{array},
$$

for which there are two four-block structures.

15.7.8 Order Invariance

If all distances computed during the execution of the 0–1 two-way joining algorithm are different, then the final clusters will not depend on the original input order of rows or columns. It is not uncommon to have several different pairs of rows or columns corresponding to the same minimum distance. The pair selected to be joined may depend on the initial input order and may effect the final clustering. It is therefore necessary to develop a procedure for settling tied minimum distances—a difficult task.

First suppose a number of different pairs of rows are involved, with no row common to the different pairs. Then all these pairs of rows are joined before any other pairs of rows or columns are considered.

Next suppose there is a row which is closest to several other rows. All rows are joined in one step, by using the modal values for the new row and creating blocks wherever a value differs from the modal value.

Finally, if a pair of rows and a pair of columns are of minimum distance, join them simultaneously. One value in the new array will be derived from four original values, where the new row and column coincide. It will be the mode of these values.

Any set of pairs of rows and pairs of columns may be handled by generalizations of the above procedures.

PROGRAMS

JOIN2 joins rows or columns, identifying disparate elements as blocks.

RDIST computes distances between rows.

CDIST computes distances between columns.

PMUT permutes array to be consistent with clusters found in JOIN2.

```
      SUBROUTINE JOIN2(A,M,N,NB,KC,KA,NR,NC,RD,CD,TH)
C.........................................................................20 MAY 1973
C.... SUCCESSIVELY JOINS ROWS AND COLUMNS,WHICHEVER CLOSEST
C     CONSTRUCTS BLOCKS FROM PAIRS OF ELEMENTS NOT IN THRESHOLD.
C     USES RDIST,CDIST
C.... M = NUMBER OF ROWS
C.... N = NUMBER OF COLUMNS
C.... A = M BY N BORDERED ARRAY ,DESTROYED BY CRUEL MANIPULATION
C.... NB = 4 BY KC BLOCK ARRAY
C.... KC = UPPER BOUND ON NUMBER OF BLOCKS, M*N IS ALWAYS SAFE
C.... KA = ACTUAL NUMBER OF BLOCKS
C.... NR = 1 BY M ARRAY, NR(I)= OLD INDEX OF ITH ROW IN REORDERED MATRIX
C.... NC = 1 BY N ARRAY, NC(I)= OLD INDEX OF ITH COLUMN IN REORDERED ARRAY
C.... CD = 1 BY N SCRATCH ARRAY
C.... RD = 1 BY M SCRATCH ARRAY
C.... TH = THRESHOLD,IF EXCEEDED, MAKES BLOCKS
C.........................................................................
      DIMENSION A(M,N),NB(4,KC),NR(M),NC(N)
      DIMENSION CD(N),RD(M)
      KB=KC
      XM=99999.
      DO 10 I=1,M
   10 A(I,1)=I
      DO 11 J=1,N
   11 A(1,J)=J
      KA=0
      MM=M-1
      NN=N-1
C.... FIND CLOSEST ROWS AND COLUMNS
      DO 22 I=2,M
   22 CALL RDIST(A,M,N,I,NR(I),RD(I),NN,TH)
      DO 32 J=2,N
   32 CALL CDIST(A,M,N,J,NC(J),CD(J),MM,TH)
   70 CONTINUE
C.... FIND CLOSEST ROWS
      DR=10.**10
      I1=2
      I2=2
      DO 20 I=2,M
      IF(A(I,1).LT.0) GO TO 20
      IF(DR.LE.RD(I)) GO TO 20
      DR=RD(I)
      I1=I
      I2=NR(I)
   20 CONTINUE
C.... FIND CLOSEST COLUMNS
      DC=DR
      J1=2
      J2=2
      DO 30 J=2,N
      IF(A(1,J).LT.0) GO TO 30
      IF(DC.LE.CD(J)) GO TO 30
      DC=CD(J)
      J1=J
      J2=NC(J)
   30 CONTINUE
      IF(I2.GE.I1) GO TO 21
      K=I2
      I2=I1
      I1=K
   21 CONTINUE
      IF(J2.GE.J1) GO TO 31
      K=J2
      J2=J1
      J1=K
   31 CONTINUE
      IF(DC.LT.DR) GO TO 60
C.... AMALGAMATE ROWS
      IF(I2.EQ.2) GO TO 80
C.... FIND CLOSEST ROW TO I1 AND I2
      A(I1,1)=-A(I1,1)
      CALL RDIST(A,M,N,I2,NR(I2),RD(I2),NN,TH)
      A(I2,1)=-A(I2,1)
      A(I1,1)=-A(I1,1)
      CALL RDIST(A,M,N,I1,NR(I1),RD(I1),NN,TH)
      II=NR(I2)
      IF(RD(I1).LE.RD(I2)) II=NR(I1)
      L1=A(I1,1)
```

294

```
      L2=-A(I2,1)
      LL=A(II,1)
      A(I1,1)=L2
      A(I2,1)=-L1
      MM=MM-1
C.... AMALGAMATE VALUES AND CREATE BLOCKS
      DO 40 J=2,N
      IF(KA.GE.KB-2) GO TO 80
      IF(A(1,J).LT.0) GO TO 40
      K=A(1,J)
      IF(AMAX1(A(L1,K),A(L2,K))-AMIN1(A(I1,J),A(I2,J)).LE.TH) GO TO 43
      Z1=AMAX1(A(L1,K),A(LL,K))-AMIN1(A(I1,J),A(II,J))
      Z2=AMAX1(A(L2,K),A(LL,K))-AMIN1(A(I2,J),A(II,J))
      IF(II.EQ.2) Z2=10.**2
      IF(II.EQ.2) Z1=10.**10
      IF(Z1.LE.TH) A(L2,K)=A(L1,K)
      IF(Z2.LE.TH) A(I1,J)=A(I2,J)
      IF(Z1.GT.TH.AND.Z2.GT.TH) A(I1,J)=A(II,J)
      IF(Z1.GT.TH.AND.Z2.GT.TH) A(L2,K)=A(LL,K)
      IF(Z1.LE.TH) GO TO 41
      KA=KA+1
      NB(1,KA)=I1
      NB(2,KA)=L1
      NB(3,KA)=J
      NB(4,KA)=A(1,J)
   41 IF(Z2.LE.TH) GO TO 40
      KA=KA+1
      NB(1,KA)=I2
      NB(2,KA)=A(I1,1)
      NB(3,KA)=J
      NB(4,KA)=A(1,J)
      GO TO 40
   43 CONTINUE
      Z=A(L1,K)
      IF(A(I1,J).GE.A(I2,J))A(I1,J)=A(I2,J)
      IF(Z.GE.A(L2,K))A(L2,K)=Z
   40 CONTINUE
C.... UPDATE CLOSEST ROWS
      DO 45 I=2,M
      IF(A(I,1).LT.0) GO TO 45
      IF(NR(I).NE.I1.AND.NR(I).NE.I2.AND.I.NE.I1) GO TO 45
      CALL RDIST(A,M,N,I,NR(I),RD(I),NN,TH)
      J=NR(I)
      IF(I.NE.I1.OR.RD(I).GE.RD(J)) GO TO 45
      NR(J)=I
      RD(J)=RD(I)
   45 CONTINUE
      DO 46 J=2,N
      IF(A(1,J).LT.0) GO TO 46
      CALL CDIST(A,M,N,J,NC(J),CD(J),MM,TH)
   46 CONTINUE
      GO TO 70
   60 CONTINUE
C.... AMALGAMATE COLUMNS
C.... FIND CLOSEST COLUMN TO J1 OR J2
      A(1,J1)=-A(1,J1)
      CALL CDIST(A,M,N,J2,NC(J2),CD(J2),MM,TH)
      A(1,J2)=-A(1,J2)
      A(1,J1)=-A(1,J1)
      CALL CDIST(A,M,N,J1,NC(J1),CD(J1),MM,TH)
      JJ=NC(J2)
      IF(CD(J1).LE.CD(J2)) JJ=NC(J1)
      L1=A(1,J1)
      L2=-A(1,J2)
      LL=A(1,JJ)
      A(1,J1)=L2
      A(1,J2)=-L1
      NN=NN-1
C.... AMALGAMATE VALUES AND CREATE BLOCKS
      DO 50 I=2,M
      IF(KA.GE.KB-2) GO TO 80
      IF(A(I,1).LT.0) GO TO 50
      K=A(I,1)
      IF(AMAX1(A(K,L1),A(K,L2))-AMIN1(A(I,J1),A(I,J2)).LE.TH) GO TO 53
      Z1=AMAX1(A(K,L1),A(K,LL))-AMIN1(A(I,J1),A(I,JJ))
      Z2=AMAX1(A(K,L2),A(K,LL))-AMIN1(A(I,J2),A(I,JJ))
      IF(JJ.EQ.2) Z2=10.**10
      IF(JJ.EQ.2) Z1=10.**10
```

```
          IF(Z1.LE.TH) A(K,L2)=A(K,L1)
          IF(Z2.LE.TH) A(I,J1)=A(I,J2)
          IF(Z1.GT.TH.AND.Z2.GT.TH)A(I,J1)=A(I,JJ)
          IF(Z1.GT.TH.AND.Z2.GT.TH) A(K,L2)=A(K,LL)
          IF(Z1.LE.TH) GO TO 51
          KA=KA+1
          NB(1,KA)=I
          NB(2,KA)=A(I,1)
          NB(3,KA)=J1
          NB(4,KA)=L1
       51 IF(Z2.LE.TH) GO TO 50
          KA=KA+1
          NB(1,KA)=I
          NB(2,KA)=A(I,1)
          NB(3,KA)=J2
          NB(4,KA)=L2
          GO TO 50
       53 CONTINUE
          Z=A(K,L1)
          IF(A(I,J1).GE.A(I,J2)) A(I,J1)=A(I,J2)
          IF(Z.GE.A(K,L2)) A(K,L2)=Z
       50 CONTINUE
C.... UPDATE CLOSEST COLUMNS
          DO 55 J=2,N
          IF(A(1,J).LT.0) GO TO 55
          IF(NC(J).NE.J1.AND.NC(J).NE.J2.AND.J.NE.J1) GO TO 55
          CALL CDIST(A,M,N,J,NC(J),CD(J),MM,TH)
          I=NC(J)
          IF(J.NE.J1.OR.CD(J).GE.CD(I)) GO TO 55
          NC(I)=J
          CD(I)=CD(J)
       55 CONTINUE
          DO 56 I=2,M
          IF(A(I,1).LT.0) GO TO 56
          CALL RDIST(A,M,N,I,NR(I),RD(I),NN,TH)
       56 CONTINUE
          GO TO 70
C.... COMPUTE NEW ORDERING OF ROWS AND COLUMNS ,EXPRESS BLOCK BOUNDARIES
       80 CONTINUE
          IF(KA.GE.KB-2) WRITE(6,1)
        1 FORMAT(36H TOO MANY BLOCKS, INCREASE THRESHOLD   )
          IF(KA.EQ.0) WRITE(6,2)
        2 FORMAT(30H NO BLOCKS, DECREASE THRESHOLD   )
          IF(KA.EQ.0.OR.KA.GE.KB-2) RETURN
C.... FIND ROW ORDER
          J=A(2,1)
          DO 84 I=2,M
          IM=M-I+2
          NR(IM)=J
       84 J=-A(J,1)
C.... FIND COLUMN ORDER
          J=A(1,2)
          DO 87 I=2,N
          IN=N-I+2
          NC(IN)=J
       87 J=-A(1,J)
          DO 89 I=2,M
          J=NR(I)
       89 A(J,1)=I
          DO 91 J=2,N
          I=NC(J)
       91 A(1,I)=J

C.... ADJUST BLOCKS
          DO 90 K=1,KA
          I1=NB(1,K)
          I2=NB(2,K)
          J1=NB(3,K)
          J2=NB(4,K)
          I1=A(I1,1)
          I2=A(I2,1)
          J1=A(1,J1)
          J2=A(1,J2)
          NB(1,K)=I1
          NB(2,K)=I2
          NB(3,K)=J1
          NB(4,K)=J2
       90 CONTINUE
          KA=KA+1
          NB(1,K)=2
          NB(2,K)=M
          NB(3,K)=2
          NB(4,K)=N
          RETURN
          END
```

296

```
      SUBROUTINE RDIST(A,M,N,I,II,DR,NN,TH)
C.........................................................20 MAY 1973
C.... USED IN JOIN2
C.... FIND CLOSEST ROW TO ROW I
C.... A = M BY N BORDERED ARRAY
C.... M = NUMBER OF ROWS
C.... N = NUMBER OF COLUMNS
C....o I = TARGET ROW
C.... II = ROW CLOSEST TO I
C.... DR = DISTANCE OF II TO I
C.... NN = NUMBER OF COLUMNS NOT AMALGAMATED
C.... TH = THRESHOLD
C.............................................................
      DIMENSION A(M,N)
      TT=TH
      IF(TT.EQ.0) TT=1
      DR=10.**10
      II=2
      LL=A(I,1)
      IF(LL.LT.0) RETURN
      DO 20 J=2,M
      IF(J.EQ.I) GO TO 20
      L=A(J,1)
      IF(L.LT.0) GO TO 20
C.... COMPUTE THRESHOLD DISTANCE
      DN=0
      DD=0
      DO 21 K=2,N
      IF(A(1,K).LT.0) GO TO 21
      KK=A(1,K)
      DIF=AMAX1(A(L,KK),A(LL,KK))-AMIN1(A(I,K),A(J,K))
      DN=DN+1
      IF(DIF.GT.TH) DIF=TT
      DD=DD+DIF
      IF(DD.GE.DR*NN) GO TO 20
   21 CONTINUE
      IF(DN.NE.0) DD=DD/DN
      IF(DN.EQ.0) DD=TH
      IF(DD.GE.DR) GO TO 20
      DR=DD
      II=J
   20 CONTINUE
      RETURN
      END
```

```
      SUBROUTINE CDIST(A,M,N,I,JJ,DC,MM,TH)
C.................................................................20 MAY 1973
C.... FIND CLOSEST COLUMN TO I
C.... USED IN JOIN2
C.... A = M BY N BORDERED ARRAY
C.... M = NUMBER OF ROWS
C.... N = NUMBER OF COLUMNS
C.... I = TARGET COLUMN
C.... JJ = CLOSEST COLUMN TO I
C.... DC = DISTANCE OF I TO JJ
C.... MM = NUMBER OF ROWS NOT AMALGAMTED
C.... TH = THRESHOLD
C..................................................................................
      DIMENSION A(M,N)
      DC=10.**10
      TT=TH
      IF(TT.EQ.0) TT=1
      JJ=2
      LL=A(1,I)
      IF(LL.LT.0) RETURN
      DO 30 J=2,N
      IF(I.EQ.J) GO TO 30
      L=A(1,J)
      IF(L.LT.0) GO TO 30
C.... COMPUTE THRESHOLD DISTANCE
      DN=0
      DD=0
      DO 31 K=2,M
      IF(A(K,1).LT.0) GO TO 31
      KK=A(K,1)
      DN=DN+1
      DIF=AMAX1(A(KK,L),A(KK,LL))-AMIN1(A(K,I),A(K,J))
      IF(DIF.GT.TH) DIF=TT
      DD=DD+DIF
      IF(DD.GE.DC*MM) GO TO 30
   31 CONTINUE
      IF(DN.NE.0) DD=DD/DN
      IF(DN.EQ.0) DD=TH
      IF(DD.GE.DC) GO TO 30
      DC=DD
      JJ=J
   30 CONTINUE
      RETURN
      END




      SUBROUTINE PMUT(B,A,M,N,NR,NC)
C.................................................................20 MAY 1973
C.... PERMUTES AN ARRAY A ACCORDING TO NR AND NC INTO AN ARRAY B.
C.... M = NUMBER OF ROWS
C.... N = NUMBER OF COLUMNS
C.... A = M BY N BORDERED ARRAY
C.... B = M BY N BORDERED ARRAY. OBTAINED FROM A BY PERMUTATION
C.... NR = 1 BY M ARRAY,NR(I)=OLD INDEX IN ITH ROW POSITION IN NEW ARRAY
C.... NC = 1 BY N ARRAY, NC(I)=OLD INDEX IN ITH COLUMN POSITION IN NEW ARRAY
C..................................................................................
      DIMENSION A(M,N),B(M,N),NR(M),NC(N)
      DO 20 I=2,M
      K=NR(I)
      DO 20 J=1,N
   20 B(I,J)=A(K,J)
      DO 40 J=1,N
      DO 40 I=2,M
   40 A(I,J)=B(I,J)
      DO 30 J=2,N
      K=NC(J)
      DO 30 I=1,M
   30 B(I,J)=A(I,K)
      RETURN
      END
```

CHAPTER 16

Simultaneous Clustering and Scaling

16.1 INTRODUCTION

In the usual case when the data consist of a number of variables measured in different scales, it is necessary to express the variables in a common scale before distances between cases may be computed. A typical *ad hoc* rescaling requires all variables to have variance one or, more generally, requires every variable to make the same average contribution to the distance.

If the variables are $V(1), V(2), \ldots, V(N)$, then a scale V will be such that $V(I) = T(V, I)$, where $T(V, I)$ is a transformation of the common scale V to the variable $V(I)$. The transformation T will be linear for interval scale variables and monotonic for ordered variables. The variance standardizing transformation would be

$$V(I) = A(I)V + B(I),$$

where $A(I)$ is the standard deviation of $V(I)$ and $B(I)$ is the mean. In Table 16.1, relationships between the votes on various questions in the U.N. General Assembly (1969–1970) are tabulated. These show the necessity of various monotonic transformations to represent the responses on a common scale. For example, the relationship between $V1$ and $V3$ is essentially that the large yes vote on $V1$ has fragmented into yes, abstain, and no votes, in about equal proportions, for $V3$. A suitable common scale would take five values, 1, 2, 3, 4, 5, with

$$T(1, 1) = T(2, 1) = T(3, 1) = 1, \quad T(4, 1) = 2, \quad T(5, 1) = 3$$

and

$$T(1, 2) = 1, \quad T(2, 2) = 2, \quad T(3, 2) = T(4, 2) = T(5, 2) = 3.$$

In other words, the five values on the common scale correspond to values of $(V1, V3)$, successively: (1, 1), (1, 2), (1, 3), (2, 3), (3, 3).

Returning to the case of interval variables, there are serious defects with the method of equalizing variances. The principal one is that the variance calculation is very much affected by the presence of outliers or other clusters in the data. What is necessary is to continue rescaling as cluster information is exposed and to use standardizing techniques, such as equalizing interquartile ranges, that are not too sensitive to outliers or other clusters.

There follows a number of algorithms, of the joining type, for simultaneously clustering cases and variables while rescaling variables. These algorithms are different according to the type of variable being rescaled. The more difficult and intricate procedures necessary for combining different types of variables have been neglected.

299

Table 16.1   Relationship Between Votes on Various U.N. Questions (1969–1970)

		V1						V2		
		1	2	3				1	2	3
	1	27	2	1			1	2	6	22
V3	2	23	3	1		V3	2	7	10	10
	3	23	0	44			3	54	13	0

SHIFT OF V1 YES TOWARD V4 NO .   REVERSAL V2 YES TO V4 NO .

		V4						V4		
		1	2	3				1	2	3
	1	11	0	0			1	2	22	41
V5	2	0	26	2		V1	2	0	1	3
	3	0	4	69			3	9	8	26

IDENTICAL QUESTIONS.   WEAK RELATIONSHIPS.

1, yes; 2, abstain; 3, no. $V1$, declare the China admission question an important question; $V2$, to make the study commission on China admission "important"; $V3$, to form a study commission on China admission; $V4$, replace last paragraphs of preamble, on South Africa expulsion from UNCTAD, by Hungarian amendment; $V5$, adopt the Hungarian amendment of paragraph 1 and 2 on South Africa expulsion.

16.2   SCALING ORDERED VARIABLES

Preliminaries.   Given two ordered variables X and Y taking values $\{X(I), Y(I), 1 \leq I \leq M\}$ on M cases, it is desired to find a scale Z, an ordered variable taking values $\{Z(I), 1 \leq I \leq M\}$, and monotonic transformations $T(Z, 1)$ and $T(Z, 2)$ of Z, such that $X(I) = T[Z(I), 1]$ and $Y(I) = T[Z(I), 2]$ with maximum frequency.

Let the values taken by X be the integers $1, 2, \ldots, N1$, let the values taken by Y be the integers $1, 2, \ldots, N2$, and let $N(I, J)$ denote the number of cases with values $X = I$ and $Y = J$. The variable Z will take values $[I(1), J(1)], \ldots, [I(K), J(K)]$, where $I(1) \leq I(2) \leq \cdots \leq I(K)$ and $J(1) \leq J(2) \leq \cdots \leq J(K)$ or $J(1) \geq J(2) \geq \cdots \geq J(K)$, and $N[I(1), J(1)] + N[I(2), J(2)] + \cdots + N[I(K), J(K)]$ is a maximum. [The transformations are $T[I(L), J(L), 1] = I(L)$, and $T[I(L), J(L), 2] = J(L)$.] The algorithm uses a maximization technique similar to dynamic programming. Let $\text{NMAX}(I, J)$ denote the maximum value of $N[I(1), J(1)] + N[I(2), J(2)] + \cdots + N[I(K), J(K)]$ subject to the constraints $1 \leq I(1) \leq I(2) \leq \cdots \leq I(K) \leq I$ and $1 \leq J(1) \leq J(2) \leq \cdots \leq J$. Then $\text{NMAX}(I, J) = N(I, J) + \max [\text{NMAX}(I, J - 1), \text{NMAX}(I - 1, J)]$. In this way, an optimal sequence increasing in I and J, connecting $(1, 1)$ to (N_1, N_2), is discovered. Similarly, discover an optimal sequence increasing in I but decreasing in J.

STEP 1.   Compute $N(I, J)$, the number of times variable X takes value I and variable Y takes value J. Set

$$\text{NMAX}(0, J) = \text{NMAX}(I, 0) = 0 \qquad (1 \leq J \leq N1, 1 \leq I \leq N2).$$

STEP 2.   For each J $(1 \le J \le N2)$, compute for each I $(1 \le I \le N1)$ $\text{NMAX}(I, J) =$ $N(I, J) + \max [\text{NMAX}(I, J - 1), \text{NMAX}(I - 1, J)]$.

STEP 3.   Set $L = N1 + N2 - 1$, $I(L) = N1$, $J(L) = N2$.

STEP 4.   By definition,

$$\text{NMAX}[I(L), J(L)] = N[I(L), J(L)] + \text{NMAX}[I(L), J(L) - 1]$$

or

$$\text{NMAX}[I(L), J(L)] = N[I(L), J(L)] + \text{NMAX}[I(L) - 1, J(L)].$$

In the first case $I(L - 1) = I(L)$, $J(L - 1) = J(L) - 1$, and in the second case $I(L - 1) = I(L) - 1$, $J(L - 1) = J(L)$. If $L = 2$, go to step 5. Otherwise, decrease L by 1 and repeat this step.

STEP 5.   Define a new variable U by $U = N2 - J + 1$ when $Y = J$. Discover the optimal monotonic relationship between X and U, following Steps 1–4. If

Table 16.2 Scaling Components of Mammal's Milk

	ANIMAL	WATER %	PROTEIN %	NMAX	JMAX	Reversed Protein NMAX	JMAX
1.	Dolphin-----	44.9	10.6	1	-	1	-
2.	Seal--------	46.4	9.7	1	-	2	1
3.	Reindeer----	64.8	10.7	2	2	2	4
4.	Whale-------	64.8	11.1	3	3	1	-
5.	Deer--------	65.9	10.4	2	2	3	3
6.	Elephant----	70.7	3.6	1	-	4	5
7.	Rabbit------	71.3	12.3	4	4	1	-
8.	Rat---------	72.5	9.2	2	6	4	5
9.	Dog---------	76.3	9.3	3	8	4	5
10.	Cat---------	81.6	10.1	4	9	4	5
11.	Fox---------	81.6	6.6	2	6	5	9
12.	Guinea Pig--	81.9	7.4	3	11	5	9
13.	Sheep-------	82.0	5.6	2	6	6	12
14.	Buffalo-----	82.1	5.9	3	13	6	12
15.	Pig---------	82.8	7.1	4	14	6	12
16.	Zebra-------	86.2	3.0	1	-	7	15
17.	Llama-------	86.5	3.9	2	16	7	15
18.	Bison-------	86.9	4.8	3	17	7	15
19.	Camel-------	87.7	3.5	2	16	8	18
20.	Monkey------	88.4	2.2	1	-	9	19
21.	Orangutan---	88.5	1.4	1	-	10	20
22.	Mule--------	90.0	2.0	2	21	10	20
23.	Horse-------	90.1	2.6	3	22	9	19
24.	Donkey------	90.3	1.7	2	21	11	22

Data ordered by water percentage.

NMAX($N1$, $N2$) is larger for U and X than for Y and X, the optimal relationship overall is increasing in X and decreasing in Y.

NOTE   If the ordered variables X and Y take a very large number of different values, the contingency table $N(I, J)$ will mostly consist of 0's and 1's and will be rather expensive to store and manipulate. Suppose that the variables X and Y take the values $\{X(I), Y(I)\}$ $N(I)$ times. Assume the data ordered so that $X(I) \leq X(J)$ if $I \leq J$, and $Y(I) < Y(J)$ if $X(I) = X(J)$ and $I < J$. The quantity NMAX(I) is the maximum value of $N[I(1)] + \cdots + N[I(K)]$ subject to

$$X[I(1)] \leq X[I(2)] \leq \cdots \leq X[I(K)] = X(I)$$

and

$$Y[I(1)] \leq Y[I(2)] \leq \cdots \leq Y[I(K)] = Y(I).$$

Compute NMAX(I) iteratively ($1 \leq I \leq M$), setting

$$\text{NMAX}(I) = \max_J [\text{NMAX}(J)] + N(I),$$

where $X(J) \leq X(I)$ and $Y(J) \leq Y(I)$ and $I \neq J$. The quantity JMAX(I) is the value of J which maximizes NMAX(J) under the above constraint. The sequence of Z values is $I(1)$, maximizing NMAX(I), then $I(2) = \text{JMAX}[I(1)]$, $I(3) = \text{JMAX}[I(2)]$, and so on.

This algorithm is applied to mammal's milk components in Table 16.2. A scale is computed with water and protein both increasing and also with water increasing and protein decreasing. The second relationship is preferred since 11 of 24 points are covered in the fitting curve. The curve is graphed in Figure 16.1.

16.3 SCALING ORDERED VARIABLES APPLIED TO U.N. QUESTIONS

The questions to be scaled are $V2$ and $V3$ as given in Table 16.1, two questions on a study commission on the China admission question.

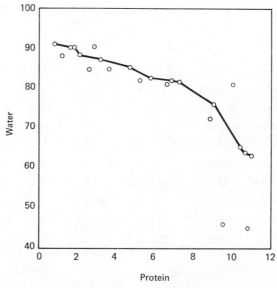

Figure 16.1 Monotonic scale for mammals' milk.

STEP 1.   In the terminology of the algorithm, $X = V3$ and $Y = V2$. Then $N(1, 1) = 2$, $N(1, 2) = 6$, $N(1, 3) = 22$, $N(2, 1) = 7$, $N(2, 2) = 10$, $N(2, 3) = 10$, $N(3, 1) = 54$, $N(3, 2) = 13$, $N(3, 3) = 0$. $NMAX(0, 1) = NMAX(0, 2) = NMAX(0, 3) = 0$, $NMAX(1, 0) = NMAX(2, 0) = NMAX(3, 0) = 0$.

STEP 2   First $NMAX(1, 1) = 2$, then

$$NMAX(2, 1) = N(2, 1) + \max [NMAX(1, 1), NMAX(2, 0)]$$
$$= 7 + 2 = 9.$$

$NMAX(3, 1) = 63$, $NMAX(1, 2) = 8$, $NMAX(2, 2) = 19$, $NMAX(3, 2) = 76$, $NMAX(1, 3) = 30$, $NMAX(2, 3) = 40$, $NMAX(3, 3) = 76$.

STEP 3.   Set $L = 5$, $I(5) = 3$, $J(5) = 3$.

STEP 4.   Since $NMAX(3, 3) = 0 + NMAX(3, 2)$, $I(4) = 3$, $J(4) = 2$. Decrease L to 4, since $NMAX(3, 2) = N(3, 2) + NMAX(3, 1)$. Therefore $I(3) = 3$, $J(3) = 1$. Similarly, $I(2) = 2$, $J(2) = 1$ and $I(1) = 1$, $J(1) = 1$. The final optimal increasing sequence is thus $(1, 1)$, $(2, 1)$, $(3, 1)$, $(3, 2)$, $(3, 3)$.

STEP 5.   Define the variable U: $U = 1$ if $V2 = 3$, $U = 2$ if $V2 = 2$, $U = 3$ if $V2 = 1$. Repeating Steps 1–4, discover the sequence $(1, 3)$, $(1, 2)$, $(2, 2)$, $(3, 2)$, $(3, 1)$, which covers 109 points. This sequence is thus preferred to the sequence increasing in I and J. The final scale is $Z = (1, 3)$, $(1, 2)$, $(2, 2)$, $(3, 2)$, $(3, 1)$ with $T[(I, J), 2] = J$. The function $T(Z, 1)$ is increasing; the function $T(Z, 2)$ is decreasing.

A number of such scales, which can be computed very quickly by hand for ordered variables taking just a few values, are given in Table 16.4.

16.4   JOINER SCALER

Preliminaries.   The data matrix $\{A(I, J), 1 \leq J \leq M, 1 \leq J \leq N\}$ is a collection of N ordered variables measured on different scales. During the algorithm's execution, pairs of the variables are joined to form new variables, pairs of cases are joined to form new cases, and a common scale for all variables is constructed.

The output consists of data clusters within which all values are equal when expressed in the common underlying scale. The data clusters $1, 2, \ldots, KD$ are determined by the corresponding row and column clusters $IR(I)$, $IC(I)$ for the Ith clusters. The tree structure of the row clusters $1, 2, \ldots, KR$ is determined by the function $JR(I)$ which is the smallest row cluster properly including cluster I. The tree structure of the column clusters $1, 2, \ldots, KC$ is determined by the function $JC(I)$, which is the smallest column cluster properly including cluster I.

STEP 1.   Set $KR = M$, $KC = N$, $JR(I) = 0$ $(1 \leq I \leq M)$, and $JC(I) = 0$ $(1 \leq I \leq N)$.

STEP 2.   Compute distances between all pairs of row clusters I, J $[1 \leq I, J \leq KR$, $JR(I) = JR(J) = 0]$ as the proportion of columns in which $A(I, K) \neq A(J, K)$, among columns for which $A(I, K)$ and $A(J, K)$ are both defined and in which $JC(K) = 0$. Let the smallest distance be DROW and the corresponding rows be IROW, JROW.

Table 16.3   Mammal's Milk

	WATER	PROTEIN	FAT	LACTOSE	ASH
Horse	90.1	2.6	1.0	6.9	0.35
Orangutan	88.5	1.4	3.5	6.0	0.24
Monkey	88.4	2.2	2.7	6.4	0.18
Donkey	90.3	1.7	1.4	6.2	0.40
Hippo	90.4	0.6	4.5	4.4	0.10
Camel	87.7	3.5	3.4	4.8	0.71
Bison	86.9	4.8	1.7	5.7	0.90
Buffalo	82.1	5.9	7.9	4.7	0.78
Guinea Pig	81.9	7.4	7.2	2.7	0.85
Cat	81.6	10.1	6.3	4.4	0.75
Fox	81.6	6.6	5.9	4.9	0.93
Llama	86.5	3.9	3.2	5.6	0.80
Mule	90.0	2.0	1.8	5.5	0.47
Pig	82.8	7.1	5.1	3.7	1.10
Zebra	86.2	3.0	4.8	5.3	0.70
Sheep	82.0	5.6	6.4	4.7	0.91
Dog	76.3	9.3	9.5	3.0	1.20
Elephant	70.7	3.6	17.6	5.6	0.63
Rabbit	71.3	12.3	13.1	1.9	2.30
Rat	72.5	9.2	12.6	3.3	1.40
Deer	65.9	10.4	19.7	2.6	1.40
Reindeer	64.8	10.7	20.3	2.5	1.40
Whale	64.8	11.1	21.2	1.6	1.70
Seal	46.4	9.7	42.0	-	0.85
Dolphin	44.9	10.6	34.9	0.9	0.53

From *Handbook of Biological Data* (1956), William S. Spector, ed., Saunders.

STEP 3.   Compute distances between each pair of columns by finding the monotonic scale which covers most rows, as in the algorithm for scaling ordered variables. (Look only at columns I, J for which $JC(I) = JC(J) = 0$, and look only at rows K for which $JR(K) = 0$ and for which $A(I, K)$ and $A(J, K)$ are both defined.) The distance between I and J is the number of rows not covered by the monotonic scale, divided by the total number of rows considered less 2. (The reason for the 2 is that two rows will always be covered.) Let the smallest distance be DCOL and the corresponding columns be ICOL, JCOL.

STEP 4.   (If the minimum of DCOL and DROW is 1, go to Step 6.) If DCOL < DROW, go to Step 5. Otherwise increase KR by 1, $JR(IROW) = JR(JROW) = KR$, $JR(KR) = 0$. For each column K $[1 \leq K \leq KC, JC(K) = 0]$, set $A(KR, K) = A(IROW, K)$ if $A(IROW, K) = A(JROW, K)$. If $A(IROW, K)$ is undefined, set $A(KR, K) = A(JROW, K)$. If $A(IROW, K) \neq A(JROW, K)$, leave $A(KR, K)$ undefined and define data clusters KD + 1 and KD + 2 by $IR(KD + 1) = IROW$, $IC(KD + 1) = K$, $IR(KD + 2) = JROW$, $IC(KD + 2) = K$. Increase KD by 2 and go to the next column cluster K. If all column clusters have been adjusted, return to Step 2.

STEP 5. Increase KC by 1, define JC(ICOL) = JC(JCOL) = JC, JC(KC) = 0. For each row cluster K [$1 \leq K \leq$ KR, JR(KR) = 0] define $A(K, \text{KC})$ to be the value in the new scale corresponding to $A(K, \text{JCOL})$ and $A(K, \text{JCOL})$] if this value is uniquely defined. Otherwise, define data clusters KD + 1 and KD + 2 by IR(KD + 1) = K, IC(KD + 1) = ICOL, IR(KC + 2) = K, JC(KD + 2) = JCOL, and increase KD by 2. Return to Step 2.

STEP 6. A single underlying scale has been constructed with monotonic functions from this scale to each original variable. Within each data cluster, consider the data values that are not included in some smaller cluster. Each such data value corresponds to a range of scale values. The intersection of these ranges is always nonempty, and this intersection range is recorded for each data cluster.

Beginning with the largest clusters and moving toward the smaller, eliminate a cluster I if the smallest cluster containing it has an intersection range which includes the intersection range for I. Otherwise, change the intersection range for I to be the smallest value in the range.

Table 16.4 Scaling U.N. Questions

V1

		1	2	3
	1	27	2	1
V3	2	23	3	1
	3	23	0	44

V2

		1	2	3
	1	2	6	22
V3	2	7	10	10
	3	54	13	0

V4

		1	2	3
	1	11	0	0
V5	2	0	26	2
	3	0	4	69

V4

		1	2	3
	1	2	22	41
V1	2	0	1	3
	3	9	8	26

Blocks are different values of constructed scale.

16.5 APPLICATION OF JOINER-SCALER ALGORITHM TO U.N. VOTES

It is natural to apply a two-way clustering algorithm to the U.N. votes (Table 16.5) because there are blocs of countries such as Bulgaria, Romania, and the USSR that vote similarly, and blocs of questions that arise from the same issue, such as "China admission," "importance of China admission," "study China admission," "importance of studying China admission."

Table 16.5 Selected Votes in the United Nations (1969–1970)

Y = YES				N = NO			A = ABSTAIN		
1	2	3	4	5	6	7	8	9	10

	1	2	3	4	5	6	7	8	9	10
1. CANADA	N	A	Y	A	N	A	A	Y	Y	Y
2. CUBA	Y	A	N	Y	Y	N	Y	A	N	N
3. MEXICO	N	Y	Y	N	N	Y	Y	A	A	Y
4. UNITED KINGDOM	N	N	Y	Y	N	A	N	A	Y	Y
5. NETHERLANDS	N	N	Y	A	N	Y	A	A	Y	Y
6. FRANCE	N	A	N	Y	A	N	A	A	Y	Y
7. SPAIN	N	A	Y	N	Y	Y	A	A	A	Y
8. PORTUGAL	A	N	A	A	A	A	N	N	Y	Y
9. POLAND	Y	Y	N	Y	A	N	Y	Y	A	A
10. AUSTRIA	N	A	A	A	A	A	A	Y	Y	Y
11. HUNGARY	Y	Y	N	Y	Y	N	Y	Y	A	A
12. CZECHOSLOVAKIA	Y	Y	N	Y	A	N	Y	Y	A	A
13. ITALY	N	A	Y	N	N	Y	A	A	Y	Y
14. BULGARIA	Y	Y	N	Y	Y	N	Y	Y	A	A
15. ROMANIA	Y	Y	N	Y	Y	N	Y	Y	A	A
16. USSR	Y	Y	N	Y	A	N	Y	Y	A	A
17. FINLAND	A	A	N	Y	A	N	A	Y	Y	Y
18. GAMBIA	N	A	Y	N	A	N	A	A	A	A
19. MALI	A	Y	N	Y	Y	N	A	Y	N	N
20. SENEGAL	A	Y	Y	A	A	A	Y	Y	N	N
21. DAHOMEY	A	Y	Y	N	Y	N	Y	Y	N	N
22. NIGERIA	N	Y	Y	N	Y	N	Y	Y	N	N
23. IVORY COAST	N	Y	Y	N	Y	N	Y	Y	A	A
Y/N/A	7/11/5	12/3/8	11/10/2	11/7/5	9/5/9	4/14/5	12/2/9	14/1/8	8/5/10	10/5/8

Columns: 1, to adopt USSR proposal to delete item on Korea unification; 2, to call upon the UK to use force against Rhodesia; 3, declare the China admission question an important question; 4, recognize mainland China and expel Formosa; 5, to make study commission on China admission important; 6, to form a study commission on China admission; 7, convention on no statutory limits on war crimes; 8, condemn Portuguese colonialism; 9, defer consideration of South Africa expulsion; 10, South Africa expulsion is important question.

Also, the questions must be rescaled. For example, "importance of China admission" and "study China admission" are similar questions translated on an underlying scale, so that some of the yes votes on "importance" become abstains on "study." The other two China questions are very similar in producing opposite votes from almost every country.

STEP 1. To initialize, set KR = 23, KC = 10, JR(I) = 0 (1 ≤ I ≤ 23), and JC(I) = 0 (1 ≤ I ≤ 10).

STEP 2. Compute the distance between all pairs of rows. For example, row 1 and row 2 match in just one vote, so the distance between Canada and Cuba is $\frac{9}{10}$. The smallest row distance (there are several, and one is chosen arbitrarily) is DROW = 0, IROW = 12 (Czechoslovakia), JROW = 16 (USSR).

STEP 3. Find the monotonic scale for columns 1 and 2 that covers most rows. This is done by using the previous algorithm of Section 16.2. The optimal scale has five values, YY, AY, NY, YA, NN, which cover 18 of 23 rows. The distance between columns 1 and 2 is therefore $1 - \frac{5}{21}$. (Note that 2 is subtracted from the 23, because a monotonic scale always covers two points for free. This becomes important in the later stages of the algorithm when just a few rows and columns remain.) Examining all pairs of columns, discover DCOL = 0 for ICOL = 9, JCOL = 10.

STEP 4. Increase KR to 24, define JR(12) = 24, JR(16) = 24, JR(24) = 0. Since rows 12 and 16 are identical, row 24 is the same as row 12. Return to Step 2, and amalgamate rows 24 and 9 to be row 25, rows 14 and 15 to be row 26, and rows 11 and 26 to be row 27. On the next return to Step 2, columns 9 and 10 are closer than any pair of rows, and Step 5 is taken.

STEP 5. Increase KC to 11, JC(9) = JC(10) = 11, JC(11) = 0. The monotonic scale takes values 1, 2, 3, 4, corresponding to the pairs (Y, Y), (A, Y), (A, A), (N, N). Note that this sequence is monotonic in both variables. All pairs of votes fall in one of these four categories, so no data clusters are formed.

STEP 4 REPEATED. The next closest pair of row or column clusters are rows 21 and 22. Set KR = 28, JR(21) = JR(22) = 28, JR(28) = 0. Define $A(28, K) = A(21, K)$ except for $K = 1$, since $A(21, 1) \neq A(22, 1)$. Define two data clusters by IR(1) = 21, IR(2) = 22, IC(1) = 1, IC(2) = 1, and increase KD to 2. The algorithm continues in this way until a single column remains and several row clusters which are a distance of 1 from each other. The data clusters at this stage are given in Table 16.6. Also all original variables are monotonic functions of the scale of the column cluster which replaced them. These column clusters are joined, pairwise, with other column clusters till a single column cluster remains. All original variables will be monotonic functions of the scale of this final column cluster, given in Table 16.7.

STEP 6. Each data cluster generates a range of scale values, the intersection of the ranges of scale values over all values in the cluster. Consider, for example, the data cluster corresponding to rows 1 and 17 and columns 3, 4, 6, 1, 2, 5. The data values which are not included in smaller clusters are A, N, A, N for row 17 and columns 6, 1, 2, 5. From Table 16.7, these correspond to ranges of final scale values, 5–8, 6-E, 4-7, 7-E. The intersection of these ranges is the value 7. Such a value is associated with every data cluster.

For some data clusters, the intersection range includes that of the next largest cluster, and the data cluster is deleted. For example, the data cluster rows 21, 22, 23 by columns 7, 8, 9, 10 has intersection range C-E which includes that of the next largest cluster, rows 3–23 by columns 3–10, intersection range C. This data cluster is deleted.

Beginning with the largest clusters, every cluster is either deleted or has its intersection range replaced by the smallest value in it. A single scale value is thus associated with each remaining cluster, as in Table 16.8. The original data is recoverable from this representation in 41 data clusters, using the scale-to-variable transformations in Table 16.7.

The clusters of countries are {Senegal}, {African bloc}, {Netherlands, Italy}, {Soviet bloc}, {fringe neutrals}, {Portugal}, and {United Kingdom}. The clusters of questions are {China questions}, {African questions}.

Table 16.6   Preliminary Data Clusters in Applying Joiner-Scaler Algorithm to U.N. Data

	3	4	6	1	2	5	7	8	9	10
20	Y	A	A	A	Y	A	Y	Y	N	N
3	Y	N	Y	N	Y	N	Y	A	A	Y
7	Y	N	Y	N	A	Y	A	A	A	Y
6	Y	N	N	N	A	A	A	A	A	A
18	Y	N	N	N	A	A	A	A	Y	Y
21	Y	N	N	A	Y	Y	Y	Y	N	N
22	Y	N	N	N	Y	Y	Y	Y	N	N
23	Y	N	N	N	Y	Y	Y	Y	A	A
5	Y	A	Y	N	N	N	A	A	Y	Y
13	N	Y	Y	N	A	N	A	A	Y	Y
19	N	Y	N	A	Y	Y	A	Y	N	N
2	N	Y	A	Y	A	Y	Y	A	N	N
11	N	Y	A	Y	Y	Y	Y	Y	A	A
14	N	Y	A	Y	Y	Y	Y	Y	A	A
15	N	Y	A	Y	Y	Y	Y	Y	A	A
9	N	Y	A	Y	Y	A	Y	Y	A	A
12	N	Y	A	Y	Y	A	Y	Y	A	A
16	N	Y	A	Y	Y	A	Y	Y	A	A
10	A	A	A	N	A	A	A	Y	Y	Y
17	Y	N	N	A	A	A	A	Y	Y	Y
1	Y	A	A	N	A	N	A	Y	Y	Y
8	A	A	A	A	N	A	N	N	Y	Y
4	Y	Y	A	N	N	N	N	A	Y	Y

16.6 THINGS TO DO

16.6.1 Running the Joiner Scaler

The algorithm assumes that given variables are obtained by monotonic transformation from some underlying scale to be discovered in the course of the algorithm. It thus produces results invariant under monotonic transformation of the variables.

Table 16.7 Common Scale for All U.N. Questions, Output of Joiner-Scaler

SCALE.	1	2	3	4	5	6	7	8	9	B	C	D	E
QUESTION 1.	Y	A	A	A	A	N	N	N	N	N	N	N	N
2.	Y	Y	Y	A	A	A	A	N	N	N	N	N	N
3.	N	N	A	A	A	A	Y	Y	Y	Y	Y	Y	Y
4.	Y	Y	A	A	A	A	A	N	N	N	N	N	N
5.	Y	Y	A	A	A	A	N	N	N	N	N	N	N
6.	N	N	N	N	A	A	A	A	Y	Y	Y	Y	Y
7.	N	N	N	N	N	N	N	N	A	A	Y	Y	Y
8.	N	N	N	N	N	A	A	A	A	Y	Y	Y	Y
9.	Y	Y	Y	Y	Y	Y	Y	Y	Y	Y	A	A	N
10.	Y	Y	Y	Y	Y	Y	Y	Y	Y	Y	Y	A	N

It is expensive to use if each variable takes many different values. The time is proportional to $M^2 K^2 N^2$, where K is the number of different values taken by each variable, averaged over different variables. In using it with continuous variables, it is suggested that you reduce the number of different values taken by each variable to between 5 and 10.

16.6.2 Monotonic Subsequences

For any sequence of length n, show that there is an increasing subsequence of length r and a decreasing subsequence of length s, such that $rs \geq n$. Thus for any n points in two dimensions, there is a monotone curve passing through at least $\sqrt{n}$.

16.6.3 Category Data

If each variable is a category variable, the results should be invariant under arbitrary one-to-one transformations of each variable. Therefore there will be an underlying scale of block values, a category scale, from which the given variables must be obtained by transformation.

In the joining algorithm, the basic problem is always the distance, the amalgamation rule, and block construction for pairs of rows or columns. The rows will be handled as usual by using matching distances and constructing blocks at the mismatches. The variables require new treatment. One simple procedure measures the distance between two variables as $\sum K(I, J)[K(I, J) - 1]/M(M - 1)$, where $K(I, J)$ counts the number of times, in M cases, that the first variable takes the value I and the second variable takes the value J. The new variable just takes the set of values (I, J) which actually occur, and no blocks are constructed when variables are joined.

16.6.4* Continuous Data

In both the monotonic data and category data approaches, the final blocks have the property that every variable within a block is constant over cases within a block. This property is not realistic in the continuous data case. It is plausible to consider either monotonic transformation or linear transformations from the block-value scale, but it is necessary that a threshold be given for each variable, such that the variable ranges within the threshold over the cases in a block.

Table 16.8 Final Data Clusters in Applying Joiner Scaler to U.N. Votes

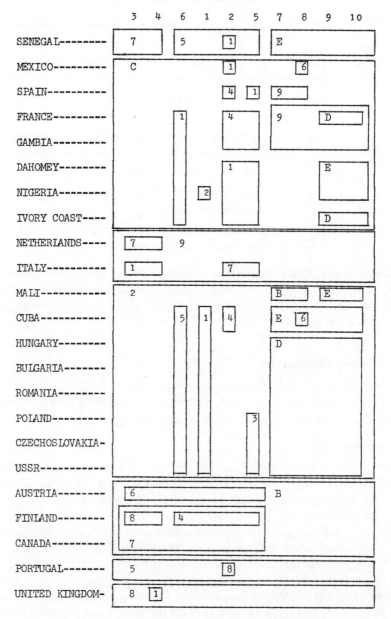

To translate this table, look at Mexico on Question 3, taking the value C . From 16.7, the value C on Question 3 is Y . Thus Mexico votes Yes on Question 3.

Original data are recovered by relating scale values to questions (Table 16.7).

In considering the linear case, pairs of cases are treated as in the range algorithm in Chapter 15, but pairs of variables must be considered freshly. Suppose that X and Y are variables taking values $[X(I), Y(I)]$ with thresholds TX and TY. A new variable Z will be constructed, connected to X and Y by

$$X = A(1)Z + A(2)$$

and

$$Y = B(1)Z + B(2).$$

There will be a threshold TZ for Z that is the minimum of $TX/A(1)$ and $TY/B(1)$. For each case I, there is a difference between the Z values $D(I) = |[X(I) - A(2)]/A(1) - [Y(I) - B(2)/B(1)]|$. Define

$$DD(I) = D(I)/TZ \qquad \text{if} \qquad D(I) \leq TZ,$$
$$DD(I) = 1 \qquad \text{if} \qquad D(I) \geq TZ.$$

Then $\sum \{1 \leq I \leq M\} DD(I)$ measures the distance between X and Y for the particular choice of scale parameters $A(1)$ and $A(2)$, $B(1)$ and $B(2)$. Of course, these must be chosen to minimize $DD(I)$. You see instantly that $B(1) = 1$, $B(2) = 0$ without loss. It is true also that the optimal choice of $A(1)$ and $A(2)$ is such that $D(I) = 0$ for two cases I. Thus the optimal values of $A(1)$ and $A(2)$ are obtained by searching over all the lines through pairs of points. (The time for a complete join of all rows and columns is thus proportional to M^3N^2.) Blocks are constructed, as in the homogeneous case, whenever a value is out of threshold with the value it is being joined to and is out of threshold with the value it is likely to be joined to next.

Complications arise later on in the algorithm, when each value becomes a range of values. For a pair of variables, the range is a rectangle with four corners. The optimal scale choice passes through corners for two cases, and so the same search procedure finds the optimal scaling.

16.6.5 Greater Generality

To handle data in which different variables are on entirely different scales, such as continuous, ordered, or category scales, it is supposed that there is an underlying block scale. All values in a block take a single block value z. For a variable I, there is a transformation $T(I, z)$ which specifies the value of variable I when the block value is z.

Thus $T(I, z)$ might be a linear transformation of z, or a monotonic transformation, or an arbitrary transformation, according to the type of variable. The problem of combining different types of variables to produce such a scale remains to be solved.

16.6.6 Median Regression

If X, Y are variables taking values $X(I)$, $Y(I)$, a median regression line of Y on X is the line $y = a + bx$, where a, b are chosen to minimize

$$\sum \{1 \leq I \leq M\} |Y(I) - a - bX(I)|.$$

Show that there is a median regression line for which $Y(I) = a + bX(I)$ for two values of I. Suppose that cases I, J are such that $Y(I) = a + bX(I)$, $Y(J) = a + bX(J)$. Suppose that for every K the lines through I, K and J, K have larger sums of absolute deviations than the line through I, J. Then the line through I, J is a median regression line.

REFERENCES

HAMMERSLEY, J. M. (1972). "A few seedlings of research." *Proceedings of the Sixth Berkeley Symposium on Mathematical Statistics and Probability*, Vol. I, pp. 345–393. U. of California, Berkeley. Considers various properties of the longest increasing subsequence of a random permutation of n integers. For example, the length L of the subsequence exceeds $\sqrt{n} - 1$, a result proved by Erdos and Szekeres in 1935. Hammersley shows that $L/\sqrt{n} \to c$ as $n \to \infty$, where $\frac{1}{2}\pi \le c \le e$. Hammersley presents a variety of nonrigorous arguments suggesting that c is close to 2.

CHAPTER 17

Factor Analysis

17.1 INTRODUCTION

Consider the correlations between physical measurements listed in Table 17.1. The correlations are relatively high within the groups head length, head breadth, face breadth, and foot, forearm, height, finger length, and somewhat smaller between the groups. An explanation for such a pattern is that all variables contain a "dimension factor," that the first group contains a "head factor," and the second group a "height factor." The general technique of representing variables as weighted sums of hypothetical factors is known as factor analysis. The principal developers and users of this technique have been psychologists, although it has been applied to every type of data.

Denote the variables to be investigated by $V(1), V(2), \ldots, V(N)$, and denote the factors by $F(1), F(2), \ldots, F(K)$. Then

$$V(I) = \sum \{1 \leq J \leq K\} B(I, J)F(J).$$

Thus each variable lies in the vector space spanned by the factors $F(1), F(2), \ldots, F(K)$. To be concrete, suppose the variables to be measured are the results of the following: $V(1)$, arithmetic; $V(2)$, geometry; $V(3)$, drawing; $V(4)$, spelling; $V(5)$, writing tests. Let the factors be the following: $F(1)$, intelligence; $F(2)$, mathematical ability; $F(3)$, spatial perception; $F(4)$, verbal ability. The equations might be

$$V(1) = 0.5F(1) + 0.3F(2),$$
$$V(2) = 0.4F(1) + 0.2F(2) + 0.3F(3),$$
$$V(3) = 0.3F(1) + 0.1F(2) + 0.4F(3),$$
$$V(4) = 0.4F(1) + 0.2F(4),$$
$$V(5) = 0.6F(1) + 0.3F(4).$$

In particular, if an individual had intelligence 20 and mathematical ability 10, the arithmetic score would be $0.5 \times 20 + 0.3 \times 10 = 13$. The matrix of coefficients $\{B(I, J), 1 \leq I \leq N, 1 \leq J \leq K\}$ is called the loading matrix, and the particular coefficient $B(I, J)$ is the loading of the variable $V(I)$ on the factor $F(J)$.

Basic operations in factor analysis are determination of the factors $\{F(I), 1 \leq I \leq K\}$ and of the loading matrix $\{B(I, J)\}$. The factors are usually interpreted by examination of the loadings of the given variables $\{V(I), 1 \leq I \leq N\}$ on them. It is apparent that the factors and loading matrix will not be uniquely determined. There are many

Table 17.1   Correlations Between Physical Measurements

		HL	HB	FB	FT	FM	HT	FL
HL	Head Length----	1.000	.402	.395	.339	.305	.340	.301
HB	Head Breadth---	.402	1.000	.618	.206	.135	.183	.150
FB	Face Breadth---	.395	.618	1.000	.363	.289	.345	.321
FT	Foot-----------	.339	.206	.363	1.000	.797	.736	.759
FM	Forearm--------	.305	.135	.289	.797	1.000	.800	.846
HT	Height---------	.340	.183	.345	.736	.800	1.000	.661
FL	Finger Length--	.301	.150	.321	.759	.846	.661	1.000

(From K. Pearson (1901). On lines and planes of closest fit to points
in space. Philosphical Magazine 559-572)

factor-analytic representations of a given set of data, and this plethora of solutions is a permanent embarrassment to the keen factor analyst. In particular, if the factors are transformed to some other set of factors by a linear transformation, the inverse transformation operates on the loading matrix, and exactly the same variables are represented by the new factors and loadings.

One way of reducing ambiguity assumes the factors have a unit covariance matrix, so that the covariance matrix $\{C(I, J), 1 \leq I, J \leq N\}$ of the variables may be written as

$$C(I, J) = \sum \{1 \leq L \leq K\} B(I, L)B(J, L).$$

The loading matrix is said to be a root of the given covariance matrix. This root remains undetermined up to a rotation of factors (since such a rotation does not change the unit covariance matrix of the factors). To determine the root, further constraints are necessary, such as orthogonality of the vectors $\{B(I, L), 1 \leq I \leq N\}$ for different L, or simple structure, which requires that many of the entries in the loading matrix be nearly zero. The simple structure requirement is formalized in a number of algorithms, which have names such as QUARTIMAX or VARIMAX.

A clustering model for the loading matrix associates each factor with a cluster of variables, the variables with nonzero loadings on the factor. This model thus requires many zeroes in the loading matrix and is a special type of simple structure model which permits interpretation of the final factors as clusters of variables.

The requirement of zero correlation between factors makes a simple model for the variable covariance matrix, but it is not a compelling assumption. It is necessary, in view of the factor-loading matrix nonuniqueness, to have even more stringent assumptions on the loading matrix if no assumptions are made about the factors. A simple clustering model of this type is that all entries in the loading matrix are 0 or 1, with each factor corresponding to a cluster of variables with unit loadings.

17.2 SPARSE ROOT ALGORITHM

Preliminaries.   A covariance matrix $\{C(I, J), 1 \leq I, J \leq N\}$ is to be approximated by the product of B and its transpose, where B contains many zeroes. The matrix B is assessed by two properties:

(i)   the sum of squares $SS(B) = \sum \{1 \leq I \leq N, 1 \leq L \leq K\} B(I, L)^2.$

(ii) the number of zeroes $Z(B)$, which equals the number of times $B(I, L) = 0$. During the maximization, it is required that the residual matrix

$$R(I, J) = C(I, J) - \sum \{1 \leq L \leq K\}\, B(I, L)B(J, L)$$

remain nonnegative definite. Thus

$$\sum \{1 \leq I \leq N\}\, R(I, I) = \sum \{1 \leq I \leq N\}\, C(I, I) - \text{SS}(B)$$

decreases as $\text{SS}(B)$ increases, and, if $\text{SS}(B)$ is close enough to $\sum \{1 \leq I \leq N\}\, C(I, I)$, the diagonal residuals will be negligible. Because R is nonnegative definite, the off-diagonal term $R(I, J)$ is less than $[R(I, I)R(J, J)]^{1/2}$.

The fitting proceeds stepwise. The first (trial) column of B is chosen to maximize $\sum \{1 \leq I \leq N\}\, B(I, 1)^2$. This column is the first eigenvector of C, and $\text{SS}(B)$ is the first eigenvalue. The row IMIN minimizes $[\sum \{1 \leq I \leq N\}\, B(I, 1)C(\text{IMIN}, I)]^2 \times [\text{SS}(B)C(\text{IMIN}, \text{IMIN})]^{-1}$. This quantity is the square of the correlation of the IMINth variable with the weighted average of the original variables, weighted by the coefficients $\{B(I, 1)\}$. If $B(\text{IMIN}, 1)$ is constrained to be zero, the quantity $\text{SS}(B)$ will be reduced by an amount proportional to this squared correlation.

The row IMIN is "removed" from the matrix C by replacing all correlations by the partial correlation with IMIN fixed and by setting all correlations involving IMIN equal to zero. A new eigenvector is computed on this adjusted matrix, a new IMIN is found least correlated with the weighted average of variables, and this variable is removed from the matrix. In this way, an eigenvector is obtained for N, $N - 1$, $N - 2, \dots, 1$ variables. That eigenvector is chosen to fit C for which eigenvalue/ (number of nonzero values) is a maximum.

The residual matrix R is computed, as follows:

$$R(I, J) = C(I, J) - B(I, 1)B(J, 1), \qquad 1 \leq I, J \leq N,$$

and the above steps are repeated on R. Eventually C is approximated by a loading matrix in which $\text{SS}(B)$ is high and the number of zeroes is also high.

STEP 1. Set $L = 1$. Initially, set $R(I, J) = C(I, J)$ ($1 \leq I \leq N$, $1 \leq J \leq N$). Set $\text{IP} = N$.

STEP 2. Let $\{X(I), 1 \leq J \leq N\}$ be the eigenvector with largest eigenvalue of the matrix C. Set $F(\text{IP}) = [\sum \{1 \leq I \leq N\}\, X(I)^2][\text{number of values } X(I) \neq 0]^{-1}$.

STEP 3. Choose IMIN to minimize $[\sum\{1 \leq I \leq N\}\, B(I, 1)\, C(\text{IMIN}, I)\}]\, [C(\text{IMIN}, \text{IMIN})]^{-1}$, with $X(\text{IMIN}) \neq 0$.

STEP 4. Compute the partial correlation matrix of C with IMIN "removed"; that is, change $C(J, K)$ to $C(J, K) - C(\text{IMIN}, J)C(\text{IMIN}, K)/C(\text{IMIN}, \text{IMIN})$, ($1 \leq J$, $K \leq N$), and finally set $C(\text{IMIN}, J) = C(J, \text{IMIN}) = 0$ ($1 \leq J \leq N$). Set $\text{IP} = \text{IP} - 1$, and, if IP remains greater than zero, return to Step 2.

STEP 5. Let $\text{IP} = \text{IMAX}$ maximize $\{F(\text{IP}), 1 \leq \text{IP} \leq N\}$. Set $B(I, L) = X(I)$ ($1 \leq I \leq N$), where $\{X(I)\}$ is the eigenvector corresponding to $F(\text{IP})$. Change $R(I, J)$ to $R(I, J) - B(I, L)B(J, L)$ ($1 \leq I, J \leq N$). Define $C(I, J) = R(I, J)$, increase L by 1, and return to Step 2, unless $L = K$.

17.3 SPARSE ROOT ALGORITHM APPLIED TO FACE MEASUREMENTS

The variables used are head length (HL), face length (FL), and face breadth (FB) with correlations between them as given in Table 17.1.

STEP 1. Initialization sets the column to be estimated, $L = 1$, remembers the covariance matrix C in R, $R(I, J) = C(I, J)$, $1 \leq I, J \leq 3$ (since C is destroyed in the next few steps), and sets IP $= 3$.

STEP 2. The first eigenvector of C is $(0.712, 0.852, 0.848)$. $F(IP) = F(3) = (0.712^2 + 0.852^2 + 0.848^2)/3 = 0.651$.

STEP 3. The squared correlation of the first variable with the linear combination of variables corresponding to the first eigenvector is

$$\frac{(0.712 \times 1.000 + 0.852 \times 0.402 + 0.848 \times 0.395)^2}{1.95^2} = 0.506.$$

The other two variables have a higher squared correlation] so IMIN $= 1$.

STEP 4. Remove IMIN $= 1$ from the covariance matrix. Then

$$C(1, 1) = C(2, 1) = C(3, 1) = C(1, 2) = C(1, 3) = 0,$$
$$C(2, 3) = 0.618 - 0.395 \times 0.402 = 0.459,$$
$$C(2, 2) = 0.838, \quad \text{and} \quad C(3, 3) = 0.844.$$

Set IP $= 2$, and return to Step 2.

STEP 2 REPEATED. The first eigenvector of C, with variable 1 removed, is $(0, 0.804, 0.809)$. Then $F(2) = (0.804^2 + 0.809^2)/2 = 0.650$.

STEP 3 REPEATED. The second variable is least correlated with the new eigenvector so IMIN $= 2$.

STEP 4 REPEATED. On removing IMIN $= 2$ from C, all entries are zero except $C(3, 3) = 0.592$. Return to Step 2.

STEP 2 REPEATED. The first eigenvector of C with variables 1 and 2 removed is $(0, 0, 0.770)$. Thus $F(3) = 0.592$.

STEP 5. The maximum value of $F(I)$ occurs at $I = 3$, $F(I) = 0.651$. The first column of B is $B(1, 1) = 0.712$, $B(2, 1) = 0.852$, $B(3, 1) = 0.848$. Change the residual matrix $R(I, J)$ $(1 \leq I, J \leq 3)$; for example, $R(1, 2) = R(1, 2) - B(1, 1)B(2, 1) = 0.402 - 0.712 \times 0.852 = -0.204$. Define $C = R$, set $L = 2$, and return to Step 2.

The sequence of values of C, R, and B are given in Table 17.2. There are not many zeroes, which indicates, perhaps, that there is not much clustering.

17.4 REMARKS ON THE SPARSE ROOT ALGORITHM

The sparse root algorithm is applied to the 7×7 physical measurements matrix in Table 17.3. The first three columns are rather satisfactory. First, the "bone length" cluster of four variables, foot, forearm, height, and finger length, appears, then the face and head breadth cluster, and then a general cluster containing all variables.

Table 17.2 Sparse Root Algorithm Applied to Head Measurements

(1)	INITIAL	HEAD LENGTH	HL	1.000	.402	.395
	CORRELATIONS	HEAD BREADTH	HB	.402	1.000	.618
		FACE BREADTH	FB	.395	.618	1.000

(2)	FIRST EIGENVECTOR			.712	.852	.848

(3)	CORRELATIONS		HL	0	0	0
	WITH HL		HB	0	.838	.459
	REMOVED		FB	0	.459	.844

(4)	EIGENVECTOR OF REDUCED MATRIX			0	.804	.809

(5)	CORRELATIONS		HL	0	0	0
	WITH HL, HB		HB	0	0	0
	REMOVED		FB	0	0	.592

(6)	EIGENVECTOR OF REDUCED MATRIX			0	0	.770

(7)	RESIDUAL MATRIX		HL	.493	-.204	-.209
	AFTER FIRST COLUMN		HB	-.204	.274	-.104
	FITTED		FB	-.209	-.104	.280

(8)	FIRST EIGENVECTOR			.702	-.286	-.303

(9)	EIGENVECTOR, HB REMOVED			.584	0	-.490

(10)	RESIDUAL MATRIX		HL	.152	-.204	.078
	AFTER SECOND COLUMN		HB	-.204	.274	-.104
			FB	.078	-.104	.040

(11)	RESIDUAL MATRIX		HL	0	0	0
	AFTER THIRD COLUMN		HB	0	0	0
			FB	0	0	0

(12)	LOADING		HL	.712	.584	-.390
	MATRIX		HB	.852	0	.524
			FB	.848	.490	-.199

The later columns reveal two problems: First, the cluster of variables in the fourth column overlaps with that in the first. This happens because it is not forbidden by the algorithm, and perhaps it should be. Second, some rather small entries appear in the later columns, such as 0.055 for head breadth in column four. These make only a trivial difference in the fit but cannot be replaced by zero because the residual matrix would then be no longer nonnegative definite.

Table 17.3   Sparse Root of Physical Measurements Data

HEAD LENGTH	0	0	.857	-.504	0	.109	0
HEAD BREADTH	0	.898	.431	-.055	0	.041	0
FACE BREADTH	0	.356	.774	.506	.027	-.123	.039
FOOT	.799	0	.438	.145	0	.339	-.175
FOREARM	.877	0	.356	0	0	0	.322
HEIGHT	.770	0	.426	0	-.408	.235	-.074
FINGER LENGTH	.814	0	.373	0	.409	-.171	0
VARIANCE	2.664	.932	2.160	.534	.335	.227	.142
CUMULATIVE VARIANCE	2.664	3.596	5.756	6.290	6.625	6.852	6.994

(MAXIMUM SUM = 7)

On the other hand, a substantial amount of variance is explained in the first three columns, 82%, and these have many zeroes. The later columns can be ignored in this case.

17.5*   ROTATION TO SIMPLE STRUCTURE

A difficulty with the "sparse root" method is the requirement of nonnegative definiteness on the residual matrix. This requirement is necessary so that the method can be applied stepwise, at each stage operating on the residual matrix from the previous stage.

An alternative constraint on the fitted loading matrix B^* is that there exists a root B such that $B^*(I, J) = B(I, J)$ whenever $B^*(I, J) \neq 0$. This condition is always met by the "sparse root" computed as above. The relaxed condition is justified as follows. Consider the data matrix $A(I, J)$.

Let $A(I, J) = \sum \{1 \leq L \leq K\} F(I, L)B^*(J, L) + E(I, J)$. Fit F and B^* with some values of B^* constrained to be zero, to minimize $\sum \{1 \leq I \leq M, 1 \leq J \leq N\} E(I, J)^2$. The columns $\{F(I, L), 1 \leq I \leq M\}$ are assumed orthogonal. For $K \geq N$, $A(I, J) = \sum \{1 \leq L \leq K\} F(I, L)B(J, L)$, where $\{F(I, L), 1 \leq I \leq M\}$ are orthogonal and B is a root of the covariance matrix with (I, J)th element $C(I, J) = \sum \{1 \leq L \leq M\} A(L, I)A(L, J)$. Thus the equation involving B^* may be written, for a particular F, as

$$A(I, J) = \sum \{1 \leq L \leq K\} F(I, L)B(J, L)$$
$$= \sum \{1 \leq L \leq K\} F(I, L)B^*(J, L) + E(I, J),$$

which implies that $E(I, J) = \sum \{1 \leq L \leq K\} F(I, L)[B(J, L) - B^*(J, L)]$. To minimize $\sum \{1 \leq I \leq M, 1 \leq J \leq N\} E(I, J)^2$, for a particular F, set $B^*(J, L) = B(J, L)$ whenever $B^*(J, L)$ is not constrained to be zero.

Thus, if a loading matrix B^* is constrained *a priori* to have certain elements zero, the optimal B^* (in the sense of a least sum of squared errors on the original data) is obtained by finding a root B of C and setting the constrained elements equal to zero. Each root B will be evaluated by the sum of its squared elements over those elements to be set zero; the optimal B^* is obtained when this sum is minimized or, equivalently, when $\sum \{1 \leq I \leq N, 1 \leq L \leq K\} B^*(I, K)^2$ is maximized. The family of roots of C must be searched. Any two of these are related by an orthogonal matrix T operating on the columns. Thus, if $B1$ and $B2$ are roots,

$$B1(I, J) = \sum \{1 \leq L \leq K\} B2(I, L)T(L, J).$$

These orthogonal matrices correspond to the orthogonal factors F, which disappear when the covariance matrix is used rather than the original data. The search thus proceeds by beginning with an arbitrary root of C and rotating it to maximize $\sum \{1 \leq I \leq N, 1 \leq L \leq K\} B^*(I, K)^2$. Stationary points occur at roots B for which

$$\sum \{1 \leq I \leq N\} B(I, J) B^*(I, L) = \sum \{1 \leq I \leq N\} B(I, L) B^*(I, J),$$

where $B^*(I, J) = B(I, J)$ whenever $B^*(I, J) \neq 0$. There may be many such points corresponding to many local optima.

Now consider the general problem of specifying the zeroes in the loading matrix. As before, all roots will be searched. Each root can be rendered sparse by setting elements equal to zero, and the error is the sum of the squares of these elements. The value of a root B is the maximum number of elements which may be set zero, with their sum of squares less than some threshold TH. Or, the value of a root B is the number of elements less in absolute value than some threshold TH.

A particular search pattern to minimize the number of elements exceeding threshold begins with an arbitrary root, rotates that pair of columns whose rotation most decreases the number of elements exceeding threshold, and continues until no pair of columns can be improved by rotation.

To optimize the rotation of each pair of columns, consider the pair of elements x, y in a particular row. Rotations with angle θ (between 0 and π) will transform these to $x \cos \theta + y \sin \theta, -x \sin \theta + y \cos \theta$. The possible values of θ will be divided into intervals in which 0, 1, or 2 elements are less than threshold. This is done for each row. An interval (or intervals) of θ values may now be discovered maximizing the number of elements within threshold. In general, the maximizing rotation will not be unique, and the smallest rotation that will achieve the maximum number of elements within threshold is chosen.

17.6 JOINING ALGORITHM FOR FACTOR ANALYSIS

Preliminaries. A covariance matrix $\{C(I, J), 1 \leq I \leq N, 1 \leq J \leq N\}$ is to be approximated by a product of loading matrices B, in which B has *simple tree structure*. This means that every column of B has constant nonzero elements (perhaps a different constant for different columns) and that the clusters of variables defined by the nonzero elements in each column form a tree. A covariance matrix C is exactly equal to a product of loading matrices of this type if and only if $-C$ is an ultrametric, that is, if and only if for every three variables $I, J, K, C(I, J) \geq \min [C(I, K), C(J, K)]$. Note that, if $-C$ is an ultrametric for one scaling of the variables, it might not be for another, so that careful scaling of the variables could improve the fit of this model.

The algorithm proceeds by finding the two variables with the largest covariance and joining them to construct a new factor whose covariance with each other variable is the weighted average of the covariances of the joined variables with that variable. The next highest covariance then indicates the next pair to be joined. This is no different from the standard distance and amalgamation procedure.

During the algorithm, clusters (or factors) $\{1, 2, \ldots, 2N - 1\}$ will be constructed. The first N clusters are the original variables. The cluster structure is recorded by the vector JT, where JT(I) is the cluster constructed by joining I to some other cluster. If clusters I and J are joined to form cluster K, then the loading in the Ith column of the loading matrix is $\{C(I, I) - \min [C(I, I), C(J, J), C(I, J)]\}^{1/2}.$

STEP 1.   Set K, the number of clusters, equal to N. For each I $(1 \leq I \leq N)$ define $WT(I) = 1$, $JT(I) = 0$. Define $B(I, I) = 1$ $(1 \leq I \leq N)$ and $B(I, J) = 0$ for all other I, J $(1 \leq I \leq N, 1 \leq J \leq 2N - 1)$.

STEP 2.   Find that pair $I \neq J$ with $JT(I) = JT(J) = 0$, such that $C(I, J)$ is a maximum.

STEP 3.   Increase K by 1. Define $JT(I) = K$, $JT(J) = K$, $C(K, K) = \min [C(I, I), C(J, J), C(I, J)]$, $WT(K) = WT(I) + WT(J)$. Define $B[L, I] = [C(I, I) - C(K, K)]^{1/2}$ whenever $B(L, I) = 1$ $(1 \leq L \leq N)$. Define $B(L, J) = [C(J, J) - C(K, K)]^{1/2}$ wherever $B(L, J) = 1$ $(1 \leq L \leq N)$. Define $B(L, K) = 1$ whenever $B(L, I)$ or $B(L, J)$ are nonzero $(1 \leq L \leq N)$. Define $JT(K) = 0$.

STEP 4.   For each L, $JT(L) = 0$, define $C(L, K) = C(K, L) = [WT(I) C(I, L) + WT(J)C(J, L)]/WT(K)$. If $K < 2N - 1$, return to Step 2.

NOTE 1.   There will be $2N - 1$ clusters, or factors, after the calculations. (Some of these may have only zero loadings and may be dropped.) The loading matrix may be reduced to an $N \times N$ matrix as follows. Begin with the smallest clusters and move to the largest. If I and J are joined to form cluster K, suppose that DI, DJ, DK are the corresponding nonzero loadings. Whenever $B(I, L) \neq 0$, set $B(I, L) = DI^2/(DI^2 + DJ^2)^{1/2}$. Whenever $B(J, L) \neq 0$, set $B(I, L) = -DJ^2/(DI^2 + DJ^2)^{1/2}$. Eliminate entirely the column $\{B(J, L), 1 \leq L \leq N\}$. Replace $B(K, L) \neq 0$ by $B(K, L) = [DK^2 + DI^2DJ^2/(DI^2 + DJ^2)]^{1/2}$. During this procedure, $N - 1$ columns are eliminated. The base for the elimination is the collinearity of the I, J, K columns when I and J are joined to form K.

NOTE 2.   This algorithm may be set up as an average joining algorithm by using euclidean distances when the variances are all unity. If $\rho(I, J)$ is the correlation between variables I and J, $D(I, J) = [1 - \rho(I, J)]/2M$ is the square of the euclidean distance between the standardized variables. The distance between clusters of variables is defined as the average distance over pairs of variables, one from each cluster. Then exactly the same sequence of joins will be obtained on the distances, as in the above algorithm.

The covariance between any two clusters is the average covariance between the variables in the two clusters. It is natural to associate a factor with each cluster equal to the average of all variables in the cluster, since the covariance between clusters is the covariance of these two factors. Of course, these factors will be oblique. Another convenient set of factors for a binary tree is obtained by associating a factor with each split into two clusters—the difference of the averages of the variables in the two clusters. These factors are oblique also, whereas the columns of the loading matrix are orthogonal.

17.7   APPLICATION OF JOINING ALGORITHM TO PHYSICAL MEASUREMENTS DATA

The correlation matrix in Table 17.1 has approximately the ultrametric structure required. For example, all the correlations in the block head length, head breadth, face breadth by forearm, finger, foot, height are approximately equal. The operations on the correlation matrix are given in Table 17.4, and the final loading matrices and correlation matrix are given in Table 17.5.

Table 17.4   Application of Joining Algorithm to Measurements Data

1. HL	1000						
2. HB	402	1000					
3. FB	395	618	1000				
4. FM	305	135	289	1000			
5. FR	301	150	321	846	1000		
6. FT	339	206	363	797	759	1000	
7. HT	340	183	345	800	661	736	1000

```
STEP 1.  JOIN FR TO FM                    STEP 2.  JOIN FMFR TO FT
   HL      1000                              HL      1000
   HB       402  1000                        HB       402  1000
   FB       395   618  1000                  FB       395   618  1000
   FMFR     303   142   305   846            FMFRFT   315   163   324   778
   FT       339   206   363   778  1000      HT       340   183   345   732  1000
   HT       340   183   345   730   736  1000
```

```
STEP 3.  JOIN FMFRFT TO HT                 STEP 4.  JOIN HB TO FB
   HL        1000                             HL        1000
   HB         402  1000                       FBHB       398   618
   FB         395   618  1000                 FMFRFTHT   321   248   732
   FMFRFTHT   321   168   328   732
```

```
STEP 5.  JOIN HBFB TO HL                   STEP 6.  JOIN HLFBHB TO FMFRFTHT
   HLFBHB     398                             HLFBHBFMFRFTHT   270
   FMFRFTHT   270   732
```

```
TREE IS  HL-FBHB---FMFR-FT--HT.
See Table 17.5 for loading matrix.
```

The pair with highest covariance are joined. New covariances are weighted averages of the old.

STEP 1.   Set K, the number of clusters, equal to 7. Define $WT(I) = 1$, $JT(I) = 0$, and $B(I, I) = 1$ for $1 \leq I \leq 7$ and $B(I, J) = 0$ for other I, J ($1 \leq I \leq 7, 1 \leq J \leq 13$).

STEP 2.   The pair FM and FR have the highest covariance, so $I = 4$, $J = 5$.

STEP 3.   Increase K to 8. Define $JT(4) = JT(5) = 8$, $C(8, 8) = 0.846$ [since $C(4, 5) = 0.846$ is less than $C(4, 4)$ or $C(5, 5)$], $WT(8) = 2$. Define $B(4, 4) = (1 - 0.846)^{1/2} = 0.392$, $B(5, 5) = (1 - 0.846)^{1/2} = 0.392$, $B(4, 8) = B(5, 8) = 1$. Define $JT(8) = 0$.

STEP 4.   Define

$$C(1, 8) = \tfrac{1}{2}[C(1, 4) + C(2, 4)] = \tfrac{1}{2}(0.305 + 0.301) = 0.303.$$

Similarly, other covariances with the new cluster or factor are defined by averaging the old. Since $K < 13$, return to Step 2. And so on.

Table 17.5   Loading Matrix and Residual Correlation Matrix Obtained by Joining Algorithm

[1] LOADING MATRIX (SQUARED COEFFICIENTS, MULTIPLIED BY 1000)

HL	0	0	0	0	0	0	0	0	0	128	0	270
HB	0	0	0	0	0	0	382	0	220	128	0	270
FB	0	0	0	0	0	0	0	382	220	128	0	270
FM	154	0	68	0	46	0	0	0	0	0	462	270
FR	0	154	68	0	46	0	0	0	0	0	462	270
FT	0	0	0	222	46	0	0	0	0	0	462	270
HT	0	0	0	0	0	268	0	0	0	0	462	270

(All rows sum to 1)

[2] REDUCED LOADING MATRIX (COEFFICIENTS MULTIPLIED BY 1000)

HL	0	0	0	0	598	391	700
HB	0	0	0	437	-409	391	700
FB	0	0	0	-437	-409	391	700
FM	278	240	181	0	0	-565	700
FR	-278	240	181	0	0	-565	700
FT	0	-365	181	0	0	-565	700
HT	0	0	-434	0	0	-565	700

[3] RESIDUAL CORRELATION MATRIX (MULTIPLIED BY 1000)

(Values in blocks average to zero)

HL	0						
HB	4	0					
FB	- 3	0	0				
FM	35	-135	19	0			
FR	31	-120	51	0	0		
FT	69	- 64	93	19	- 19	0	
HT	70	- 88	75	68	- 71	4	0

Consider also the construction of oblique factors from the binary tree. One such set of factors is the differences between the averages of the variables in pairs of clusters joined during the algorithm. Thus,

$$F(8) = [V(4) - V(5)]\sqrt{\tfrac{1}{2}},$$
$$F(9) = [\tfrac{1}{2}V(4) + \tfrac{1}{2}V(5) - V(6)]\sqrt{\tfrac{2}{3}},$$
$$F(10) = [\tfrac{1}{3}V(4) + \tfrac{1}{3}V(5) + \tfrac{1}{3}V(6) - V(7)]\sqrt{\tfrac{3}{4}},$$
$$F(11) = [V(2) - V(3)]\sqrt{\tfrac{1}{2}},$$
$$F(12) = [\tfrac{1}{2}V(2) + \tfrac{1}{2}V(3) - V(1)]\sqrt{\tfrac{2}{3}},$$
$$F(13) = [\tfrac{1}{3}(V(1) + V(2) + V(3)) - \tfrac{1}{4}(V(4) + V(5) + V(6) + V(7))]\sqrt{\tfrac{12}{7}},$$
$$F(14) = [V(1) + V(2) + V(3) + V(4) + V(5) + V(6) + V(7)]\sqrt{\tfrac{1}{7}}.$$

The multiplicative constants $\sqrt{\frac{1}{2}}$, etc., ensure that the sum of the squares of the coefficients is unity. The method of definition guarantees orthogonality between coefficients for different factors. The coefficient matrix is thus easily inverted to

$$V(1) = -\sqrt{\tfrac{2}{3}}\, F(12) + \tfrac{1}{3}\sqrt{\tfrac{12}{7}}\, F(13) + \sqrt{\tfrac{1}{7}}\, F(14),$$

$$V(2) = \sqrt{\tfrac{1}{2}}\, F(11) + \tfrac{1}{2}\sqrt{\tfrac{2}{3}}\, F(12) + \tfrac{1}{3}\sqrt{\tfrac{12}{7}}\, F(13) + \sqrt{\tfrac{1}{7}}\, F(14),$$

$$V(3) = -\sqrt{\tfrac{1}{2}}\, F(11) + \tfrac{1}{2}\sqrt{\tfrac{2}{3}}\, F(12) + \tfrac{1}{3}\sqrt{\tfrac{12}{7}}\, F(13) + \sqrt{\tfrac{1}{7}}\, F(14),$$

$$V(4) = \sqrt{\tfrac{1}{2}}\, F(8) + \tfrac{1}{2}\sqrt{\tfrac{2}{3}}\, F(9) + \tfrac{1}{3}\sqrt{\tfrac{3}{4}}\, F(10) - \tfrac{1}{4}\sqrt{\tfrac{12}{7}}\, F(13) + \sqrt{\tfrac{1}{7}}\, F(14),$$

$$V(5) = -\sqrt{\tfrac{1}{2}}\, F(8) + \tfrac{1}{2}\sqrt{\tfrac{2}{3}}\, F(9) + \tfrac{1}{3}\sqrt{\tfrac{3}{4}}\, F(10) - \tfrac{1}{4}\sqrt{\tfrac{12}{7}}\, F(13) + \sqrt{\tfrac{1}{7}}\, F(14),$$

$$V(6) = -\sqrt{\tfrac{2}{3}}\, F(9) + \tfrac{1}{3}\sqrt{\tfrac{3}{4}}\, F(10) - \tfrac{1}{4}\sqrt{\tfrac{12}{7}}\, F(13) + \sqrt{\tfrac{1}{7}}\, F(14),$$

$$V(7) = -\sqrt{\tfrac{3}{4}}\, F(10) - \tfrac{1}{4}\sqrt{\tfrac{12}{7}}\, F(13) + \sqrt{\tfrac{1}{7}}\, F(14).$$

This representation is similar to the reduced loading matrix in Table 17.5, but it is the columns of the loading matrix, rather than the factors, that are orthogonal. The factors will be oblique in general, even if the simple tree structure model holds.

If the factors obtained above *are* orthogonal and are normalized to have a variance of unity, then the loading matrix is a root of the covariance matrix with orthogonal columns, so that the columns of the loading matrix are just the eigenvectors of the covariance matrix. A possible use of these factors, then, is as a first approximation to the eigenvectors of a matrix. The suggested procedure would be to construct a tree, transform to the oblique factors $F(1), \ldots, F(14)$ by using the above rotation, then use a standard technique on the covariance matrix of the F's that would hopefully be substantially more nearly diagonal.

A similar use of these factors occurs in regression. It is desired to predict a variable Y from the variables $V(1), \ldots, V(N)$. The standard stepwise technique finds the variable $V(I)$ that best predicts Y, a second variable $V(J)$ such that $V(I)$ and $V(J)$ best predict Y, and so on. A difficulty of this approach is that, if there are three or four variables highly correlated with $V(I)$, they will not appear in the regression formulas although they are nearly as good predictors as $V(I)$. Suppose, instead, that the factors $F(I)$ are used in the stepwise fitting. Any variables that are highly correlated will be grouped together in the factors and so will appear simultaneously, as averages, in the regression equations.

17.8 THINGS TO DO

17.8.1 Running the Factor Analysis Algorithms

All the algorithms operate on a covariance matrix and are aimed at discovering clusters of variables rather than clusters of cases. The techniques are not appropriate if there is a substantial clustering of cases, since this will not be visible in the covariance matrix. For 1000 cases and 50 variables, preliminary analyses on the covariance matrix may reveal, say, 10 significant factors, and a more detailed analysis may then be performed on the 1000 cases by 10 factors.

Any tree on the variables generates a loading matrix in which each factor corresponds to a cluster of variables. A binary tree on N variables generates a loading matrix on N factors, in which each factor corresponds to the split at each node into two clusters of variables. (See Table 17.6, Indian caste measurements, for a trial data set.)

Table 17.6   Indian Caste Measurements

	SH	ND	NH	HL	FB	BB	HD	NB
STATURE---------------	5849	1774	1974	2698	2173	2891	1412	2103
SITTING HEIGHT-------		2094	2170	2651	3012	2995	2069	1182
NASAL DEPTH----------			2910	1537	1243	1575	1308	1139
NASAL HEIGHT----------				1758	1139	1852	1735	0438
HEAD LENGTH----------					2270	2792	1982	1930
FRONTAL BREADTH------						4930	4461	1831
BIZYGOMETIC BREADTH--							5407	2729
HEAD BREADTH---------								1413
NASAL BREADTH--------								

[From C. R. Rao (1948), The utilization of multiple measurements in problems of biological classification, *J. Roy. Stat. Soc. B* **10,** 159–193.] The correlations (by 10,000) are computed within 22 caste groups containing between 67 and 196 individuals.

17.8.2   Direct Factoring of a Data Matrix

Consider the representations

$$A = FB,$$

where A is an $M \times N$ data matrix, F is an $M \times K$ factor matrix, and B is an $N \times K$ loading matrix. The arrays F and B are row and column factors of A. Many factor analyses ignore or assume away structure in F in order to concentrate on the more malleable loading matrix B. For example, if the columns of F are assumed uncorrelated, then B is a root of the covariance matrix of A.

Since important interactions between the clustering of variables and the clustering of cases are gladly expected, it is necessary to have factor analysis models that will reveal two-way clustering. Let the Kth cluster be a submatrix of A or a block. Define $F[I, K] = 1$ if case (row) I is in the Kth cluster, and define $F(I, K) = 0$ otherwise. Define $B(J, K) = S(J, K)$ if variable (column) J is in the Kth cluster, and define $B(J, K) = 0$ otherwise. The values of F and B are not defined analogously because it is anticipated that variables will be constant within a cluster, but not necessarily cases, since the variables may be measured on different scales.

Then $A = FB = \sum \{1 \leq K \leq L\} C(K)$, where $C(K)$, the Kth cluster, is zero except on a submatrix within which the variables are constant. For example,

$$\begin{bmatrix} 3 & 6 & 2 & 6 \\ 3 & 11 & 3 & 2 \\ 2 & 8 & 3 & 2 \end{bmatrix} = \begin{bmatrix} 1 & 3 & 0 & 0 \\ 1 & 3 & 0 & 0 \\ 0 & 0 & 0 & 0 \end{bmatrix} + \begin{bmatrix} 0 & 0 & 0 & 0 \\ 0 & 5 & 1 & 2 \\ 0 & 5 & 1 & 2 \end{bmatrix} + \begin{bmatrix} 0 & 0 & 0 & 6 \\ 0 & 0 & 0 & 0 \\ 0 & 0 & 0 & 0 \end{bmatrix} + \begin{bmatrix} 2 & 3 & 2 & 0 \\ 2 & 3 & 2 & 0 \\ 2 & 3 & 2 & 0 \end{bmatrix}.$$

Note that there are no overlapping constraints, which may make the blocks difficult to understand in large arrays. (A similar decomposition is available if all values within a block are equal.)

The model may be fitted in stepwise fashion. At each stage, an optimal block is fitted to the residual matrix. Given the rows of this block, any variable is placed in the block for which the square of the mean by the number of rows exceeds a given threshold. Given the variables in a block, each row is added or deleted from the block (as in K means) according to its distances from the block mean and the nonblock mean.

REFERENCES

HARMAN, H. H. (1967). *Modern Factor Analysis*, University of Chicago Press, Chicago. This is a really excellent book, very clearly written. One method which discovers clusters (in a backhand way) is the *centroid* method (p. 171). The first factor is $F(1) = -1/N \sum \{1 \leq I \leq N\} V(I)$. The second factor is

$$F(2) = -1/N \sum \{1 \leq I \leq N\} E(I)[V(I) - \alpha(I)F(1)],$$

where $V(I) - \alpha(I)F(1)$ is orthogonal to $F(1)$ and $E(I)$ is a pattern of plus and minus ones chosen, in an *ad hoc* way, to maximize the variance of $F(2)$. The third factor is an average of residual variables orthogonal to $F(1)$ and $F(2)$, and so on.

The patterns of plus and minus ones, at least for the first few factors, often conform to clusters obtained by other methods.

A second method involving clusters is the multiple-group method, which partitions the variables and defines factors equal to the averages of the variables in each group. A method of constructing the clusters is discussed on p. 119—the two variables with highest correlation are grouped together, then the variable with highest average correlation to these two is added to the group, and so on, until the average correlation has "a sharp drop" with the addition of a new variable. Then a new group is begun, and the process continues until all the variables are grouped.

HORST, P. (1965). *Factor Analysis of Data Matrices*, Holt, Rinehart and Winston, New York. This is a general text which may be used as a guide to the extremely extensive literature. This particular book has many algorithms explicitly described and an array of supporting Fortran subroutines. Some quotes will be used to show the relation between factor analysis and classification. On p. viii, "Factor analysis has had its most popular appeal as a device for generating plausible and acceptable taxonomies in disciplines where these have been confused and unsatisfactory.... Labels seem to play a fundamental role in providing emotional security for all human beings, including scientists, and therefore the taxonomic function of factor analysis procedures probably needs no justification. However, it is in parsimonious description of natural phenomena that factor analysis has more fundamental and universal significance."

In the "group centroid" method, discussed on p. 148, a loading matrix is specified, and then the factors are computed by regression on the original variables. The loading matrix "consists of binary vectors each of which serves to group the variables into subsets according to unit elements in the vector.... It is not necessary that the unit elements in the vectors of the binary matrix be mutually exclusive.... Harman, H. H. (1960) ("Factor Analysis" in *Mathematical Methods for Digital Computers*, H. S. Wilf and A. Ralston (eds.), Wiley, New York) has suggested that a preliminary cluster analysis of the correlation matrix may provide a basis for the grouping of the variables. However, since cluster analysis procedures are, in general, not objective, it is perhaps better to depend on some *a priori* or even arbitrary hypothesis concerning the way variables should be grouped in setting up the binary matrix."

In order to approximate a given loading matrix by one with a given pattern of zeroes, Horst assumes that the fitted loading matrix will be constant or zero in the columns and seeks that orthogonal transformation of the original matrix which most closely approximates a loading matrix of this form (p. 415).

TRYON, R. C., and BAILEY, D. E. (1970). *Cluster Analysis*, McGraw-Hill, New York. This book was published shortly after the death of R. C. Tryon, a pioneer in clustering from the factor analysis side. (Tryon wrote a monograph "Cluster Analysis" in 1939.) In the foreword Charles Wrigley states: "Tryon spent a sabbatical year at the University of Chicago in the later 1930's. He grasped the point, as many others at Chicago must have done, that similar tests would have high correlations between them and that clusters of related tests could therefore be identified, without the labour of a centroid factor analysis, by direct search of the correlations. Thus cluster analysis, as originally conceived by Tryon, was a poor man's factor analysis."

Part of the purpose of this book is to describe a large package of computer programs called the BCTRY system. Tryon's general approach to clustering begins with a factor analysis of the variables identifying similar groups of variables and then continues by clustering cases, using new variables, each representing a group of the old.

Two types of clustering take place—clustering of variables, and then clustering of cases. The clustering of variables proceeds as follows: To construct the first cluster, a pivot variable is found maximizing the "index of pivotness," the variance of the squares of the correlation coefficients of all variables with this variable. A measure of similarity between variables is defined, the "proportionality index":

$$P(I, J) = (\sum \{1 \leq K \leq N\} \, \rho(I, K)\rho(J, K))^2$$
$$\times (\sum \{1 \leq K \leq N\} \, \rho^2[I, K] \sum \{1 \leq K \leq N\} \, \rho^2(J, K))^{-1},$$

where $\rho(I, K)$ is the correlation between variables I and K. The variable with largest proportionality to the pivot variable is added to the cluster. The variable with the highest mean proportionality to the two already added is added provided its proportionality with the first two exceeds 0.40. Similarly, for the fourth variable. For more than four variables, additional variables are added "if their mean index of proportionality is within twice the range of the indexes of proportionality among the four first-selected variables, and if all of the indexes of proportionality of the variable and the previously selected variables are greater than 0.81."

The clustering of cases proceeds principally through a K-means-type algorithm in which initial cluster means are guessed, each object is assigned to whichever mean it is closest to, then all means are recomputed, then the objects are reassigned, and so on.

PROGRAMS

SPARSE   finds root of covariance matrix containing many zeroes.
FIRST   finds first eigenvector.
FIND   finds eigenvector with many zeroes.
REMOVE   computes partial covariances.

```
      SUBROUTINE SPARSE(C,R,N,X,B,KK)
C.................................................................20 MAY 1973
C.... APPROXIMATES C BY ROOT B WHICH HAS MANY ZEROES. USES FIRST,REMOVE,FIND,OUT
C.... C = N BY N BORDERED COVARIANCE MATRIX, DESTROYED.
C.... N = NUMBER OF ROWS
C.... KK = NUMBER OF FACTORS, TRY KK = N
C.... X = 1 BY N SCRATCH ARRAY
C.... R = N BY N RESIDUAL MATRIX,C-BB'.
C.... B = N BY KK LOADING MATRIX
C.................................................................................
      DIMENSION C(N,N),R(N,N),X(N),B(N,KK)
      DIMENSION S(100)
      DATA XL/4HLOAD/
      B(1,1)=XL
      WRITE(6,1)
    1 FORMAT(18H COVARIANCE MATRIX )
      CALL OUT(C,N,N)
      SS=0
      DO 50 J=2,N
      R(J,1)=C(1,J)
      B(J,1)=C(1,J)
      R(1,J)=C(1,J)
   50 SS=SS+C(J,J)
      DO 20 KR=2,KK
      B(1,KR)=0
      DO 30 I=2,N
      DO 30 J=2,N
   30 R(I,J)=C(I,J)
      CALL FIND(C,N,X,B(1,KR))
      DO 31 I=2,N
      DO 31 J=2,N
   31 C(I,J)=R(I,J)-B(I,KR)*B(J,KR)
   20 CONTINUE
      WRITE(6,3)
    3 FORMAT(16H RESIDUAL MATRIX)
      CALL OUT(R,N,N)
      WRITE(6,2)
    2 FORMAT(33H SPARSE ROOT OF COVARIANCE MATRIX)
      CALL OUT(B,N,KK)
      DO 60 K=2,KK
      S(K)=0
      DO 61 J=2,N
   61 S(K)=S(K)+B(J,K)**2
   60 CONTINUE
      WRITE(6,4) SS,(S(K),K=2,KK)
    4 FORMAT(29H VARIANCES, TOTAL AND FACTORS/
     *6G10.4,2X,5G10.4/(10X,5G10.4,2X,5G10.4))
      S(1)=0
      DO 62 K=2,KK
   62 S(K)=S(K)+S(K-1)
      DO 63 K=2,KK
   63 S(K)=100.*S(K)/SS
      WRITE(6,5)(S(K),K=2,KK)
    5 FORMAT(31H PERCENTAGE CUMULATIVE VARIANCE/(10X,5G10.4,2X,5G10.4))
      RETURN
      END
```

```
      SUBROUTINE FIRST(C,N,X)
C..........................................................20 MAY 1973
C.... COMPUTES FIRST EIGENVECTOR OF SUBMATRIX OF C OVER ROWS I WHERE X(I).NE.0
C.... C = POSITIVE DEFINITE ARRAY
C.... N = NUMBER OF ROWS
C.... X = N BY 1 INDICATOR VARIABLE ON INPUT, EIGENVECTOR ON OUTPUT
C..........................................................................
      DIMENSION C(N,N),X(N)
      DIMENSION Y(100)
      TH=.0005
      ICNT=0
      SN=0
      DO 20 I=1,N
      IF(C(I,I).GT.TH) GO TO 22
      DO 21 J=1,N
      C(I,J)=0.
   21 C(J,I)=0.
   22 CONTINUE
      IF (X(I).NE.0.) X(I)=C(I,I)**0.5
   20 CONTINUE
   10 CONTINUE
      SN=0.
      DO 50 I=1,N
   50 SN=SN+X(I)**2
      SN=SN**0.5
      DO 51 I=1,N
   51 IF(SN.NE.0) X(I)=X(I)/SN
      SYY=0.
      SXY=0.
      DO 30 I=1,N
      Y(I)=0.
      IF(X(I).EQ.0.) GO TO 30
      DO 40 J=1,N
      IF (X(J).EQ.0.) GO TO 40
      Y(I)=Y(I)+C(I,J)*X(J)
   40 CONTINUE
      SXY=SXY+X(I)*Y(I)
      SYY=SYY+Y(I)**2
   30 CONTINUE
      IF(SXY.LT.0.) SXY=0.
      E=SXY**0.5
      SYY=SYY**0.5
      ERR=0.
      DO 60 I=1,N
      IF(SYY.NE.0) Y(I)=Y(I)/SYY
      ERR=ERR+(X(I)-Y(I))**2
   60 X(I)=E*(1.2*Y(I)-0.2*X(I))
      ICNT=ICNT+1
      IF (ICNT.GT.20) RETURN
      IF (ERR.GT.TH**2) GO TO 10
      RETURN
      END
```

```
      SUBROUTINE FIND(C,N,X,Y)
C.......................................................................20 MAY 1973
C...... FINDS BEST EIGENVECTOR FITTING TO C, MINIMIZING EIGENVALUE/NON-ZERO VALUE
C.... C = N BY N BORDERED COVARIANCE
C.... N = NUMBER OF ROWS
C.... X = SCRATCH VARIABLE
C.... Y = N BY 1 FITTED VARIABLE
C.......................................................................
      DIMENSION C(N,N),X(N),Y(N)
      DO 20 I=1,N
   20 X(I)=1.
      X(1)=0
      CMAX=0.
   50 CALL FIRST(C,N,X)
      XS=0.
      SS=0.
      DO 30 I=1,N
      IF (X(I).EQ.0.) GO TO 30
      XS=XS+1.
      Z=X(I)**2
      SS=SS+Z
   30 CONTINUE
      XMIN=10.**10
      DO 31 J=1,N
      IF(X(J).EQ.0.) GO TO 31
      XX=0.
      DO 32 I=1,N
   32 XX=XX+X(I)*C(I,J)
      XX=XX**2/(SS**2*C(J,J))
      IF (XX.GT.XMIN) GO TO 31
      XMIN=XX
      IMIN=J
   31 CONTINUE
      IF(XS.NE.0) CC=SS/XS
      IF(CC.LE.CMAX) GO TO 33
      CMAX=CC
      DO 40 I=1,N
   40 Y(I)=X(I)
   33 CONTINUE
      CALL REMOVE(C,N,X,IMIN)
      X(IMIN)=0.
      IF (XS.GT.1.) GO TO 50
      RETURN
      END




      SUBROUTINE REMOVE(C,N,S,I)
C.......................................................................20 MAY 1973
C.... COMPUTES PARTIAL CORRELATIONS ON ARRAYC, REMOVING VARIABLE I.
C.... C = N BY N BORDERED COVARIANCE MATRIX
C.... N = NUMBER OF VARIABLES
C.... S = INDICATOR VARIABLE,ONLY NON-ZERO ROWS CONSIDERED
C.... I = INDEX OF VARIABLE TO BE REMOVED.
C.......................................................................
      DIMENSION C(N,N),S(N)
      TH=.0005
      IF (C(I,I).LT.TH) GO TO 50
      IF (S(I).EQ.0.) RETURN
      DO 20 J=1,N
      IF (J.EQ.I) GO TO 20
      IF (S(J).EQ.0.) GO TO 20
      DO 30 K=1,N
      IF (S(K).EQ.0.) GO TO 30
       C(J,K)=C(J,K)-C(I,J)*C(I,K)/C(I,I)
   30 CONTINUE
   20 CONTINUE
   50 CONTINUE
      DO 40 J=1,N
      C(J,I)=0.
   40 C(I,J)=0.
      RETURN
      END
```

329

CHAPTER 18

Prediction

18.1 INTRODUCTION

A base data matrix $\{A(I, J), 1 \leq I \leq M, 1 \leq J \leq N\}$ is assumed. A new variable $\{A(I, O), 1 \leq I \leq M\}$ is observed on the given cases. A new case $\{A(O, J), 1 \leq J \leq N\}$ is observed on the given variables. How can the missing value $A(O, O)$ be estimated?

The standard regression approach would predict $A(I, O)$ as a linear function $\sum \{1 \leq J \leq N\} C(J)A(I, J)$ of the base variables. This linear function is then applied to the new case to predict the value $A(O, O)$ by $\sum \{1 \leq I \leq M\} C(J)A(O, J)$. Of course, if the new case is quite different from the base cases, the prediction may be quite wrong (and the error estimate much too small), so that an underlying similarity between the new case and the base set is necessary for the extension to be valid. A typical assumption is that all $\{A(I, J), 1 \leq J \leq N\}$ for $O \leq I \leq M$ are randomly sampled from an M-dimensional multivariate normal. In piecewise fitting, the set of all possible cases is partitioned into clusters, and a different predicting function is used within each of the clusters. Essentially the clusters are used as a means of generalizing the fitting function. For example, suppose that a variable $Y(I)$ is to be fitted by a linear function of $X(I)$ $(1 \leq I \leq M)$. The usual least-squares fit is

$$y = a + bx,$$

where

$$a = \bar{Y} - b\bar{X},$$

$$b = (\sum \{1 \leq I \leq M\} Y(I)[X(I) - \bar{X}])(\sum \{1 \leq I \leq M\} [X(I) - \bar{X}]^2)^{-1},$$

$$\bar{X} = \sum \{1 \leq J \leq M\} \frac{X(I)}{M},$$

and

$$\bar{Y} = \sum \{1 \leq I \leq M\} \frac{Y(I)}{M}.$$

This fit could be computed separately for $X(I) \leq 0$ and $X(I) > 0$. The fitting equation would be

$$y = a(1) + b(1)x, \qquad x \leq 0$$
$$y = a(2) + b(2)x, \qquad x > 0$$

where $a(1)$ and $b(1)$ are computed from those $\{X(I), Y(I)\}$ with $X(I) \leq 0$, and $a(2)$ and $b(2)$ are computed from those $\{X(I), Y(I)\}$ with $X(I) > 0$. Notice that not only the given cases are clustered but the *set of all possible cases* is clustered, so that new cases can be inserted in the fitting equations.

The clustering used in piecewise fitting is chosen to optimize some measure of the fit within each of the clusters. For example, in the above line fitting problem, suppose that the possible clusters were $x \leq c$ and $x > c$. For each c, there will be a sum of squared deviations of $Y(I)$ from its predicted value. As c increases, this residual sum of squares will change only as c passes through $X(I)$. So $c = X(I)$ $(1 \leq I \leq M)$ are the only values which need be tried.

For N variables, there are two types of approaches to piecewise fitting—the first via mixtures, and the second via the automatic interaction detection (AID) technique of Sonquist and Morgan.

First, as in discriminant analysis, suppose the variable to be predicted is an integer variable $\{A(I, O), 1 \leq I \leq M\}$. When $A(I, O) = L$, $\{A(I, J), 1 \leq J \leq N\}$ is randomly sampled from a multivariate normal with mean $\{E(L, J), 1 \leq J \leq N\}$ and covariance matrix Σ. A new case $\{A(O, J), 1 \leq J \leq N\}$ is used to predict $A(O, O)$. Let the probability be $P(L)$ that the new case is from the Lth multivariate normal. The minimum expected number of misclassifications occurs if $A(O, O) = L$ is predicted, where L maximizes

$$-\tfrac{1}{2}[A(O, I) - E(L, I)] \, \Sigma^{-1} (I, J)[A(O, J) - E(L, J)] + \log P(L).$$

Thus N-dimensional space is divided into a number of convex polytopes in each of which a different prediction of $A(O, O)$ is made. In practice, E, Σ, and P must be estimated from the data matrix $\{A(I, J), 1 \leq I \leq M, 1 \leq J \leq N\}$; for example, $E(L, J)$ is the average of $A(I, J)$ over those cases for which $A(I, O) = L$.

For $\{A(I, O), 1 \leq I \leq M\}$ continuous, assume that each vector of length $N + 1$, $(A(I, O), \{A(I, J), 1 \leq J \leq N\})$ is a sample from a mixture of K multivariate normals, where the Lth has mean $[E(L, O), E(L, 1), \ldots, E(L, N)]$ and variance Σ and sample I comes from the Lth with probability $P(L)$. The expected value of $A(I, O)$, given $\{A(I, J), 1 \leq J \leq N\}$ and that the Ith sample lies in the Lth population, is $F(L) + \sum \{1 \leq J \leq N\} C(J)A(I, J)$. Note that the coefficients $C(J)$ depend on Σ but not on L, since the covariance matrix does not change between clusters. Given $\{A(O, J), 1 \leq J \leq N\}$, let $P(L, O)$ be the probability that the Oth sample lies in the Lth cluster. If the sample were always assigned to the cluster of highest probability, the sample space would be divided into polytopes as in the discriminant analysis case. But for predicting a continuous variable, the correct prediction is the expectation of $A(O, O)$, $\sum \{1 \leq L \leq K\} P(L, O)F(L) + \sum \{1 \leq J \leq N\} A(O, J)C(J)$. The quantities E, Σ, and P are estimated, as in the mixtures algorithm, by maximum likelihood.

The difference in the prediction of $A(O, O)$, due to the clustering, is carried in the constant terms $F(L)$ in the various predicting equations within the various clusters. In particular, if the $F(L)$ do not much differ, it will not much matter what cluster the Oth observation lies in.

The other approach to multivariate prediction is the AID technique of Sonquist and Morgan (see the references in Chapter 14). This constructs clusters of cases by splitting, using the variables one at a time. At each split a cluster is divided into two clusters of cases, according as $A(I, J) > C$ or $A(I, J) \leq C$. That variable J and that constant C are chosen which minimize the sum of squared deviations within clusters of the variable to be predicted. Thus the final AID clusters are of especially simple form—parallelipipeds consisting of all cases $\{I \mid D(J) < A(I, J) \leq C(J), 1 \leq J \leq N\}$. These clusters are easy to interpret and compute. Because of its simple, one-at-a-time treatment of variables, the AID technique can also accommodate ordered or category variables. An ordered variable generates a split into $\{I \mid A(I,J) \leq C\}$ and $\{I \mid A(I,J) > C\}$. A category variable is temporarily converted into an ordered variable by ordering its categories according to the mean value of $A(I, O)$ within each of the categories.

A possible drawback of the AID approach is the simplicity of the final clusters. It may be that convex shapes other than parallellepipeds are appropriate. A generalization has a first split according to $\sum \{1 \leq J \leq N\} B(J)A(I,J)$ not greater than, or greater than, C. Later splits will also be according to some linear combination of the variables rather than a single variable. Each set of coefficients $\{B(J)\}$ and constant C is evaluated according to the sum of squared deviations of $\{A(I, O)\}$ from cluster means generated by the split. Determining the optimal $\{B(J)\}$ is not a simple procedure (especially when compared to the AID technique), and "local optimization" approximations are necessary.

In a third approach, it is assumed that a tree of convex clusters has already been computed by using the data matrix $\{A(I,J), 1 \leq I \leq M, 1 \leq J \leq N\}$. By using analysis of variance techniques, clusters in the tree are eliminated if the cluster mean of $\{A(I, O)\}$ does not differ significantly from the mean of the next including cluster. The tree is thus reduced to a smaller tree from which predictions are made. The advantage of this approach is that the original clustering may be applied to a number of different variables to be predicted.

18.2 VARIANCE COMPONENTS ALGORITHM

Preliminaries. A tree of clusters $1, 2, \ldots, K$ is given with its structure specified by an ancestor function F, which for each cluster I $(I < K)$ specifies $F(I) > I$, the smallest cluster containing I. For the largest cluster K, $F(K) = K$.

A variable X takes values $\{X(I), 1 \leq I \leq M\}$ on the objects at the ends of the trees. It is fitted to the tree by using a variance components model. A random variable $X(I)$ is associated with each cluster I $(1 \leq I \leq K)$, such that the $X(I) - X[F(I)]$ for I $(1 \leq I < K)$ are independent normal variables with mean 0 and variance $V[F(I)]$. The quantity $X(I)$ is a "mean value" associated with cluster I, and the quantity $V(I)$ is a variance component associated with cluster I $(M < I \leq K)$. The variance of the value at the Ith object is the sum of the variance components corresponding to clusters including the Ith object. If a new object is classified into the tree, into cluster I, its X value is estimated by $X(I)$. The variance components are a guide in reducing the tree, with the cluster I eliminated if $V[F(I)]$ is sufficiently small.

A threshold T is used that is injected into the probability model by assuming a uniform prior distribution of V and X and a threshold observation $X(I) - X[F(I)] = \sqrt{T}$ for each $F(I)$. The log posterior density of X and V is given by

$$\text{LPD} = C - \tfrac{1}{2}\sum \{1 \leq I < K\} \frac{\{X(I) - X[F(I)]\}^2}{V[F(I)]}$$

$$- \tfrac{1}{2}\sum \{M < I \leq K\} [N(I) + 1] \log V(I) - \tfrac{1}{2}\sum \{M < I \leq K\} \frac{T}{V(I)}.$$

The maximum posterior density values of X and V are determined iteratively, first by getting optimal V's for a given X and then by getting optimal X's for given V's. The posterior density is surely increased at every step.

STEP 1.   *Initialization.* Let NC(I) be the number of L with $F(L) = I$. Let $X(I)$ be the average of $X(L)$ over L with $F(L) = I$, defined successively for $M < I \leq K$. Let $V(I) = T$ ($M < I \leq K$).

STEP 2.   *Update X.* The value of $X(I)$ which maximizes the log posterior density is, for $M < I < K$ in turn,

$$X(I) = \left(\sum \{F(L) = I\} \frac{X(L)}{V(I)} + \frac{X(J)}{V(J)} \right) \left(\frac{NC(I)}{V(I)} + \frac{1}{V(J)} \right)^{-1},$$

where $J = F(I)$. Thus $X(I)$ is a weighted average of X values on the adjoining clusters.

$$\text{Set } X(K) = \sum \{F(L) = K\} \frac{X(L)}{NC(K)}.$$

STEP 3.   *Update V.* Change $V(I)$ ($M < I \leq K$) to

$$V(I) = [(\sum \{F(L) = I\} [X(I) - X(L)]^2) + T][N(I) + 1]^{-1}.$$

STEP 4.   *Compute log posterior density* by LPD $= - \frac{1}{2}\sum\{M < I \leq K\}$ [NC(I) + 1] log $V(I)$. If the increase from its previous value is less than 0.01, stop. Otherwise, return to Step 2.

NOTE.   To eliminate clusters, remove all clusters I with $V[F(I)] < T$, and repeat the algorithm on the reduced tree.

18.3 VARIANCE COMPONENTS ALGORITHM APPLIED TO PREDICTION OF LEUKEMIA MORTALITY RATES

Consider the leukemia mortality data in Table 18.1. Suppose the 1967 rate for Switzerland were unknown. The regression approach to its prediction would find a linear function of, say, the three rates in 1964, 1965, 1966 that best predicts the 1967 rate for countries other than Switzerland. This formula would be then extended to Switzerland.

The clustering approach constructs a tree of countries by their rates in 1964, 1965, 1966. The 1967 rates and variance components are then computed for every cluster in the tree. Switzerland is classified into the tree according to its 1964, 1965, 1966 rates, and its 1967 rate is predicted from the estimated rate for the smallest cluster containing it.

The tree, given in Table 18.2, is constructed by using the adding algorithm. Switzerland, if added to the tree, forms a new cluster of two countries with Belgium.

STEP 1.   First NC(I) = 0 ($1 \leq I \leq M = 16$). Then, NC(17) = 2 because cluster 17 contains 2 objects. Because the tree is a binary tree, NC(I) = 2 for every cluster I ($17 \leq I \leq 31$). The initial $X(17)$ is the average of $X(1) = 34$ and $X(2) = 29$, so $X(17) = 31.5$. Then cluster 18 has 17 and 3 as its maximal clusters, so $X(18) = (31.5 + 33)/2 = 32.25$, and so on. The initial threshold is $T = 1$, the smallest variance component which is judged noticeable.

Table 18.1   Mortality Rates per Million from Leukemia Among Children Aged 0–14 Years, Various Countries, 1956–1967

From Spiers (1972). "Relationship at age of death to calendar year of estimated maximum leukemia mortality rate." *HSMHA Health Rep.* **87**, 61–70.

	1956	1957	1958	1959	1960	1961	1962	1963	1964	1965	1966	1967
AUSTRALIA	34	33	43	44	38	38	39	34	36	35	37	34
AUSTRIA	27	35	40	45	39	34	37	39	35	39	37	29
BELGIUM	41	29	31	40	39	39	47	40	34	29	34	30
CANADA	36	35	32	42	35	35	35	38	36	37	33	33
DENMARK	33	44	46	45	35	48	43	49	56	39	39	38
FINLAND	43	41	34	32	37	28	28	42	43	38	32	41
FRANCE	44	41	41	42	40	36	40	36	36	34	34	31
GERMAN FED. REP.	35	33	35	36	35	35	34	38	36	38	34	39
HUNGARY	28	26	31	31	29	27	27	34	36	35	29	35
ISRAEL	38	44	45	28	61	33	32	28	29	36	30	30
JAPAN	26	25	29	28	30	29	32	32	31	33	33	31
NETHERLANDS	32	37	39	39	39	31	41	37	40	42	36	33
NORTHERN IRELAND	17	25	27	39	29	27	22	31	33	26	30	32
NORWAY	45	44	36	31	44	44	56	31	36	37	50	40
PORTUGAL	20	28	21	34	30	30	30	29	35	37	40	36
SCOTLAND	27	31	32	30	35	30	31	26	29	28	29	25
SWEDEN	33	45	46	45	44	37	44	31	39	35	37	38
SWITZERLAND	42	46	44	47	38	35	45	42	33	31	36	31

STEP 2.   The updating of $X(17)$ is based on the X values for clusters 1, 2, 18, which immediately adjoin it; that is, $F(1) = F(2) = 17$, $F(17) = 18$. Thus,

$$X(17) = \frac{34 + 29 + 32.25}{3} = 31.75.$$

Other X values are updated similarly, except for the largest cluster $X(31) = [X(16) + X(30)]/2 = (38 + 37.35)/2 = 37.67$.

STEP 3.   Change $V(17)$ to

$$V(17) = \frac{(34 - 31.75)^2 + (29 - 31.75)^2 + 1}{2 + 1} = 4.54$$

$$= \frac{[X(1) - X(17)]^2 + [X(2) - X(17)]^2 + T)}{NC(17) + 1}.$$

Similarly adjust other variance estimates.

STEP 4.   Compute log posterior density by $-\frac{1}{2} \times 3 \times \log 4.54 - \frac{1}{2} \times 3 \times \log 1.15 - \cdots = -18.534$. Return to Step 2, and continue updating X and V. At each stage increase the log posterior density. When this increase is less than 0.01, the algorithm stops. The optimum log posterior density is -13.053.

The final estimates of X and V are given in Table 18.3. Notice that several clusters have very small variance components, which justifies the reduction of the tree. For example, the largest cluster, 31, has a variance component of 0.2 because there is very

Table 18.2   Tree Using Adding Algorithm, with Euclidean Distances, on the Variables 1964, 1965, 1967, Leukemia Mortality Data (Table 10.1)

COUNTRY	1967 RATE
1. AUSTRALIA--------- / - I - I - I - I - I - I - I	34
2. AUSTRIA----------- / - / I I I I I I	29
3. CANADA------------ / - - - / I I I I I	33
4. SWEDEN------------ / - - - - - / I I I I	38
5. NETHERLANDS------- / - I I I I I	33
6. FINLAND----------- / - / - - - - - / I I I	41
7. BELGIUM----------- / - - - I - - - I I I I	30
8. NORTHERN IRELAND-- / - I I I I I I	32
9. SCOTLAND---------- / - / - / I I I I	25
10. FRANCE------------ / - I - I - I I I I I	31
11. GERMANY----------- / - / I I I I I I	39
12. HUNGARY----------- / - - - / I I I I I	35
13. JAPAN------------- / - I I I I I I	31
14. ISRAEL------------ / - / - - - / - / - / I I	30
15. NORWAY------------ / - - - - - - - - - - - / I	40
16. DENMARK----------- / - - - - - - - - - - - - - /	38

TREE STRUCTURE FOR 31 CLUSTERS

I	F(I)	I	F(I)	I	F(I)	I	F(I)
1	17	9	22	17	18	25	27
2	17	10	24	18	19	26	27
3	18	11	24	19	21	27	28
4	19	12	25	20	21	28	29
5	20	13	26	21	29	29	30
6	20	14	26	22	23	30	31
7	23	15	30	23	28	31	31
8	22	16	31	24	25		

little difference between its subcluster means 37.9 and 38. These subclusters should therefore be eliminated. The algorithm is repeated on the reduced tree, which eliminates all subclusters where the variance component is less than $T = 1$. The reduction step is again performed, and the final reduced tree is given in Table 18.4.

If Switzerland was classified into this tree, it falls with Belgium into the large cluster Belgium–Israel with relatively low rates. The predicted value for Switzerland in 1967 would be 32.1 with a variance of 1.8. The actual value is 31.

The final reduced tree has only a few clusters in it, which may be briefly summarized. The first division is into four clusters: Australia–Finland, Belgium–Israel, Norway,

Table 18.3 Maximum Posterior Density Estimates of Means and Variances by Applying Variance Components Algorithm to Leukemia Mortality

The first value in each block is the mean; the second value is the variance.

```
OBJECT    X VALUE   ------------------------------------------------
  1         34      I32.9I33.0I35.4I35.4I34.4I37.9I37.9I
  2         29      I 2.9I 0.2I 2.2I 0.2I 1.6I 2.8I  .2I
  3         33         I   I    I    I    I    I    I
  4         38              I   I    I    I    I    I
  5         33      I35.5          I    I    I    I    I
  6         41      I 6.3          I    I    I    I    I
  7         30          I30.4      I31.7I    I    I    I
  8         32      I30.2I 0.2     I  .5I    I    I    I
  9         25      I 5.3I     I    I    I    I    I
 10         31      I34.7I34.7I32.0I    I    I    I    I
 11         39      I 5.5I  .2I 1.7I    I    I    I    I
 12         35         I   I    I    I    I    I    I
 13         31      I30.6    I    I    I    I    I    I
 14         30      I  .3   I    I    I    I    I    I
 15         40                             I    I    I
 16         38                                  I    I
```

Table 18.4 Reduced Tree, Small Variance Components Eliminated

```
                         1967 RATE    --------------------
  1. AUSTRALIA---------      34        I32.7I36.3I36.6I
  2. AUSTRIA-----------      29        I 1.5I 2.6I 1.9I
  3. CANADA------------      33        I   I    I    I
  4. SWEDEN------------      38            I    I    I
  5. NETHERLANDS-------      33            I    I    I
  6. FINLAND-----------      41            I    I    I
  7. BELGIUM----------       30        I32.1     I    I
  8. NORTHERN IRELAND--      32        I 1.8     I    I
  9. SCOTLAND---------       25        I         I    I
 10. FRANCE-----------       31        I         I    I
 11. GERMANY----------       39        I         I    I
 12. HUNGARY----------       35        I         I    I
 13. JAPAN------------       31        I         I    I
 14. ISRAEL-----------       30        I         I    I
 15. NORWAY-----------       40                  I    I
 16. DENMARK----------       38                       I    I
```

336

Denmark. The last two have high death rates. The Belgium–Israel cluster has a low rate, and within Australia–Finland, Australia–Canada has a relatively low rate.

18.4   ALTERNATIVES TO VARIANCE COMPONENTS ALGORITHM

There are two separable problems in the variance components approach—estimation of the means and estimation of the variance components. The technique used for estimating the means given the variances is just the least-squares fit of the model, $X(I) - X[F(I)]$, uncorrelated with variance $V[F(I)]$. The normality assumption is not necessary to justify this procedure.

The variance components themselves are more difficult to estimate. Another method of estimating these is to use the raw cluster variances

$$\sum \{F(L) = I\} \frac{[X(L) - X(I)]^2}{N(I) - 1},$$

where $X(I)$ is the mean of $X(L)$ with $F(L) = I$, defined recursively for I increasing $(M < I \leq K)$. The expectation of this quantity is $V(I)$ plus other variance components of clusters contained in I, and thus an unbiased estimate of $V(I)$ is available. This procedure was not used, principally because of the problem of negative estimates of variance components, but it seems plausible and is certainly simpler than the iterative procedure.

18.5   AUTOMATIC INTERACTION DETECTION

Preliminaries.   There is a base data set $\{A(I, J), 1 \leq I \leq M, 1 \leq J \leq N\}$ and a variable to be predicted $\{A(I, O), 1 \leq I \leq M\}$. The variables are all assumed real valued, but an extension to category-valued variables will be discussed later.

The algorithm considers splits of the M cases into two clusters defined by the variable J, such that the first cluster consists of those cases I with $A(I, J) \leq C$ and the second cluster consists of those cases I in which $A(I, J) > C$. The variable J and the constant C are chosen to minimize the sum of the squared deviations of $\{A(I, O), 1 \leq I \leq M\}$ from the cluster means.

During the algorithm, the clusters are numbered $1, 2, \ldots, $ KC. The cluster K splits into two clusters $L1(K)$ and $L2(K)$, where $L2(K) > L1(K) > K$. The number of cases in cluster K is $NC(K)$. The average value of $A(I, O)$ over cases in cluster K is $AVE(K)$. The between-cluster sum of squares at cluster K is

$$SSQ(K) = \frac{\{AVE[L2(K)] - AVE[L1(K)]\}^2}{1/NC[L2(K)] + 1/NC[L1(K)]}.$$

At each stage, whichever cluster has the smallest value of $SSQ(K)$ is split, and the two subclusters $L1(K)$ and $L2(K)$ become available for further splitting.

The splitting will continue until KC clusters are obtained. If KC is chosen somewhat larger than the number of clusters expected, the later splits may be rejected after examination of the tree, for example, by using the variance components algorithm.

STEP 1.   Let $NP(I)$ denote the cluster into which object I is assigned. Initially $NP(I) = 1$ $(1 \leq I \leq M)$. Set $LL = 1$, $KK = 1$.

STEP 2.   Find the optimal split of cluster LL. For cluster LL and for variable J $(1 \leq J \leq N)$, define $I(1), I(2), \ldots, I(L)$ to be the cases with $NP(I) = LL$ reordered so that $A[I(1), J] \leq A[I(2), J] \leq \cdots \leq A[I(L), J]$.

Find the maximum value for $1 \leq K < L$ of $ST(J) = [B(K) - B(L)]^2 LK/(L - K)$, where $B(L) = \sum \{1 \leq K \leq L\} A[I(K), O]/L$. Let $D(J) = A[I(K), J]$ if K maximizes $ST(J)$.

The maximum value of $ST(J)$ over J $(1 \leq J \leq N)$ is $SSQ(LL)$, the reduction in the sum of the squared deviations of $\{A(I, O), 1 \leq I \leq M\}$ due to splitting cluster LL optimally. Correspondingly, $J(LL)$ is the maximizing J, and $C(LL)$ is the split point $D(J)$.

STEP 3.   Find the maximum value of $SSQ(K)$ over all clusters K with $NP(I) = K$ for at least one I. If cluster K has maximum SSQ, set $L1(K) = KK + 1$, $L2(K) = KK + 2$, and increase KK by 2. For each I $[1 \leq I \leq M, NP(I) = K]$, define $NP(I) = L1(K)$ if $A[I, J(K)] \leq C(K)$, and $NP(I) = L2(K)$ if $A[I, J(K)] > C(K)$. If $1 + KK \geq$ KC, stop.

STEP 4.   Perform Step 2 for $LL = L1(K)$, then for $LL = L2(K)$, and then return to Step 3.

18.6.   APPLICATION OF AID ALGORITHM TO LEUKEMIA MORTALITY

The problem is again a regression problem: to predict the 1967 death rate from the rates in 1964, 1965, 1966. The AID algorithm operates similar to a stepwise regression, except that the base variables are converted to 0–1 variables before being used for prediction.

STEP 1.   Initially, there is only one cluster, $KK = 1$. Set $NP(I) = 1$ $(1 \leq I \leq 16)$ $LL = 1$.

STEP 2.   This step finds the best split of a given cluster. If $J = 1$, the cases $I(1)$, $I(2), \ldots, I(16)$ are 10 (Israel), 15 (Scotland), 11 (Japan), . . . , 5 (Denmark). This is the ordering of the cases by the first variable 1964. Thus the death rate for Israel in 1964 is 36, the minimum, and the death rate for Denmark is 56, the maximum.

The quantity $[B(K) - B(L)]^2 LK/(L - K)$ is to be computed for each K $(1 \leq K < L)$; $B(L)$ is the average over all 16 countries, 33.4, and $B(1)$ is the average over the first country, $B(1) = 30$. Thus, for $K = 1$,

$$ST(J) = (30 - 33.4)^2 16 \times \tfrac{1}{15} = 14.5.$$

Similarly,

$$B(2) = 27.5, \qquad ST(J) = (33.4 - 27.5)^2 16 \times \tfrac{2}{14} = 87.5.$$

The maximum value of 168.3 is achieved at $K = 6$, $D(1) = 35$. The split is thus according as the 1964 rate is less than or equal to 35, or greater than 35.

The other variables attain a maximum reduction of 109.9 and 64.5, respectively, so the best split is $J(1) = 1$ with $C(1) = 35$. The reduction is $SSQ(1) = 168.3$.

STEP 3.   The only cluster K with $NP(I) = K$ for some I is $K = 1$. Thus cluster 1 is split, $L1(1) = 2$, $L2(1) = 3$, $KK = 3$. If $I = 1$, set $NP(I) = 1$, and, since $A(1, 1) = 36 > C(1) = 35$, $NP(I) = 3$. If $I = 2$, $NP(I) = 1$, and since $A(2, 1) = 35 \leq C(1) = 35$, $NP(I) = 2$. In this way, all cases will be assigned to one or other of clusters 2 and 3.

STEP 4. Find the optimal split for cluster 2 by returning to Step 2.

STEP 2 REPEATED. With LL = 2, $J = 1$, there are only six cases with NP$(I) = 2$. Ordered by the first variable, 1964 rate, these are 10, 15, 11, 13, 3, 2. The value of ST(1) is then 12.5. The best split occurs for the third variable, so SSQ(2) = 24.3, $J(2) = 3$, and $C(2) = 29$.

Continuing with Step 2, LL = 3, the best split occurs with variable 2, SSQ(3) = 30.0, $J(3) = 2$, and $C(3) = 34$.

STEP 3. The only clusters that need be examined are 2 and 3. For these, $K = 3$ minimizes SSQ. Set $L1(3) = 4$, $L2(3) = 5$, increase KK to 5. Cases with NP$(I) = 3$ are changed to NP$(I) = 5$, except for case 7, where $A(7, 2) = 34$, so NP$(I) = 4$.

The splitting is stopped after seven clusters are obtained, with the next split being of cluster 2, variable 3. The splits and cluster averages are given in Table 18.5. It will be seen that the first split gives a rather satisfactory division into high and low rates.

Table 18.5 Aid Algorithm Applied to Leukemia Mortality

1, 1964 rate; 2, 1965 rate; 3, 1966 rate.

		1967 RATES	ESTIMATES (MEANS)
SCOTLAND---------- I3<29I1<35I		25	25.0
AUSTRIA----------- I3>29I I		29	30.4
BELGIUM----------- I I I		30	30.4
ISRAEL------------ I I I		30	30.4
JAPAN------------- I I I		31	30.4
NORTHERN IRELAND-- I I I		32	30.4
FRANCE------------ I2<34 I1>35I		31	31.0
AUSTRALIA--------- I2>34I I		34	37.0
CANADA------------ I I I		33	37.0
DENMARK----------- I I I		38	37.0
FINLAND----------- I I I		41	37.0
GERMANY----------- I I I		39	37.0
HUNGARY----------- I I I		35	37.0
NETHERLANDS------- I I I		33	37.0
NORWAY------------ I I I		40	37.0
SWEDEN------------ I I I		38	37.0

A new state—say, Switzerland—is classified according to its rates in 1964, 1965, 1966. Since Switzerland in 1964 has a rate of 33, it is classified into the low group. Next, in looking to 1966, it is classified high. Thus the predicted value for Switzerland is 30.4. The real value is 31.

Portugal would be classified into the same group—predicted value 30.4, real value 36. The reliability of the method (as with many clustering methods) should be assessed cautiously by the within-cluster sum of squares, since this has been minimized in the algorithm.

18.7 REMARKS ON THE AID ALGORITHM

The AID algorithm offers a simple clustering alternative to linear regression. The algorithm may be used with discrete or continuous base variables.

Suppose the base variables $\{A(I, J), 1 \leq I \leq M, 1 \leq J \leq N\}$ are 0–1 variables. Two clusters are associated with the Jth variable—the cluster of cases where the variable is zero and the cluster where the variable is one. That variable is first chosen which maximizes the between-cluster sum of squares (or equivalently, the correlation squared) for the predicted variable $\{A(I, O), 1 \leq I \leq M\}$. Suppose this variable is the Jth. A number of new variables are now defined, based on the first one fitted, just as in stepwise regression. The Kth variable becomes two new category variables. The first of these—say, X—is defined as

   (i) $X(I) = 0$ if $A(I, J) = 0, A(I, K) = 0,$

  (ii) $X(I) = 1$ if $A(I, J) = 0, A(I, K) = 1,$

 (iii) $X(I) = 2$ if $A(I, J) = 1.$

The second of these Y, is defined as

   (i) $Y(I) = 2$ if $A(I, J) = 0,$

  (ii) $Y(I) = 0$ if $A(I, J) = 1, A(I, K) = 0,$

 (iii) $Y(I) = 1$ if $A(I, J) = 1, A(I, K) = 1.$

Each variable divides the cases into three clusters, and that variable is found which has the largest between-cluster sum of squares. A number of new variables are now defined, each with four categories, obtained by dividing one of the three present case clusters according to whether or not one of the original variables is zero or one.

The technique used then is to construct a finer and finer partition, at each stage having a number of variables available which generate partitions even finer than the present partition. The best of these trial partitions is the next partition of cases.

The technique can obviously be generalized to category variables taking more than two values. If the category variables take very many values, any one variable may use up too many degrees of freedom. A way out of this is to construct 0–1 variables from the category variables at each stage. The best 0–1 variables (ones generating best between-cluster sums of squares) are obtained by ordering the categories according to the mean of $A(I, O)$ within each category, and then generating $N-1$ 0–1 variables by splitting consistent with this order, where N is the number of categories.

For continuous base variables these also are converted to 0–1 variables by considering a number of 0–1 variables of the type

$$Z(I) = 0 \quad \text{if} \quad A(I, J) \leq C,$$
$$Z(I) = 1 \quad \text{if} \quad A(I, J) > C.$$

One of the defects of the AID algorithm is that it considers a single variable at a time. This makes the results easy to interpret, but, if several variables cluster about equally well, only one of these may define the split and the others may never appear. There is a similar "arbitrary selection" problem in stepwise regression. A possible cure is to consider more general ways of constructing 0–1 variables, to be used in the fit, from many variables. For example, for two variables define

$$X(I) = 0 \quad \text{if} \quad A(I, J)B(J) + A(I, K)B(K) \leq C$$
$$X(I) = 1 \quad \text{if} \quad A(I, J)B(J) + A(I, K)B(K) > C.$$

The number of 0–1 variables of this type is $M(M - 1)/2 + 1$, versus $M - 1$ for the 0–1 variables based on a single continuous variable. So this approach certainly complicates the computing.

To handle many variables, it is suggested that an initial splitting variable be found, then find one which combines with it to split well, then find one which combines with the new variable (a linear combination of the first two), and so on, until no significant improvement takes place. Then begin over within the two clusters obtained in the first step.

The AID algorithm takes $O(M \log MNK)$ calculations (where M is the number of cases, N is the number of variables, and K is the number of clusters) for continuous variables. For 0–1 variables, by clever sorting, the calculations are $O(MN)$. The technique is thus appropriate for very large data sets.

The above procedure for handling many variables is $O(M^2N^2)$, which is much more expensive. Approximate optimization of the K-means type will reduce the computation time.

The AID algorithm is remarkable in its ability to handle continuous and category variables, but it should be noted that the method is to reduce all variables to 0–1 variables. The method operates somewhat differently on the continuous than on the category variables. In particular, a category variable taking many possible values is more likely to be chosen early in the fitting than a 0–1 or continuous variable, because the many-valued category variable generates many 0–1 variables.

18.8 THINGS TO DO

18.8.1 Running Prediction Algorithms

The AID algorithm is to be used in somewhat the same circumstances as stepwise regression; it is much more able to cope with clusters in the cases and with non-linear relationships between variables. It generates a simple interpretable prediction rule for new observations. (Use the city crime data in Table 18.6.)

The variance components algorithm fits a continuous variable to a given tree and is just a generalization of standard variance components fitting for partitions. The analogous procedure for category data is the minimum-mutation method. Thus this algorithm might be used to summarize a number of variables in terms of their average values at a few nodes of the tree. Predictions of missing values are made from these fitted average values.

18.8.2 By Rows or Columns

The matrix

$$\begin{bmatrix} 3 & 1 & 1 \\ 1 & 2 & 5 \\ 6 & 2 & * \end{bmatrix}$$

has a missing value denoted by an asterisk. Show that the following procedures are equivalent:

(i) Using the first two rows, predict the third column as a linear combination of the first two. Extend this prediction to the third row.

Table 18.6   City Crime

[From *The Statistical Abstract of the United States* (1971), Bureau of the Census, U.S. Department of Commerce, Grossett and Dunlap, New York.] Each variable, except if specified otherwise, is per 100,000.

CITY	POPN (1000's)	%WHITE CHANGE, INNER CITY 1960-1970	BLACK (1000's)	MURDER	RAPE	ROBBERY	ASSAULT	BURGLARY	AUTO THEFT
ANAHEIM	1420	50.8	39	2.7	21.9	94	103	1607	377
BALTIMORE	2071	-21.4	501	13.2	34.9	564	396	1351	701
BOSTON	2754	-16.5	151	4.4	14.8	136	95	1054	984
BUFFALO	1349	-20.7	118	5.7	13.7	145	111	862	448
CHICAGO	6979	-18.6	1306	12.9	25.4	363	233	830	708
CINCINATTI	1385	-17.2	156	6.4	16.8	120	107	912	348
CLEVELAND	2064	-26.5	343	14.5	18.7	288	132	826	1208
DALLAS	1556	14.2	261	18.4	41.0	206	338	1581	577
DETROIT	4200	-29.1	780	14.7	31.1	649	223	1986	758
HOUSTON	1985	25.5	399	16.9	27.1	335	183	1532	741
LOS ANGELES	7032	4.7	1026	9.4	50.0	307	371	1981	945
MIAMI	1268	13.5	196	15.6	17.0	427	421	1858	781
MILWAUKEE	1404	-10.4	116	3.8	8.3	53	63	499	405
MINNEAPOLIS	1814	- 7.9	50	2.6	16.8	179	89	1198	615
NEW YORK	11529	- 9.3	2080	10.5	19.9	665	286	1821	947
NEWARK	1857	-36.7	363	9.5	20.5	333	182	1315	667
PATERSON	1359	- 9.1	83	2.6	5.7	99	63	815	491
PHILADELPHIA	4818	12.9	873	9.3	15.2	173	123	754	534
PITTSBURGH	2401	-18.0	176	4.4	14.0	145	108	696	537
ST. LOUIS	2363	-31.6	388	14.8	34.4	280	203	1458	840
SAN FRANCISCO	3110	-17.2	535	8.3	42.9	348	226	2164	957
SAN DIEGO	1358	17.2	106	4.1	19.4	93	106	1011	428
SEATTLE	1422	- 8.5	85	4.4	23.8	164	114	1976	547
WASHINGTON	2861	-39.4	737	11.4	23.0	504	232	1433	767

  (ii)   Repeat (i) with rows and columns interchanged.

  (iii)   Estimate the missing value by requiring zero determinant.

18.8.3   Block Models

If any array contains a number of blocks and all values within a block have a specified range, a missing value will be predicted to lie in the range of the block containing it. For a large block it might be reasonable to assume the missing value has the same distribution as the distribution of values in the block.

This technique works well for filling in missing values, but it is less compelling as a prediction technique because a new case must be fitted into the blocks of the array before predictions can be made. These predictions are like the AID predictions—if $0.3 < V1 < 0.5$ and $0.2 < V2 < 0.7$ and $0.5 < V3 < 0.9$, then predict $0.3 < V4 < 0.6$.

18.8.4   Median Fits

In fitting a continuous variable to the nodes of a tree, it is plausible to minimize $|X(I) - X[T(I)]|$ summed over all nodes of the tree, where $T(I)$ is the ancestor of node I. The values of X at the ends of the tree are assumed given. The algorithm given in Section 13.8.6 computes this median fit exactly.

VARCO   variance components model fitted to a given tree.

AID   splits data to best predict a target variable, basing each split on a predicting variable.

AIDOUT   outputting program for AID.

REL   splits data to best predict a given variable, with split based on another variable.

```
      SUBROUTINE VARCO(M,K,NT,TH,X,XN,V,XLL,Y)
C..............................................................20 MAY 1973
C....      GIVEN A TREE, AND A SET OF VALUES ON ITS ENDS,THIS PROGRAM FINDS
C      VARIANCE COMPONENTS AND MEANS ASSOCIATED WITH EACH NODE.  THE PROBABILITY
C      MODEL ASSUMES THAT X(L)-X(K) IS NORMAL WITH MEAN ZERO AND VARIANCE V(L)
C      WHERE K IS ANY CLUSTER AND L IS THE SMALLEST ONE INCLUDING IT.
C....  M = NUMBER OF OBJECTS AT END OF TREE, PLUS ONE FOR DUMMY FIRST OBJECT.
C....  K = NUMBER OF CLUSTERS OR NODES OF TREE, K.GT.M.
C....  NT = 1 BY K VECTOR, NT(I).GT.I IS ANCESTOR OF I, EXCEPT FOR ROOT K,NT(K)=K
C....  TH = THRESHOLD,PRIOR EXPECTATION OF (X(L)-X(K))**2
C....  X = K BY 1 VECTOR, FOR I.LE M, X(I) = INPUT VALUE AT END OF TREE
C          FOR I.GT.M, X(I) = ESTIMATED MEAN AT NODE I.
C....  XN = K BY 1 VECTOR, XN(I) = NUMBER OF NODES WITH ANCESTOR I.
C....  V = 1 BY K VECTOR,V(I) = VARIANCE COMPONENT AT I.
C....  XLL = LOG LIKELIHHOD AT EACH STEP
C....  Y = K BY 1 SCRATCH VECTOR
C..................................................................................
      DIMENSION NT(K),X(K),V(K),XN(K),Y(K)
      MM=M+1
      KK=K-1
C..... INITIALIZE MEANS,VARIANCES,COUNTS AT EACH NODE
      DO 10 I=MM,K
      X(I)=0.
   10 V(I)=TH
      DO 11 I=2,K
   11 XN(I)=0.
      DO 12 I=2,KK
      J=NT(I)
      IF(I.GT.M) X(I)=X(I)/XN(I)
      X(J)=X(J)+X(I)
   12 XN(J)=XN(J)+1.
      X(K)=X(K)/XN(K)
      WRITE(6,1)
    1 FORMAT(22H1OBJECT ANCESTOR VALUE  )
      WRITE(6,2)(I,NT(I),X(I),I=2,M)
    2 FORMAT(2I5,5X,G12.5)
      ITER=20
      XLL=-10.**10
      DO 40 IT=1,ITER
      WRITE(6,3) XLL
    3 FORMAT(14HOLOGLIKELIHOOD ,G12.5)
      WRITE(6,4)
    4 FORMAT(44H NODE ANCESTOR COUNT      MEAN      VARIANCE  )
      WRITE(6,5) (I,NT(I),XN(I),X(I),V(I),I=MM,K)
    5 FORMAT(2I5,G12.2,2G12.5)
C..... UPDATE MEANS
      DO 20 I=MM,K
   20 Y(I)=0.
      DO 21 I=2,KK
      J=NT(I)
      IF (I.GT.M) X(I)=(Y(I)/V(I)+X(J)/V(J))/(XN(I)/V(I)+1./V(J))
   21 Y(J)=Y(J)+X(I)
      X(K)=Y(K)/XN(K)
C..... UPDATE VARIANCES
      DO 30 I=MM,K
   30 V(I)=0.
      DO 31 I=2,KK
      J=NT(I)
   31 V(J)=V(J)+(X(J)-X(I))**2/XN(J)
C..... COMPUTE LOG LIKELIHOOD
      XOLD=XLL
      XLL=0.
      DO 32 I=MM,K
      V(I)=(V(I)+TH)/(1.+XN(I))
   32 XLL=-(1.+XN(I))*0.5*ALOG(V(I))+XLL
      IF(XLL.LT.XOLD+.00001) RETURN
   40 CONTINUE
      RETURN
      END
```

```
      SUBROUTINE AID(A,M,N,K,TH,NB,SP,KA,JP)
C.......................................................................20 MAY 1973
C.... SPLITS DATA TO BEST PREDICT COLUMN JP,BASING EACH SPLIT ON SOME OTHER COL.
C.... USES AIDOUT,REL
C.... A = M BY N BORDERED ARRAY
C.... M = NUMBER OF ROWS
C.... N = NUMBER OF COLUMNS
C.... JP = VARIABLE TO BE PREDICTED
C.... K = MAXIMUM NUMBER OF CLUSTERS
C.... TH = THRESHOLD, STOP SPLITTING WHEN THRESHOLD EXCEEDS SSQ REDUCTION
C.... NB = 4 BY K ARRAY DEFINING BLOCKS
C          NB(1,K) = FIRST ROW IN CLUSTER
C          NB(2,K) = LAST ROW OF CLUSTER
C          NB(3,K) = VARIABLE USED IN SPLIT
C          NB(4,K) = VARIABLE IN SPLIT
C..... SS = 5 BY K ARRAY DEFINING FOR EACH CLUSTER THE SPLITTING VARIABLE, THE
C          SPLITTING CONSTANT, THE CLUSTER MEAN, THE SSQ DUE TO SPLIT
C.... KA = ACTUAL NUMBER OF CLUSTERS
C.......................................................................
      DIMENSION A(M,N),NB(4,K),SP(6,K)
C.... INITIALIZE BLOCK ARRAY
      KA=0
      IL=2
      IU=M
      JM=JP
      IS=1
   80 IF(KA.GT.K-2) GO TO 90
C.... DEFINE BLOCKS
      KA=KA+1
      NB(1,KA)=IL
      NB(2,KA)=IS
      NB(3,KA)=JM
      NB(4,KA)=JM
      KA=KA+1
      NB(1,KA)=IS+1
      NB(2,KA)=IU
      NB(3,KA)=JM
      NB(4,KA)=JM
C.... DEFINE SPLITTING PARAMETERS FOR NEW BLOCKS
      KL=KA-1
      DO 43 KK=KL,KA
      LL=IL
      LU=IS
      IF(KK.EQ.KA) LL=IS+1
      IF(KK.EQ.KA) LU=IU
      IF(LL.GT.LU) GO TO 43
      SP(3,KK)=0
      DO 40 J=2,N
      IF(J.EQ.JP) GO TO 40
      NN=LU-LL+1
      CALL REL(A(LL,JP),A(LL,J),NN,C,SSQ)
      IF(SSQ.LT.SP(3,KK)) GO TO 40
      SP(1,KK)=J
      SP(2,KK)=C
      SP(3,KK)=SSQ
   40 CONTINUE
      S0=0
      S1=0
      S2=0
      DO 42 I=LL,LU
      S0=S0+1
      S1=S1+A(I,JP)
   42 S2=S2+A(I,JP)**2
      SP(4,KK)=S0
      SP(5,KK)=S1/S0
      SP(6,KK)=S2/S0-(S1/S0)**2
   43 CONTINUE
C.... FIND BEST BLOCK TO SPLIT
      SM=0
      DO 10 KK=2,KA
      IF(SP(3,KK).LE.SM) GO TO 10
      KM=KK
      SM=SP(3,KK)
   10 CONTINUE
      IF(SM.LT.TH) GO TO 90
C.... REORDER DATA ACCORDING TO BEST SPLIT
      IL=NB(1,KM)
      IU=NB(2,KM)
```

344

```
            JM=SP(1,KM)
            DO 20 I=IL,IU
            DO 20 II=I,IU
            IF(A(I,JM).LE.A(II,JM)) GO TO 20
            DO 21 J=1,N
            C=A(I,J)
            A(I,J)=A(II,J)
         21 A(II,J)=C
         20 CONTINUE
            SP(3,KM)=-SP(3,KM)
            DO 31 I=IL,IU
            IF(A(I,JM).GT.SP(2,KM)) GO TO 32
         31 CONTINUE
         32 IS=I-1
            GO TO 80
         90 CONTINUE
                CALL AIDOUT(A,M,N,NB,SP,KA,TH,JP)
            RETURN
            END




      SUBROUTINE AIDOUT(A,M,N,NB,SP,KA,TH,JP)
C....................................................................20 MAY 1973
C....PRINTS OUTPUT OF SUBROUTINE AID.  SEE AID FOR ARGUMENTS.
C...................................................................................
      DIMENSION NB(4,KA),SP(6,KA),A(M,N)
      WRITE(6,1) JP,TH
    1 FORMAT(51H1CLUSTERS OBTAINED IN SPLITTING TO PREDICT VARIABLE,I5,
     */16H USING THRESHOLD  ,G12.6)
      WRITE(6,3)
    3 FORMAT(
     *79H FIRST    LAST   SPLIT    SPLIT       SSQ          COUNT
     *MEAN      VARIANCE        ,/,
     *45H OBJECT OBJECT  VARIABLE  POINT    REDUCTION
      DO 20 K=2,KA
      JS=SP(1,K)
      IL=NB(1,K)
      IU=NB(2,K)
   22 WRITE(6,2) A(IL,1),A(IU,1),A(1,JS),(SP(I,K),I=2,6)
    2 FORMAT(3(2X,A5),5X,5G12.5)
   20 CONTINUE
      RETURN
      END
```

```
      SUBROUTINE REL(U,V,N,C,SSQ)
C.............................................................20 MAY 1973
C.... USED IN AID FOR PREDICTING A VARIABLE U BY SPLITTING ON A VARIABLE V.
C.... U = N BY 1 ARRAY, TO BE PREDICTED
C.... V = N BY | ARRAY,USED IN PREDICTION
C.... N = NUMBER OF ELEMENTS OF ARRAY
C.... C = SPLIT POINT
C.... SSQ = SUM OF SQUARES IN U AFTER SPLIT
C..........................................................................
      DIMENSION U(N),V(N)
      DIMENSION X(500),Y(500)
      DO 10 I=1,N
      X(I)=U(I)
   10 Y(I)=V(I)
C- - - - - - - - REORDER X AND Y VECTORS BY Y VECTOR
      DO 20 I=1,N
      DO 20 J=I,N
      IF (Y(I).LE.Y(J)) GO TO 20
      C=X(I)
      X(I)=X(J)
      X(J)=C
      C=Y(I)
      Y(I)=Y(J)
      Y(J)=C
   20 CONTINUE
C- - - - - - - - FIND BEST SPLIT
      SSQ=0.
      AVE=0.
      DO 30 I=1,N
   30 AVE=AVE+X(I)
      XN=N
      AVE=AVE/XN
      A=0.
      DO 31 I=1,N
      IF (I.EQ.N) GO TO 31
      A=A+X(I)
      XI=I
      B=(AVE-A/XI)**2*XN*XI/(XN-XI)
      IF (Y(I).EQ.Y(I+1)) GO TO 31
      IF(B.LE.SSQ) GO TO 31
      SSQ=B
      C=Y(I)
   31 CONTINUE
      RETURN
      END
```

Index